THE OCEAN OF TRUTH

THE OCEAN OF TRUTH

A Personal History of Global Tectonics

H. W. MENARD

PRINCETON UNIVERSITY PRESS

PRINCETON, NEW JERSEY

All Rights Reserved
Library of Congress Cataloging in Publication Data
will be found on the last printed page of this book
ISBN 0-691-08414-9

This book has been composed in Linotron Times Roman
Clothbound editions of Princeton University Press books
are printed on acid-free paper, and binding materials
are chosen for strength and durability

Printed in the United States of America by
Princeton University Press
Princeton, New Jersey

To the memory of Teddy and Harry, the two Maurices, and Bruce, and the good health of those who remember.

MEMORIAL TRIBUTE TO H. W. MENARD

This work represents the confluence of two of Bill Menard's great passions in life: plate tectonics and the history of science. It is the story as only he could tell it, with wit, charm, and the insight that comes from a full participant who can, nevertheless, step back to analyze the events chronicled.

Bill Menard came to the forefront of marine geology in the early 1950s and led the field through the most exciting three decades of submarine research. He has authored or coauthored more than 115 scientific books and articles, of which even the early papers are widely referenced today. Bill obtained a B.S. degree from California Institute of Technology in 1942. After three years as a Naval photo interpreter and air intelligence officer in the western Pacific during World War II, during which time he earned the Bronze Star and a Navy Commendation ribbon, he returned to Caltech for an M.S. in 1947, continuing on to a Harvard Ph.D. in 1949. Bill spent the next six years working for the Naval Electronics Laboratory in San Diego before becoming Professor of Geology at Scripps Institution of Oceanography. He took leave from Scripps on two occasions to serve in government positions, first in the Office of Science and Technology under the Johnson administration and later as the tenth Director of the U.S. Geological Survey under President Carter.

Bill's earliest scientific accomplishments include his accurate morphological descriptions of the Pacific ocean basins and rises, including the discovery of the great Pacific fracture zones (one of which bears his name) and the Hawaiian swell, all before the plate-tectonic revolution. His knowledge of the Pacific was a key factor in putting together the sea-floor spreading history of the eastern Pacific in a series of publications coauthored with illustrious students such as Roger Larson, Tanya Atwater, Jean Francheteau, and Clem Chase. Bill participated in or led about 30 oceanographic expeditions and scientifically sampled 50 islands representing almost every major chain in the Pacific. His contributions did not go unrecognized by his colleagues. He received the Penrose Medal from the Geological Society of America, the Shepard Medal from the Society of Economic Paleontologists and Mineralogists, the Bowie Medal from the American Geophysical Union, and membership in the National Academy of Sciences.

The recollections in this book transport me back to my days of Camelot in a chart room overlooking the Pacific, where science seemed so easy and a glow of excitement for problems yet to be solved surrounded Bill and those few privileged enough to work with him. His students agree that Bill was a tough teacher. He always insisted on having the facts at hand to back up new hypotheses, and he never let us gloss over discrepancies or inconsistencies. He was always the first to reject even his own theories if new data proved them wrong, so that in spite of the passing years his science, and his mind, never aged.

Henry William Menard, pioneer of deep-sea exploration, patriot of the plate-tectonic revolution, and patriarch of modern marine geomorphology, died on February 9, 1986. From his hospital bed he labored over the galley sheets for this book until exhaustion and pain consumed him. The eternal optimist, he spoke with me of plans for its sequel and of our upcoming expedition to the Marquesas Islands, knowing full well that his cancer was terminal. At his memorial service in La Jolla, Tom Jordan recalled some lines from one of Bill's favorite poets, Robert Penn Warren, with which I bid him farewell:

> To stare at the stars that remain, with eyes shut in dark, pressing an ear
> To sand, cold as cement, to apprehend,
> Not merely the grinding of shingle and sea-slosh near,
> But the groaning miles of depth where light finds its end.
>
> Below all silken soil-slip, all crinkled earth-crust,
> Far deeper than ocean, past rock that against rock grieves,
> There at the globe's deepest dark and visceral lust,
> Can you hear the *groan-swish* of magma as it churns and heaves?*

Marcia McNutt

* Reprinted, with permission, from Robert Penn Warren, "Youthful Truth-Seeker, Half-Naked, At Night, Running Down Beach South of San Francisco,"*Being There: Poetry 1977-1980.* © 1980 by Random House, Inc.

CONTENTS

ILLUSTRATIONS

PREFACE

For most of us except historians, it is fatal to look back. Thus I began this book tentatively, with only a sketchy outline and no commitment to complete it. I felt that the revolution in the earth sciences merited a history that had not been written, but it was by no means evident that I should be the one to write it. Still, I had participated in that revolution, and most of the other participants were friends or at least colleagues, so I wrote or talked to some of them and with their encouragement began to write during a sabbatical leave.

Many of those who generously helped me are specifically acknowledged in the text. I particularly appreciated the help of Allan Cox, who provided copies of his extensive correspondence while writing his book *Plate Tectonics and Geomagnetic Reversals*. Deborah Day, archivist of the Scripps Institution of Oceanography, helped me work through the Revelle papers, and Ann Lewis of Lamont-Doherty Geological Observatory provided access to the Institutional collection of photographs. Marie Tharp gave me copies of her exceptional collection of newspaper stories about Bruce Heezen. Walter and Judy Munk kindly allowed me to select copies from the lively photographs they acquired for the halls of the Institute of Geophysics and Planetary Physics in La Jolla.

Thomas Jordan and Richard Hey read the first few chapters of the initial draft of the manuscript, and their comments influenced further writing. The first half of the manuscript was read at sea by most of the scientific party of the Zed expedition, and discussions by Hey, Tanya Atwater, and Susan Dixon enlivened the watches. The entire manuscript was thoughtfully reviewed by E. L. Hamilton, whose memory of the Midpac expedition was sharper than mine. Duncan Agnew read the manuscript in several stages, and his scholarly comments have been incorporated into the narrative.

My initial goal was to cover the period from 1900 to 1980, and I acquired much useful information from some of the scientists who became prominent in the 1970s. The coverage now extends, however, to the late sixties, and these scientists do not appear in this account. I apologize to them for an inadvertent deception about when I might publish their interesting material about that decade.

Writing can be educational, and when the sabbatical leave came to an end I began to appreciate Harry Hess's practice of writing papers on airplanes. Consequently most of Chapter 16 was written while flying to and from a meeting in Strasbourg, and the last four chapters were written at sea between Kodiak and Honolulu on the Zed expedition. Considering that the entire manuscript was written longhand, and the air and sea were not always calm, my debt to Mary Beth Hiller who put it into a central computer word processor for editing is manifest.

As always in the past 38 years, I have had the encouragement and support of my wife, Gifford Merrill Menard.

THE OCEAN OF TRUTH

Chapter 1

THE DUST ON HESS'S SLATE

Revolutions are painful but they become necessary when the old order no longer makes sense. That was true in geology after World War II. There was no agreement about the most fundamental properties of the earth. Was it cooling, heating, or staying at a constant temperature? Was it contracting, expanding, or retaining its size? Did the interior convect? Did continents drift? No geologist knew for sure. Nor, with the tools and techniques of the past, was there any way to know. Only a scientific revolution could solve the stagnation in geology, but where would it start and when? As it happened it started at sea.

With the Midpac expedition of 1950 the new marine geology was fairly launched and the coming tectonic revolution became inevitable. The first campaign of the revolution had been fought to a stalemate. The dream of drifting continents, which sprang fully formed from the brain of Lt. Alfred Wegener during a long recuperation from war wounds in the First World War, was untestable with the geological tools available before World War II. The revolution sputtered along for decades with episodic sniping from the academic trenches. In 1950 the first generation of the new marine tools was assembled: echo sounders, magnetometers, temperature probes, explosion seismometers, piston corers, and dredges. Within five years they would be adequate to discover the magnetic anomalies that take the measure of all the world, the great fracture zones that offset them, and the median rift where the sea floor is born. Meanwhile the cadres and tools for other phases of the revolution were being assembled. It would also be five years before the "paleomagicians"[1] began to revive continental drift and double that time before the conjunction of precise isotopic dating and extensive magnetic sampling proved that the earth's magnetic field reverses frequently. Nonetheless paleomagnetic studies were about to begin. So was the new seismology—triggered by efforts to detect clandestine tests of nuclear weapons—that would verify transform faults and track the plunging lithosphere. All was ready to begin to end the revolution, but the campaign would be long. The wunderkinder, Morgan, McKenzie, and Parker, who would invent plate tectonics, were still schoolboys. Of the principal characters in this history, probably only Teddy Bullard, who learned it long before when he himself was a schoolboy, knew of Euler's theorem in 1950. Ultimately it was not geology or physics but a theorem proposed by the Swiss mathematician Leonhard Euler two centuries earlier that provided the linchpin of plate tectonics.

Both the written and the orally reported memory of the development of scientific ideas can be remarkably misleading. Generally an idea evolves because new data require a revision. Then, to a working scientist, in contrast to a philosopher or historian, the old idea has no value and is erased from memory. After a decade or two a vigorous creative scientist might sometimes be questioned about an idea he has long since abandoned and must ask the questioner why he ever entertained the idea. For example, I once pointed out to Harry Hess that what he was telling me was incompatible with what he had published a scant two years earlier. Having forgotten, he expressed some surprise but more indignation, claiming he was in no way responsible for ideas made obsolete by new data. Thus the only certain way to know what Hess wrote was to read it rather than to ask him. On the other hand scientific papers are rarely explicit about the broad questions to which the described research is related. The basic purpose is assumed to be obvious, so to understand why most papers are written we must put them in the context of current theories. In this respect the author of the paper can be of some help.

The most difficult fact to establish is what background information a scientist had when he started his research. The paper itself ordinarily leads one to believe that the author was one day reading a scientific journal and it occurred to him to conduct some research based on his reading. The citations credit, and references identify, all the previous work that the scientist had digested before he began his own contribution. However, few research scientists seem to function in this way. I, for example, tend to get ideas while walking along the beach in La Jolla, and if I cannot think of a test I brood about it. If I can test, I immediately do so by whatever means, crude or refined, are at hand. Barring an immediate test, I might phone a colleague to outline the problem and ask if she had solved it or knew of a solution—thereby saving me the effort. I can't think of a time when I went to a library seeking a solution to a hypothesis. I have queried many productive scientists in rapidly advancing fields, and they all seem to work in this way. In part this is because, within their own specialties, they know most of the literature. I suspect, however, that the main reason for this behavior is that it is more exciting to conceive and test an idea than to read about one. Perhaps another reason is that scientists are not always easy to understand; "it is easier to rediscover Gibbs than to read him."[2]

When a scientist publishes, the circumstances usually change. The whole fabric of science consists of linkages to past work, and so after the fact the researcher repairs to the library to find what has been published on a subject. Often one scientist (A) realizes that a bit of apparently isolated research published by another scientist (B) is part of a newly significant body of work. Scientist B of course did not realize this, but A dutifully cites B, and later a third scientist (C) cites B as the real discoverer of A's work.

Another problem for a historian of science arises from what Robert K. Merton calls the "Matthew effect,"[3] citing the Gospel according to St. Matthew:

> For unto every one that hath shall be given, and he shall have abundance; but from him that hath not shall be taken away even that which he hath.

The effect presumably applies to all kinds of activities that involve public recognition. When a distinguished actor is reduced by age to a secondary role, we still associate his name with a movie when we have forgotten that of the young star. In science the Matthew effect creates much confusion and, as with aspiring actors, makes it difficult for a young scientist to achieve recognition. For example, an established scientist (E) often receives a reprint from a young scientist (Y) who is unknown to him but who cites his work. Often Y, a graduate student, is the first author of a paper that has as a second author his professor (P) who is well known to E because they work on the same subjects. That is in fact why Y is citing E's work. The established scientist will remember the paper because it is his subject, but what is he to do with the reprint? If E files it alphabetically under "Y" he is in danger of losing track of it because he will not remember Y's name. Thus it is filed under "P." The problem for the historian of science is that the established scientist who is likely to be interviewed may innocently remember the research and the noted second author and completely forget the first author who did most of the work. The effect is so strong for Nobel Prize winners that some of them stop publishing jointly with young colleagues even though they collaborated in the research.

The Matthew effect sometimes has the curious result that the more famous scientists who were established in a field when a discovery was made may be less reliable sources for a historian than the students who came after the discovery. The students are assigned papers to read, and thus they know the names and order of the authors. As to the established scientists, an example of the Matthew effect will appear in Chapter 5 in connection with the discovery of fracture zones.

At best it is difficult to reconstruct the evolution of one idea from another, but most scientists try to lay down some sort of *pro forma* trail in the form of references. They at least go to the library after they finish the research and cite the papers as though they had read them in advance. Therein lies normal polite behavior. To fail to cite known work is dishonest. To tell the truth—namely, that the research was independent of the cited references—smacks of an unseemly concern for priorities. Indeed, if the library shows that the idea and tests have already been published, the proper procedure is to shrug it off and wait for another idea. Some of the most fruitful and imaginative scientists, however, further cloud the evolution of ideas by disdaining the postresearch visit to the library. Albert Einstein was particularly casual about reading the literature even at the start of his career. Consider the following excerpt translated from a paper in 1906 when he was 27:

> It seems to me to be in the nature of the subject, that what is to follow might already have been partially clarified by other authors. However, in view of the fact that the questions under consideration are treated here from a new point of view, I believed I could dispense with a literature search which would be very troublesome for me, especially since it is to be hoped that other authors will fill this gap, as was commendably done by Herr Planck and Herr Kaufmann on the occasion of my first paper on the principle of relativity.[4]

Few scientists expect the likes of Max Planck to tidy up for them, but more than one of the principals in the tale that follows had an attitude about the literature not unlike Einstein's, and that poses problems. Harry Hess was always generous with acknowledgments of critiques and discussions and praise for his colleagues' work, but his formal citations were rare. It was my impression that he wrote most of his papers on trains and trans-Atlantic airplanes and cited mainly from memory.[5] His attitude in his seminal "Drowned Ancient Islands of the Pacific" is an interesting variation on Einstein's reason for not visiting the library:

> Since it is difficult to discuss any theory of origin of guyots against the background of misconception and ill-founded theories which at present confound geologic literature on ocean basins and the Pacific Basin in particular, the writer proposes to wipe the slate clean and start on a new basis.[6]

Guyots are the drowned ancient islands. Hess here gives a splendid illustration of how complex the evolution of ideas can be. He did not cite Darwin's *Structure and Distribution of Coral Reefs* of 1842, but it is hardly conceivable that he had not read that classic book and accepted its proposed origin of atolls by submergence. Yet Darwin notes an elementary corollary that if volcanoes sink in the tropics to become atolls, they must also sink in high latitudes to become drowned rocky platforms. Hess had merely confirmed Darwin, but he apparently had wiped the message off the slate. Can we be sure that he had no residual memory? E. L. Hamilton, as careful a scholar as conscience can conceive, read Darwin and made no note of this point.[7] I edited a reprint of Darwin's book[8] and wrote about guyots for 30 years before noting the point. We all must have read it, but this crucial argument in one of the best known books by one of the real giants of science had no influence on thinking about guyots for over three decades. Or did it?

The evolution of scientific ideas is further complicated by the frequency of independent discoveries in science. These are almost inevitable simply because of the nature of science. At any given time, published data and interpretations are ready to lead to further advances in science. The only question is who will first identify the advance. Considering the delays in publication, several people may be expected to make simultaneous discoveries before the work of one of them receives recognition. Scientists engaged in research and publication rarely read outside the current journals, so any idea that is lost from the current literature is more apt to be rediscovered than not.

The existence of this phenomenon is well known because of a few famous examples. Newton and Leibniz independently discovered the calculus, Darwin and Wallace independently discovered evolution, and so on. However, the frequency of the phenomenon is less appreciated. Merton, the pioneering sociologist of science, calls independent discoveries ''multiples''; they may be simultaneous or spread out and may involve many people.[9] He shows, for example, that the hypothesis of the existence of independent multiples in science and technology was itself independently discovered by at least 18 people between 1828 and 1922 when it came into general recognition. The discoverers included Macaulay, Francis Galton, Friedrich Engels, George Sarton, and Albert Einstein; Benjamin Franklin did not consider himself a discoverer because he thought everyone was familiar with the subject.

The unpublished notebooks of Cavendish and Gauss are full of findings that were later independently discovered and published by others. The phenomenon is so common that when Gauss, aged 18, discovered the method of least squares, it seemed so obvious to him that he assumed it was known. It was left to others to discover it three more times. It is hardly surprising that scientists have adopted institutional expedients to deal with the problem of priority. Since the seventeenth century, sealed and dated manuscripts have been deposited with scientific societies. Early minutes of the Royal Society state that

> when any fellow should have a philosophical notion or invention, not yet made out, and desire that the same sealed up in a box might be deposited with one of the secretaries, till it could be perfected, and so brought to light, this might be allowed for the better securing inventions to their authors.[10]

From the sixteenth through the nineteenth centuries, the curious practice also existed of reporting discoveries in anagrams. The more conventional presentation of an idea or observation in a short abstract has been widely used for centuries. It is especially popular now because abstracts of talks at professional meetings are published much faster than journal articles. Merton concludes,

> These and comparable expedients all testify that scientists, even those who manifestly subscribe to the contrary opinion, in practice assume that discoveries are potential multiples and will remain singletons only if prompt action forestalls the later independent discovery.[11]

Just so, the absence of prompt action by Harry Hess changed the concept of sea-floor spreading from a potential to an actual multiple discovery by Dietz. Just so, the failure of Bruce Heezen to publish his ideas on a world-girdling rift for three years caused the discovery to be shared with Maurice Ewing. Just so, the fact that Jason Morgan gave the first talk on plate tectonics in 1967 is obscured by the fact that his discovery came after he had already submitted an abstract on a different subject. Ideas develop rapidly during a scientific revolution. Discoveries are in the air. Many occur simultaneously.

Biographical memoirs provide another useful source of historical information, and in some respects the historian of science is relatively fortunate. A sizable fraction of the scientists of historical interest are elected to the Royal Society of London, the U.S. National Academy of Sciences, or some other national equivalent. In due course they become the subjects of biographical memoirs 10 or 20 pages long, usually including complete bibliographies. Moreover, all scientists are members of professional societies such as the Geological Society of London or of America, and these societies also publish such memoirs of prominent members. In earlier times they were lengthy, but as membership has grown they have shrunk to little more than half-page obituaries. Even so, there is rarely a problem in determining vital statistics and professional achievements of scientists who have died.

Biographical memoirs are labors of love in that the author, ideally as distinguished as the subject, presumably would prefer to be pursuing research. Obviously no one would undertake such a task who was not prepared to praise the deceased. The tendency to immortalize personal virtues is only natural under the circumstances, but unfortunately it extends to professional activities as well. An example will illustrate the distortion that this may introduce in the historical record and show how, even during the deliberations of composition, the best memories can be selective, and the best minds must judge what should be printed.

Reginald Aldworth Daly (1871-1957) was as eminent as a geologist can be: Sturgis Hooper Professor of Geology at Harvard, member of the National Academy of Sciences, president of the Geological Society of America, and recipient of its Penrose Medal—the most prestigious award in American geology. I am a great admirer of Daly, who although retired still graced the Harvard campus when I was a graduate student. He was astonishingly versatile. He deduced, in a flash after a decade of cogitation, how magma works its way upward through the crust by stoping; he elaborated the implications of isostasy for the strength of the earth; he showed how the swinging sea level of glacial times determined the geography of atolls; he demonstrated that submarine canyons could be cut underwater by turbidity currents; and so on, as we shall see. Daly was honored with biographical memoirs[12] that describe his many achievements and how they were based on his very extensive geological mapping and travel. For example, they record a "protracted visit to the southern hemisphere" in 1920-1921 on the Shaler Memorial expedition. He spent two months mapping the isolated islands of St. Helena and Ascension. (Daly was fond of remarking that Napoleon would have enjoyed his stay at the former more if he had been a geologist.) He went from the Atlantic islands to South Africa where he was accompanied, at times, by Charles Palache, G.A.F. Molengraaff, and A. L. du Toit. Their main objective was the Bushveld complex of igneous rocks, but they must have seen much of South African geology and discussed its implications. Molengraaff already believed that Wegener was right about continental drift and that South America, Antarctica, Australia, and India had formerly been joined to South Africa and had been scoured by the same glaciers in Permian time. Du Toit was on the way to becoming the chief disciple of Wegener's theory. We can easily speculate on one of the subjects of their discussions in the evenings on the veldt.

The biographical memoirs deal sparingly with the next decade of Daly's life. They mention that among his many books *Our Mobile Earth* was published in 1926 and that it is based on a series of lectures at the Lowell Institute. One memoir states that in 1922 after the Shaler Memorial expedition, "Daly's research and interest now entered a new phase" and then leaps to 1932 when Daly initiated the Harvard program in experimental high-pressure research on rocks and minerals.[13]

Our Mobile Earth was a comprehensive exposition of the evidence for and existence of continental drift. It differed from Wegener's ideas largely in the driving mechanism for drift, but the differences were enough for the book to be reviewed in Germany under the title (in German) *New American Theory of the Origin of Continents and Oceans.*[14] The contents of the book and Daly's early commitment to continental drift are unmentioned not only in the lengthy biographical memoirs but also in his biography in the *Encyclopedia Britannica* or anywhere else that I can find. The biographical memoirs were published in 1958 and 1960. Sea-floor spreading was confirmed by the symmetry of magnetic anomalies in 1966. One wonders what might have been written if Daly had lived another decade.

Regardless of the volume or character of the biographical information about deceased scientists, it cannot solve a basic problem for the historian. Most scientists who ever lived are still alive. Derek Price demonstrated this by a simple calculation.[15] The output of scientific papers for the last 200 years is known to be about 6 million, and the average output per scientist is about three papers per year. Thus the number of scientists is known. Analysis shows that the number of papers (and scientists) has doubled every 15 years for two centuries. Let us say a scientific career lasts 45 years. At the end of a career, of each eight scientists who ever lived, four will have started 15 years ago, two started 30 years ago, and only one of the remaining two was active 45 years ago. Thus now or at any given time 87.5% of all scientists who ever lived are alive. In rapidly expanding fields such as molecular biology or particle physics or marine geology the doubling time is only from four to five years.[16] In such circumstances the number of scientists expands from 1 to 1,024 in 45 years, so 99.9% of them are alive at a given time.

Harry Hess is gone. So are Bullard and Ewing. Wegener died on the Greenland ice and Heezen in the abyss; but most of the people involved in this history are still alive. Biographical information about living scientists who have distinguished themselves in published research is abundant. Moreover, working scientists correspond with others similarly inclined, exchange reprints, and talk by telephone or at professional meetings. They go to sea or spend weeks in the field with colleagues and with students who grow up to be colleagues. I have survived almost eight doubling periods of marine geology and tectonics, so I have been fortunate enough to know almost everyone involved in the recent scientific revolution. My involvement may present some problems for this history. The principal problem is that my perception of what is important may not be correct. Even if it is, a risk exists that the emphasis will favor those activities that I know best because I participated in them. In addition, I shall be focusing on some people with whom I interacted frequently and shall neglect equally important people with whom I did not converse so often. A less involved historian might present a more even treatment. I tell it as I remember it. If I know of anyone else who has remembered it differently, I shall say so. Moreover, I have taken every opportunity to question my friends, colleagues, and professional acquaintances about their memories or understanding of the events recounted here. This has not been a process of passive recording but rather of pointed, iterative discussion. We have compared our ideas on who did what, when, and why, and why someone did not do something at a logical time, and so on. My correspondents have been generous—even indulgent—in their responses, and I believe that we have all had some surprises as a consequence.

As for units and names, I shall discuss measurements or observations in the units current at the time they were made. In contrast I usually abandon the geographical, geological, and geophysical terminology of the past in favor of a uniform use of current terminology. Thus I hope to avoid the tedious confusion that arises as a consequence of simultaneous discovery or the evolution of ideas.

I shall often indulge myself in the slightly eccentric use of the word "geology" to be equivalent to the more fashionable term "earth science." Half a century ago "geology" was a generic term, equivalent to "physics" or "chemistry," that embraced a wide range of specialties. Some geologists, however, began to define geology very narrowly. Even in the nineteenth century the greatest American geologist, Grove Karl Gilbert, had difficulty in having his work recognized as geology.[17] The premier geomorphologist of the time, William Morris Davis,

> did not criticize the reigning organization of geology, as history, but he attacked its refusal to accept other than paleontologic and stratigraphic criteria. He campaigned unceasingly for the legitimization of "physiographic evidence," and the reluctance of conservative geologists to accept the new data was a major reason why he, and Gilbert, often labeled themselves geographers rather than geologists.[18]

Thus it was that physicists and chemists who studied the earth tended not to be accepted as insiders. Instead they became separate breeds and called themselves geophysicists and geochemists. Geologists who worked at sea became oceanographers. After World War II, the application of modern physics, chemistry, and mathematics to the study of the earth burgeoned. Caltech, for example, deliberately abandoned vertebrate paleontology and replaced it with isotopic geochemistry. A problem then arose; the outsiders actually refused to become identified as insiders. In order to recruit the new generation of scientists, one geology department after another cravenly renamed itself a department of earth sciences or, even worse, a department of earth and planetary sciences. I hope this trend has peaked and I am merely leading the way toward the restoration of "geology" to its rightful place as a generic term.

I am used to thinking of most of the people who appear in this book by the names or nicknames used in addressing them or referring to them at the time. I use these names in the way that seems natural to me with the hope that the reader will feel immersed in the story rather than embarrassed or confused by familiarities.

I shall be mentioning honors that are awarded to research scientists to the extent that they are related to this history. This might lead the reader to assume that the scientists were directing their activities with the aim of winning the honors. A few of them were. Certainly, James Watson looked on the search for the structure of DNA as a race for the Nobel Prize.[19] Moreover, I remember thinking in passing that somewhere associated with the spectacular fracture zones should be a Penrose Medal. I was right, but I did not surmise who would win it. Ordinarily, however, whether because of uncertainty, modesty, or even shyness, scientists are not thinking of any personal rewards that may derive from their work. The work itself—basic research—is enough.

Research scientists are as single-minded as test pilots or surfers in deciding what counts. Research is what counts. If a research scientist tells his colleagues that he has accepted an appointment as Assistant Secretary of State for Science, they will furrow their brows and ask, "But won't that interfere with your research?" And if, in fact, the scientist at that instant is still engaged in research, he will think "It's only four years, but will I be up to real work afterwards?"

Research is not only what counts; it is all that counts. One indication of a false research laboratory is that it closes on weekends. A false scientist takes holidays. A true research scientist does not even know there is a holiday.

Research is what counts. Holding office in a professional society is not research. Neither is advising government agencies or writing articles for popular magazines. Neither is administering scientific organizations. I had occasion to remark to Dallas Peck that I kept seeing on the walls of our buildings pictures of former Chief Geologists of the U.S. Geological Survey but not of former Directors. He said he believed that the Chief Geologists were more distinguished. I naturally agreed. Most of the Chief Geologists had had famous achievements in research, the Directors less so. Both of us were then administrators, he as Chief Geologist and I as Director, but we could remember what counts.

Consider Joseph Henry of Henry's law. After a professorship at Princeton he was elected the first Secretary of the Smithsonian Institution. He helped found the Weather Bureau. He established the first system for free publication and international distribution of American scientific results. He served as Chairman of the Lighthouse Board, organized scientific support for the Union in the Civil War, and was a prime mover in the founding of the National Academy of Sciences and the American Association for the Advancement of Science. The Congress directed the publication of a *Memorial to Joseph Henry* when he died. And yet, what does the true researcher think of all this?

> He might have been the first and was certainly not the last scientist to sacrifice his opportunity for research in order to serve the United States Government as an administrator.[20]

Research consists of creating some new information that is published in a refereed scientific journal. A hobby is not research. Research is an addiction much stronger than a hobby, so true researchers do not worry about being distracted.

I have said not a word about the quality or wide importance or practical significance of research. These are not really considerations. It is the fact of research that counts. "I cannot," Harold Urey remarked, "talk with a man who is not an expert on something." By that he meant that any man who is actively engaged in research is therefore an expert in some subject, however narrow. Being an expert he knows about the statistical significance of data, uncertainties, errors, the procedures of research, the necessity for critical review—the whole business of being a scientist, regardless of the subject. Harold himself won the Nobel Prize for discovering deuterium in 1934, and he had the ability that distinguishes the best scientists. He could identify important problems that were ripe for solution. After that it was merely a matter of concentration. At a memorial ceremony, a former student, now a member of the National Academy of Sciences, Jim Arnold, told a story. Harold met a colleague at lunch time on the grounds of the University of Chicago in the forties. They chatted about research. At the conclusion Harold asked whether the colleague had chanced to notice the direction in which Harold was walking when they met. When asked why, he said, "Well, if I was coming from that direction it means I was going to lunch, and if from the other direction it means I have had it."[21]

The importance of a problem can usually be assessed by the breadth of the applicability of the answer. Thus, the discovery of radiocarbon dating by Willard Libby had broad utility in archaeology and geology and merited the Nobel Prize. Fortunately, a little luck can always metamorphose the distinguished work of a contented specialist into something unexpectedly fertile. Just so, John West Wells studied fossil corals for decades and one day became the cynosure of geophysicists because his observations of growth rings on Devonian corals could determine whether the tidal friction of the moon was slowing the rotation of the earth. He was elected to the National Academy within the year.

Most research is relatively narrow, has little application to research in other fields, and thus is basically of little importance. A distinction then arises between the quality of research and the assessment of that quality. The chances that quality will be appreciated as exceptional tend to depend on the reputation of the experts in the narrow field, or the visibility of the individual scientist in the right places, or a variety of social factors such as an appreciation of perseverance against odds.

The importance of the reputation of one's fellow experts is readily apparent. In a narrow field (and at the research level what field is not?) only fellow experts are capable of judging one's work. If no one outside the field pays any attention to the experts, whence comes the recognition of important work? One possibility is that a scientist has the right connections and thus his work obtains visibility outside a special research field. The importance of visibility is shown by the frequency of awards. A scientist will do superb research for decades with minimal recognition and then win a string of medals and honorary degrees in a year or two. The problem is how to obtain the first visibility among influential scientists. In this respect, nothing is better than the old school connection, knowing one's fellow graduates who will later grant scientific awards. It was not by chance that at one time most of the geologists in the National Academy were formerly together at Yale. They naturally tended to read and recognize the work their classmates did in later life. The extent of these connections can hardly be exaggerated because only a few universities have the great graduate schools that train most research scientists.

Although it is research that counts, scientists, or at least geologists, feel a warm glow when the virtues of perseverance and courage against odds are also recognized. For many decades, J. Harlan Bretz tried to prove that the scablands of eastern Washington and Oregon were produced by a giant flood caused by the collapse of a glacial dam across the Columbia River. His contemporaries were contemptuous. He fought a long lonely battle, writing paper after paper, and was mostly ignored until at last it was clear he was right and he won the Penrose Medal at age 97. If Alfred Lothar Wegener had been alive in 1966, aged only 86, he would have received standing ovations at every meeting where the symmetrical magnetics were discussed.

I should pause to recognize the existence of the wrong connections. Be he ever so brilliant, a chap who is openly engaged in felonious activities may be passed by when the awards committee meets. The wretch who does the unspeakable—become a public personality prematurely—certainly will be bypassed. Regarding popularity, timing is of the essence. The true research scientist sees no point in popularity. He is not ungenerous, however, if a Linus Pauling or a Harold Urey becomes a public personality. He assumes they did their best to avoid it. After all, it must interfere with their research. But woe betide the upstart who takes to popular lecturing and political endorsements before becoming a tried and true research scientist. He must actually enjoy that sort of thing. Thus it is that most of the scientists best known to the public are not the ones assessed generously by their colleagues.

I should pause again to mention style. Among any mutually recognized group of professionals style is everything, and in science it is intuition, boldness, modesty, brilliance, elegance, and wit. P.A.M. Dirac conceiving antimatter in order to balance his equations had style. Style leads to stories. Who, among those present at the time, could ever forget when Walter Munk was talking to a small seminar at Scripps about a century of ideas on the rotation of the earth? He wrote a long equation on the blackboard, described its ancient and honorable antecedents and the history of its use by eminent scientists, and began "And for over a century no one . . ." when Carl Eckart spoke from the floor. "Walter, the sign in the third term is wrong." Walter continued smoothly, ". . . but Carl Eckart ever realized that the sign in the third term is wrong." That was style twice over. For the most part scientists appreciate elegance, meaning "pleasing by ingenious simplicity," in the identification and solution of scientific problems. Brute force may solve a problem and win a Nobel Prize without being elegant. When a speaker at a national meeting of mathematicians writes an equation on a blackboard and the audience bursts into applause, that is an elegant solution. That is style.

The creation of ideas, the discovery of new relations, these are lonely pursuits whether in art, literature, or science. They are doubly lonely, for the creator always wonders if the creation is meritorious—and who is to judge? The critics. No wonder creators hate critics. Let two poets, the most sensitive of creators, speak for all:

A man must serve his time to every trade
Save censure—critics all are ready made.
—Lord Byron, *English Bards and Scotch Reviewers*

They have tried their talents at one or the other and have failed; therefore
they turn critics.
—Samuel Taylor Coleridge, *Lectures on Shakespeare and Milton*

How the artist and the playwright must envy the physicist whose work is criticized, or rejected, by working creative scientists—criticized, moreover, before publication. Thereafter, the scientist usually has little to fear of criticism
except for the silent but deadly one of having his work ignored. Fortunately,
again, scientific research is the one creative activity in which a large number of
people can make a living. Thus there is an excellent chance that someone will
read the work. But will it be used? Will it be worth the effort? Will it be important enough to be incorporated in the ever-expanding structure of science?
The answer may be long in coming, but it is determinable because the paper,
if used, will be cited.

In fact, most papers probably are not cited. The 16,224 publications of
3,078 geologists have been correlated with the citations in the prestigious *Bulletin of the Geological Society of America* from 1888 to 1969.[22] Fully 78% of
the geologists were never cited. All that work and no citations. In contrast,
only 2% received 65% of the 6,646 citations. This detailed study shows that
citations are numerous but rare enough to be a form of critical praise. It also
shows that a disproportionally small fraction of scientists are well known and
are the likely candidates for further signs of approval and acclaim.

A few scientists receive medals from professional societies, are elected to
honorary academies, and occupy endowed chairs. The perception of the honor
by the recipient and colleagues often has little or nothing to do with money—
if indeed any money is involved. What counts is the open demonstration that
one's research is meritorious in the eyes of those who count. Even more important is the fact that an award implies that the new winner merits association
with the pantheon of previous winners. Thus in many ways the oldest awards
are the most desirable. How sweet to be the Lucasian Professor at Cambridge
knowing that Sir Isaac Newton was your predecessor. How Charles Darwin
must have appreciated the Copley Medal of the Royal Society knowing that it
had been awarded a century before to Captain James Cook for his research on
the prevention of scurvy. When Teddy Bullard won the Wollaston Medal of the
Geological Society of London, he became linked with the previous winners including Darwin, Murchison, Lyell, Dana, Suess, and Gilbert. When Maurice
Ewing received the Penrose Medal from the Geological Society of America,
he followed Daly, Coleman, Willis, Vening-Meinesz, Holmes, Hess, and

Tuzo Wilson. I include these long lists because all these people and many more appear in this book. It should be remembered that whatever their attitudes about continental drift or plate tectonics at various times, they were not critics of the sort detested by Byron and Coleridge. In their own fields they were leaders in research. Harold Urey could have talked to any of them. They all counted. They all knew what counts.

Chapter 2

THE DREAM OF A GREAT POET

A sound almost imperceptible, so slight, so little different from silence itself, of continents *en marche*, which slowly, oh very slowly, as great pontoons floating on the calm waters of a port, or as great icebergs borne by the polar currents, are drifting toward the equator.

—Pierre Termier, Director, Geological Survey of France
Annual Report of the Smithsonian Institution, 1925

Whereas a palace revolution is recognizable by the gunfire, a scientific revolution is recognizable by the language, whether it be in newly coined words or in the symbols of a new mathematics. Data and hypotheses are expertly discussed for decades in an expanding special vocabulary. Then one day a scientific revolution occurs, and the hypotheses, vocabulary, and experts are swept away. New experts appear among the few hardy survivors of the old, and they generate the vocabulary that characterizes the new theory.

The revolution that included continental drift, sea-floor spreading, and plate tectonics was no exception, but the timing was odd. The innovator Alfred Wegener, who had every right and opportunity, coined not a word, but that is not unusual. It is the followers who need new words. Barring a spirited try by S. Warren Carey[1] with his *oroclines* and the like, the new vocabulary did not arise until *transform fault, spreading center, plate, subduction, hotspot*, and so on revolutionized terminology in the 1960s—half a century after Wegener mercifully spared us. The vocabulary of the early decades of discussion of continental drift was novel not because it was special to the subject but because it was exceptional for science at the time. Flowery phrases, denunciations of method and motives, accusations of bias and advocacy are not the language of twentieth-century science. They can be revealing, however, to the historian of science. The history of these early decades has been well studied, and more books and papers will follow, so it is treated only in the abstract here.[2]

THE CONTRACTING EARTH

As the nineteenth century approached its end, the science of geology rested on a reasonable, satisfactory, and generally accepted theory of global tectonics.[3] The paleontologist cleaning trilobites, the mineralogist measuring crystal angles, or the field geologist hammering rocks could feel as confident about the foundations of the earth sciences as they feel at present. The theory was different from what it is now and it was wrong, but that made it no less satisfactory. Facts that were inconsistent merely provided the problems that needed to be solved. In short, the earth is very old; it was initially a hot liquid sphere and the interior was gravitationally stratified into a dense core and less dense mantle; a crust formed, and as the earth continues to cool and contract this crust is subjected to compression; it adjusts to the decreasing size of the mantle by fracturing into blocks that subside from time to time; the higher blocks are eroded, shedding sediment into the lower ones; in time, perhaps by chance, some of the highs become lows and the lows are exalted.

This theory was in pleasing accord with the grand concepts of contemporary cosmology and was capable of an integrated solution to diverse geological observations. Layers of sedimentary rock 10-12 km thick are exposed in mountain ranges. They accumulated in subsided blocks that then became elevated. Earthquakes in the Basin and Range Province in the western United States produce steeply dipping faults that bound many mountain ranges. Naturally so; they are the boundaries of still subsiding blocks. In contrast, many faults in the Alps, the Scottish Highlands, and the Appalachians are almost horizontal and stacked one above another. The horizontal movement in each range is enormous and certainly extends from tens to hundreds of kilometers. Crustal shortening is further confirmed by folding at the peripheries of the overthrust regions. These facts are wholly explained by the contraction theory because, inevitably, parts of the crust must override other parts in order to fit on the shrinking interior. Indeed, for this is the nature of normal science; one need merely sum the measurable shortening shown by all the folds and thrust faults in order to calculate the crustal shortening and, thus, the radial shortening.

The stratigraphers and structural geologists were not the only ones content with this theory. The paleontologists found countless occurrences of identical or closely related fossil species of land and shallow water organisms in places such as Africa and South America that are separated by broad expanses of deep water. This was readily explained if parts of the deep oceans had formerly been continents connecting the fossil localities. They were called "land bridges," and as paleontological sampling advanced, they proliferated—to growing embarrassment. For example, see Charles Darwin's letter to A. R. Wallace of 5 June 1876 regarding the latter's book *Geographical Distribution*, which Darwin was reading:

> The point which has interested me most, but I do not say the most valuable point, is your protest against sinking imaginary continents in a quite reckless manner, as was stated by Forbes, followed, alas, by Hooker, and caricatured by Wollaston and [Andrew] Murray![4]

Darwin was writing a private letter but he became a public figure, and it was published in 1887 in the *Life and Letters*, edited by his son Francis. Thus it may not be coincidence that A. P. Coleman (1852-1939) used similar terminology in 1916 in his Presidential Address to the Geological Society of America:

> other geologists especially paleontologists . . . display great recklessness in rearranging land and sea [for] the convenience of a running bird, or of a marsupial afraid to wet its feet. . . .[5]

In addition to uneasiness about the number of land bridges, there was a growing belief that continents and ocean basins are permanent features of great antiquity. This was not viewed as necessarily incompatible with the contraction paradigm, but it implied that if a continental land bridge had existed it should be detectable as anomalous crust in the deep sea. The belief became incompatible if the principle of isostasy was accepted, but the conclusive proofs of isostasy were yet to come. Nonetheless, the principle was known, and the growing number of measurements of gravity were demonstrating its reality. If the level of a continent or ocean basin is a function of its density, it is not possible to push a land bridge down to the level of the deep sea floor or to elevate the sea floor to become a land bridge. Earth scientists, not thinking of themselves as members of a community with common interests, preserved the contracting-earth theory in the normal way by specialization. The geophysicists talked about isostasy and the paleontologists about intercontinental similarities of fossil assemblages.

WEGENER—THE EXPLORER AND METEOROLOGIST

Alfred Lothar Wegener (1880-1930) was a very trusting man. When he read about isostasy he believed that the geophysicists were right and that continents cannot be changed into ocean basins. When he read about fossils he believed that the paleontologists were right and that land species in Africa had had a way to reach South America. What is truly remarkable is that, unlike any contemporary, he managed to believe these things simultaneously. Guided by the obvious match between the Atlantic shorelines, it was but a step to continental drift.

Wegener was born in Berlin and studied at Heidelberg and Innsbruck.[6] During his undergraduate years he was active in mountain climbing and skiing and interested in meteorology, particularly in the new techniques of aerology. He took his doctorate in astronomy at Berlin in 1905 with a thesis on "The Alphonsine tables for the use of modern computers," meaning people. In 1906 he returned to aerology in an extraordinary way. He and his brother Kurt, in the service of the Prussian Aeronautical Observatory, entered and won the

Gordon Bennett Contest for Free Balloons. They set a world record by staying aloft for 52 hours, breaking the old mark by 17 hours.

Wegener was rewarded in 1906 with an appointment as meteorologist on the Danish Danmark expedition led by Mylius-Erichsen to northeastern Greenland. Thus began the lifelong dedication to exploring Greenland that would raise his hopes for a proof that continents drift, and that would eventually leave him dead on the ice and his theory in ruins. He spent two winters on the first expedition and made journeys of hundreds of miles by dog sled, investigating glaciology and astronomy as well as meteorology.

In 1908 Wegener began an academic interlude of four years at Marburg as a popular lecturer and demonstrator on meteorology and astronomy. He had time to work up some of his Greenland observations and publish a book on the physics of the atmosphere before returning to Greenland.

The second expedition of 1912-1913 was a logical continuation of the first in the normal style of scientific exploration in which each field effort raises more questions than it settles. His co-leader, Captain J. P. Koch, was a friend from the 1906-1908 expedition. He wintered on the high Greenland glacier, continued his studies of weather and ice, and ultimately made the first scientific crossing, 700 miles, of the glacial core of the island.

Only 33 years old, Wegener had already distinguished himself with extraordinary feats of ballooning and glacial exploration, published a book on meteorology, and was launched on a career of research and teaching. Shortly after returning from Greenland he married Else Köppen, daughter of the famous meteorologist Wladimir Köppen, who originated the standard classification of climates that is still in use. Little more than a year later, he was called up as an officer in the general mobilization that accompanied the guns of August. He was wounded twice but remained in the army to the end of the war. On one long convalescent leave he wrote a book to which we shall return. His military duties and his wounds, however, could not wholly stifle his drive to do scientific work. He published from two to seven scientific papers every year from 1914 to 1920 as well as *The Origin of Continents and Oceans* in 1915.[7] Not much can be said for trench warfare, but static fronts make for circumstances that allow creative activities much more frequently than the mobile fighting of World War II. Of course they rarely compare with the opportunities of a naval officer, such as Harry Hess, who had not only a clean office that floated around with him but also an echo sounder to discover guyots.

Wegener returned to Marburg after the war but shortly succeeded his father-in-law as Director of the Meteorological Research Department of the Marine Observatory in Hamburg. There he was able to pursue science with his customary vigor and made contributions on a wide range of geophysical subjects. In 1924 he published, jointly with Köppen, a book on the geological history of climatic change that spanned his interests as much as anything could. Despite his growing fame, he was rejected more than once for a regular chair at a German university on the predictable grounds that his interests were not confined to a single academic discipline.[8] In 1924 a professorship in meteorology and geophysics was created for him at Graz in Austria. There he remained until April 1929 at which time he sailed for a summer reconnaissance in Greenland. The purpose was to find and prepare sites for an ascent by heavy equipment to the ice cap in 1930-1931.

Wegener departed on his fourth and last expedition on 1 April 1930. He had organized and prepared a full-scale geophysical expedition with the complex logistics of polar exploration. The work progressed—from catching and drying shark meat for the dogs to maintaining the propellor-driven sledges that he was introducing to Greenland. The first road on the island was completed, the under-ice bases outfitted for the winter, and the scientific work underway. Wegener broke free at last from his administrative duties on the coast, and with two companions he took provisions to the Eismitte Station at 3,000 m elevation. It was a hard struggle because, in late autumn, the snowfall was heavy en route and the temperature $-50°C$ when he arrived on 30 October. Georgi, who was there, describes the "joy shining in his face" as he inquired about their work and plans for the winter.[9] Two days later "on his 50th birthday, fully fit and with rested dogs" Wegener and one companion, a Greenlander, Rasmus Villumsen, set off for a relatively quick and easy trip downhill to the coast. They never arrived.

On 7 May contact between the coast and Eismitte Station was reestablished, and it was realized that Wegener had not wintered in either place. His body was found almost exactly halfway between the camps. His skiis were upright in the snow, three meters apart, with a broken ski pole between them. Under the snow, his clothed body was lying between sewn sleeping bag covers, covered with furs. His body showed no ill effects of hunger or exposure. It was deduced that he had had a heart attack in camp, perhaps from overexertion caused by his practice of skiing beside the running dogs rather than being a load in a sledge. The search party buried him as they had found him in the snow and built a monument of snow blocks above. Traces of a sled and dogs led a few miles to the west, but the body of the faithful Villumsen was never found.[10]

Thus in heroic circumstances ended the life of a famous balloonist, geophysicist, meteorologist, and polar explorer. Biographical memoirs appeared in 1931 in the *Gerlands Beiträg zur Geophysik, Petermanns Mitteilungen*, the *Meteoroligische Zeitschrift*, and doubtless in the popular press, thus bringing to an appropriate close one phase of a highly successful career. As to the other phase, development of the theory of continental drift, the story was rather different at his death. He was not memorialized in the annual address of the president of the Geological Society of London in 1931, although the passing of Pierre Termier, who had ridiculed him, was. But then the Society had honored the Director of the Geological Survey of France before. Wegener's death was hardly even mentioned in the "Deceased" section of "Personals" in the *Geologische Rundschau*; the obituary merely stated, "'Prof. Dr. Alfred Wegener (Graz) the founder of the theory of continental drift, on his expedition to Greenland."[11]

WEGENER AND CONTINENTAL DRIFT

What went wrong? Why was Wegener's great contribution to geology slighted? We now turn back to that phase of his career devoted to continental drift. Perhaps we should begin with the theory of continental drift even before Wegener became interested. As mentioned earlier, once someone shows that an idea is important or that a hypothesis explains certain published data and the corollaries are important, it is easy to go to the library and discover that other people already had similar ideas or hypotheses. This is such a popular pastime following most scientific revolutions as to suggest that these discoveries are inevitable. They can be ignored, but they invariably crop up when one stumbles upon them in the older literature. Clearly, when published, they were not considered important; usually, in fact, they are not quite the same as the new idea, but they exist. In sum, it had not escaped notice that the margins of the South Atlantic match, and this can be explained if Africa and South America had once been joined. Moreover, the Paleozoic geology on the two sides of the North Atlantic is strikingly similar—and runs out to sea. The amount of informal speculation that the trans-Atlantic congruences generated is suggested by the fact that the distinguished British geologist, E. B. Bailey, lectured on continental drift to a nontechnical audience at the Old Vic theater in 1910.[12]

Even within standard geological literature, W. A. Pickering noted the match in 1907,[13] but to suggest how the rest of the paper influences acceptance of any part of it he attributed the separation to the origin of the moon. Likewise, F. B. Taylor (1860-1938) after publishing 41 consecutive papers on surficial geology of the northeastern United States turned his interests to mountain building by continents drifting toward the equator.[14] His speculative discourse was similar to and earlier than Wegener's first publication on drift. Thus in the 1920s and 1930s in the United States at lest, continental drift was called the "Taylor-Wegener hypothesis." Taylor himself was disturbed by this coupling because the two hypotheses are not identical, and he wrote in a private letter, "Wegener was a young professor of meteorology. Some of his ideas are very different from mine and he went much farther in his speculations."[15] Those who generate hyphenated authorships for scientific theories might take note.

Wegener, functioning like a normal research scientist, apparently knew nothing of these publications when he began to think about continental drift.

His friend J. Georgi says that another friend, W. Wundt, personally recalls that Wegener pointed out the congruence between the African and South American coasts to him in 1903 before Pickering or Taylor published.[16] This may actually be true because the match is obvious (Fig. 2 A) and we remember different things; Wegener himself says, however, that he noticed the congruence first in 1910 but dismissed it as improbable.[17] He learned of the paleontological evidence of former land connections across the Atlantic in 1911 and talked about continental displacement in January 1912. He was soon to leave for Greenland, and like any sensible scientific explorer he was apparently more concerned about the loss of his research than about the loss of his life that might occasion it. In any event he left his first manuscript on continental drift with a scientific journal that published it in two parts while he was in Greenland. It was soon summarized in the *Geologische Rundschau*, which gave him such a short obituary 18 years later. Thus immediately it was widely available in Europe and among leading research geologists throughout the world. There were few geological journals then, and almost every prominent research geologist read English, French, and German in order to understand global geology. Even in 1947 I had to pass proficiency examinations in French and German as a doctoral requirement in geology at Harvard. It goes without saying that the recommended reading for geological vocabulary was from Edward Suess's *Das Anlitz der Erde* and its French translation, rather than Wegener's German original and French translation that said that Suess had missed the point. In those days the other students and I regularly perused the *Geologische Rundschau*. Fortunately the plate tectonics revolution was to be in English.

During his long convalescence from wounds in 1915, Wegener had enough time—and determination—to compose a book of 94 pages, *Die Entstehung der Kontinente und Ozeane*, which was published in the same year. This style of publication was not unusual at the time in a less competitive world. Today, books are published too slowly to be suitable media for revolutionary ideas. Instead they are abstracted to 900-word articles in *Nature*, like the one announcing the structure of DNA by Watson and Crick with cryptic sentences such as "It has not escaped our notice that the specific pairing we have postulated immediately suggests a possible copying mechanism for the genetic material."[18]

Wegener published a second, expanded edition in 1920 and a third in 1922, which was translated into English, French, Russian, and Spanish. The fourth revised edition of 1929 was reprinted in German in 1962 and translated into English in 1966. Meanwhile, after Alfred Wegener's death, his brother Kurt issued a fifth revised edition in 1936. Although a remarkable record, it hardly matches the five printings in one year of the first edition of Darwin's *Origin of Species*.

What does *The Origin of Continents and Oceans* say? I shall discuss and compare the third (1922) and fourth (1929) editions, the former because it was the center of the controversy and the latter because it was Wegener's last response to his critics.

The third edition begins with an introduction by John Evans, C.B.E., F.R.S., President of the Geological Society of London. It is unusual because Evans did not believe in many aspects of continental drift but wanted to assure that Wegener should be "allowed to state his case in his own way." This was hardly necessary by the time of the fourth edition and was omitted.

In the fourth edition, Chapter 2, Wegener states the nature of the drift theory and compares it with older theories, highlighting their weaknesses. Faunal connections indicate former land bridges where oceans now exist. Folding and faulting of mountains indicate large-scale horizontal movement of crustal blocks. The blocks are in isostatic balance. The simplest explanation of all these observations is continental drift. The transoceanic continental margins in many places are congruent.

The next five chapters present supporting evidence for drift. Chapter 3, on geodetic observations, has been moved up from being Chapter 7 in the third edition. This is becausse of the critical importance of such evidence in Wegener's version of continental drift. It was, in fact, the only evidence related to his theory that he was personally involved in collecting. This in itself would have made it important to him. Some earth scientists believe in God and some in Country, but all believe that their own field observations are without equal and they adjust other data to fit them. Moreover, geodetic observations gave Wegener's theory a unique power; to his mind

> Compared with all other theories of similarly wide scope, drift theory has the great advantage that it can be tested by accurate astronomical position-finding. If continental displacement was operative for so long a time, it is probable that the process is still continuing, and it is just a question of whether the rate of movement is enough to be revealed by our astronomical measurements in a reasonable period of time.[19]

What was the rate? Wegener believed that the youngest glacial moraines in Greenland and Europe had been connected (Fig. 1 B) and had drifted apart, so the North Atlantic had opened at an average rate of tens of meters per year for the last 50,000 years. His estimate was 1,000 times too fast, so his tests were doomed to fail, but he believed that repeated observations of longitude in Greenland and Europe had a chance of proving the reality of drift. When his first ideas on drift were emerging, Wegener recalled that longitude had been determined in 1823 and 1870 in the area of his expedition of 1906-1908, and that it had been remeasured during the expedition. He wrote to the cartographer of the expedition, J. P. Koch, who was to be his co-leader on the next expedition in 1912-1913. The letter explained the reason for Wegener's inquiry and asked if the field observations had been worked up. They had not, but Koch made some preliminary calculations. We can imagine the excitement in Marburg when he confirmed a shift in longitude of the expected order. Wegener had fulfilled the dream of every scientist. He had conceived a spectacular hypothesis, identified a quantitative test, and had it confirmed! From then on, from 1910, he must have been almost certain he was right. Later, critics would accuse him of unscientific methods because he selected only one of several existing hypotheses to explain a certain set of geological data and then claimed the data supported continental drift. Even if this were true, it would hardly be surprising if he thought that he already had quantitative and experimentally repeatable proof of drift.

But did he? The longitudes were based on observations of the moon, or lunars, which was the method used by Captain Cook but which has long since been replaced by chronometers and radio signals to time the astronomical observations. Wegener's doctorate, however, was in astronomy, and he certainly understood the errors that might arise. In any event the problem was identified, and the new methods of observation were capable of confirmation if Wegener's estimated rate of 10-30 m/yr was correct. In 1922, the Danish Survey Organization made new modern observations and repeated them in 1927. The fourth edition has a quote from a then unpublished report on the results. They were all Wegener could hope for; the modern instruments indicated a displacement of 36 m/yr. Ecstasy. The amount was nine times the mean error. "The result is therefore proof," said Wegener, but he added a cautionary note about a "most improbable" error that could invalidate the result. In any event it was wrong.

Chapter 4 is concerned with geophysical observations confirming isostasy and the fundamental differences between continents and ocean basins that prevent one from changing into the other. It is directed toward the conclusion that the continental crust is stronger than the oceanic crust and thus can drift through it. This was wrong.

In Chapter 5, at last, Wegener discusses geological evidence. The trends of geological structures, such as mountain ranges, are broken at the Atlantic margins but fit together if the effect of drift is removed. The mountains of the Cape of Good Hope are an example. The late Pleistocene moraines of the North Atlantic are another of his examples (alas). The Mid-Atlantic Ridge consists of continental debris thinned by stretching. Much of this was wrong.

Chapter 6 concerns paleontology. The evidence of former land connections is compelling; "we are justified as counting as favorable to drift theory all biological facts which imply that at one time unobstructed land connections lay across today's ocean basins."[20] This would be hotly contested.

Chapter 7 shows that drift gives a relatively simple explanation of many puzzling features of paleoclimatology; on this subject, at least, Wegener was an expert. What is most notable, the powerful evidence for late Paleozoic glaciation in South Africa, South America, India, and Australia can be explained by a single continental glacier when the congruent continental margins are fitted together. Futhermore, as shown in Chapter 8, Paleozoic glaciation requires polar wandering. Not only do the continents drift relative to one another, but the whole crust drifts relative to the pole of rotation.

Chapter 9 addresses the displacement forces. Because the center of mass of a continent does not coincide with the center of buoyancy, rotation of the earth produces a force directed toward the equator, a "Polfluchtkraft." This causes east-west mountain ranges by compression. Solar and lunar tidal forces exert a pull on continents that causes them to drift westward at rates that vary with their size. This causes the opening of the Atlantic. The westward drift compresses the west side of South America and produces the Andes. The curve of the tip of South America is a consequence of the two types of forces. Wegener did not propose these forces; he merely adopted them for his purposes from the published literature. The astute American geophysicist, W. D. Lambert, confirmed the existence of the Polfluchtkraft and that it would produce a continental rotation if the earth has a classical viscosity and will yield to the smallest force if applied long enough. On the other hand, the amount or nature of the viscosity was not established, and Wegener quotes P. S. Epstein as confirming the certainty of continental drift caused by the Polfluchtkraft.[21] Unfortunately Epstein did this by assuming that Wegener's drift velocity of 33 m/yr is correct, which gave him a viscosity five orders of magnitude too small according to present values. Wegener's mechanism was wrong.

In the final two chapters, Wegener assumes continental drift and moves on to its corollaries, the geological effects on land and sea. He discusses transgressions, regressions, different types of faulting and folding, and the origin of submarine canyons, fjords, island arcs, and the depth of the ocean; then the history and evolution of the Afar triangle, the Gulf of California, Indonesia, the entire continental crust, the Atlantic basin, the Seychelles, Fiji, Madagascar, and India. A page here, a paragraph there, and Wegener, right or wrong, presents enough ill-documented ideas to enrage almost every specialist who was not already enraged by the chapters on geology, paleontology, and paleoclimatology.

All in all, Wegener offered a theory that was wrong in many aspects but that had an extraordinary power for synthesizing the highly diverse facts collected by a century of field geology and integrating the whole with the new data of geophysics. As early as 1911 he wrote to his father-in-law, Köppen, "I don't believe that the old ideas have more than a decade to live."[22] What went wrong?

EARLY RESPONSE TO CONTINENTAL DRIFT

"Wegener's suggestion that the continents have drifted apart has met with acceptance in Europe on the part of scientists whose views we cannot but receive with respect and reasonable consideration" and the suggestion was entitled to a fair test, wrote Bailey Willis.[23] The theory of continental drift as proposed by Wegener received such a test. What went wrong was that the basic framework failed at several points and the whole structure collapsed. It would be difficult to claim that the test was not fair. An impressive number of the leading earth scientists of the 1920s wrote about continental drift and participated in symposia or informal discussions at professional society meetings. The British Association had a lengthy discussion as did the Royal Geographical Society; the Washington Academy of Sciences and the Geological Society of Washington had a joint symposium. The American Association of Petroleum Geologists had a lengthy symposium. There was a discussion at an annual meeting of the Geological Society of America. The *American Journal of Science* published a series of papers amounting to a symposium. Papers on the subject appeared by H. S. Washington, F. B. Taylor, W. D. Lambert, J. W. Evans, Charles Schuchert, Chester Longwell, Pierre Termier, Beno Gutenberg, R. A. Daly, A. P. Coleman, Harold Jeffreys, R. T. Chamberlin, G.A.F. Molengraaff, William Bowie, Joseph White, and E. W. Berry—as diverse a group of distinguished individuals as ever examined a geological hypothesis. Moreover, they were not all opposed; Daly, at Harvard, and Gutenberg, the future head of the Seismological Laboratory at Caltech, supported continental drift, and several others were neutral.

The fairness of the test was not flawed, to my mind, by an assault from the old establishment or, as was then occurring, by an attack on nationalistic grounds. Some of the attackers were old; in 1925 Coleman was 73, Willis 68, Schuchert 67, and Termier 66; but Washington was 58, Berry 50, and the implacable Jeffreys only 34. Longwell, reasonable and agnostic on this issue, was 38. On the supportive side, Wegener himself was 45, Daly 54 (and just elected to the NAS), and Gutenberg 36. Everyone except Termier and Wegener was in or would be in the Royal Society or National Academy, and thus both age and honors were divided. As to anti-German reactions, they were not unheard of in the aftermath of the Great War. The theory of relativity was attacked by an American physicist on the grounds that it was German science. The attack, however, was isolated, and I cannot find any positive statement regarding a similar attack on Wegener. Nonetheless, the possibility of anti-German thoughts cannot be wholly eliminated. Bailey Willis referred to "abilities of the Germans, such as composing fairy tales" when talking about continental drift after World War II.[24] Not long after in a discussion at a meeting of the Geological Society of America he referred to Beno Gutenberg, whom he had known or known of for two decades, as "Herr Gutenmeyer."[25]

When Wegener's ideas burst upon the entire geological world in 1922-1923, geologists on the whole were receptive. At the Royal Geographical Society meeting in 1923, G. W. Lamplugh said of Wegener, "we should like him to be right." At the end of the meeting, the society president, the Earl of Ronaldshay, remarked that "geologists, as a whole, regret profoundly that Professor Wegener's hypothesis cannot be proved to be correct." In short, the scientists of the time were behaving normally. They were excited by new and imaginative ideas, but such ideas are cheap and they awaited proof. Fortunately, they reasoned, the redeterminations of longitude would provide proof within a decade.

Meanwhile, experts examined the elements of the theory touching on their expertise. The young Harold Jeffreys, emerging as the geophysicist of the age, demolished Wegener's driving mechanism in 1923 and 1924.[26] Both the Polfluchtkraft and Westwandering forces are real but too small by three orders of magnitude to have any effect. From then on Wegener's version of continental drift was criticized as mere empiricism without theoretical support. As to the details of his geological, paleontological, and paleogeographical evidence, Wegener was a sitting duck. Philip Lake said of Wegener, "Whatever his own attitude may have been originally, in his book he is not seeking truth; he is advocating a cause and is blind to every fact and argument that tells against it."[27] Schuchert notes how easy it is for Wegener to make all facts fit his hypothesis because "he generalizes on the generalizations of others" and regards the "correlation of formations by geologists as dealing with 'relatively trifling differences of time.' "[28] E. W. Berry objected to Wegener's method because "it is not scientific, but takes the familiar course of an initial idea, a selective search through the literature for corroborative evidence, ignoring most of the facts that are opposed to the idea. . . ."[29] He indicated that "American geological philosophers" also followed this course. A careful reading of *The Origin of Continents and Oceans* makes it very difficult not to accept these comments as just. The book is much closer to Velikovsky's *Worlds in Collision* or Heyerdahl's *Aku Aku* than it is to normal science. As I have said, I believe that Wegener probably was convinced of the correctness of his hypothesis when his quantitative prediction of a change in longitude apparently was confirmed in 1910. Thereafter, it was only necessary to seek confirming data among the welter and confusion of apparently conflicting geological facts. The fact that the core of a hypothesis is correct does not mean that the scientific method was used to prove it.

R. A. Daly, who believed in continental drift, was particularly concerned by Wegener's method: "So obvious are his logical inconsequences and his failure properly to weigh ascertained facts that there is danger of a too speedy rejection of the main idea involved."[30] The basic arguments for drift were so attractive that Wegener might well have achieved more modest goals, but he wanted it all. He could have emphasized the simple beauty and integrating power of the theory without insisting that it offered the *only* explanation for any *particular* observation. It did not, and the experts quickly proved it. He could have begun by saying an empirical theory can be useful and acceptable without anyone knowing why continents drift. Certainly, half a century later, plate tectonics would be accepted as proved before the cause of plate motion was known. He could have said that repeated measurements of longitude might prove that drift is very fast, but if not they might not be accurate enough to prove that continents are not drifting slowly. He could have focused on gaining acceptance of the basic theory instead of casually tossing off the explanation of every puzzle that was vexing geologists. But he did not, and not even the great prestige of Reginald Aldworth Daly could rescue the theory of continental drift from Wegener's hubris.

The death blow to continental drift as proposed by Wegener, in my opinion, was the failure of the long-heralded test, the quantitative test that Wegener proclaimed was unique to his hypothesis. The Danish geodesists repeated their determination of longitude on exactly the same spot, using radio time signals, in 1927, 1936, 1938, and 1948. They had long since discarded the less precise measurement of 1922 for purposes of comparison. The values were essentially identical, and the Danes concluded that the measurements "do not indicate a variation in the longitude."[31] It is hard to see how a theory, correct or not, could be more discredited. The supporting data proved to be flawed, the driving mechanisms proved to be absurd, and the critical test proved to be a wretched failure.

Continental drift continued to have its supporters in Europe and the southern hemisphere, but in North America it was dormant and temporarily became literature instead of science. The trend developed early. When Daly talked favorably about drift in 1922, A. P. Coleman remarked that "Professor Daly shows a poetic imagination."[32] In 1925, Termier wrote, "The theory of Wegener is to me a beautiful dream, the dream of a great poet. One tries to embrace it and finds that he has in his arms but a little vapor or smoke; it is at the same time both alluring and intangible."[33] Longwell, the respected arbiter of continental drift for over 30 years, concluded in 1944, "The concept of drifting continents will be strengthened only by establishing a body of incontrovertible evidence in its favor; not by reiteration of diffuse and qualitative arguments. . . ."[34]

The incontrovertible evidence would be of wholly new types when it finally arrived in the 1960s. Meanwhile, when Harry Hess circulated the first draft of his theory of sea-floor spreading in 1960 he called it "an essay in geopoetry."

Chapter 3

I KEEP MY SHIPS AT SEA

The plate tectonics revolution in the late 1960s had extraordinary impact and overwhelming momentum. Its rapid acceptance was owing in no small degree to the availability of large quantities of pertinent field observations that could rapidly confirm predictions. This chapter is a history of the creation of the principal organizations that collected the data and how the work was done. The period of concern is 1946-1968. It spans Maurice Ewing's move to Columbia University, Roger Revelle's reshaping of Scripps, the invention of a new system of funding, the acquisition of ships, and the development and use of new systems for collecting and processing data.

To a remarkable degree, it all began with Richard Montgomery Field (1885-1961) "whose vision and enthusiasm started the bandwagon of marine geology on its triumphant course."[1] Field collected no data, led no expeditions, and founded no institutions, but he had an incredible ability to select and inspire the men who would. Field and his fellow committee member William Bowie convinced the young Teddy Bullard, Maurice Ewing, and Harry Hess that the future lay at sea. Bullard much later called him "a major eccentric,"[2] and Hess described him as "brilliant and erratic";[3] I first heard of him from R. S. Dietz who called him by a widespread nickname, "Tricky Dick." He was a native of Massachusetts and received a B.A. and Ph.D. from Harvard. His thesis was in paleontology. He worked successively at MIT, Anaconda Copper at Butte, and Brown before joining Princeton in 1923. There he remained until his retirement in 1950. He organized the famous Princeton International Summer School of Geology and Natural Resources, which was all done by specially modified train. Harry Hess took the course as an undergraduate at Yale. Field's interest in research waned, and "his boundless energy was devoted to broader and more imaginative scientific investigations."[4] A later biographer might have made the same statement about Hess—with the difference that Hess had some creative reserves left to do research as well as to influence it. Of course these efforts by Field did not count as research, but he was so good at them that in 1954 he won the Bowie Medal, which was awarded by the American Geophysical Union for "outstanding contributions to fundamental geophysics and unselfish cooperation in research." The essence of his enormous contribution is evident in the *Report of Committee on Geophysical and Geological Study of*

Oceanic Basins submitted to the American Geophysical Union in 1933. Field had conceived of this committee and was Chairman. The members included such notables as Bowie, Walter H. Bucher, J. A. Fleming, and J. B. Macelwane. The AGU would name a medal or award in honor of each of these geophysicists. Among the geologists were D. W. Johnson, T. W. Vaughan, the longtime Director of Scripps, and F. P. Shepard, the rising star. Important scientific problems were identified, but their solution required a new approach:

> in order that these problems may be attacked with the greatest degree of enthusiasm and efficiency, it is necessary to have a closer cooperation between geophysicists and geologists than has heretofore been achieved through previously existing institutional or cooperative agencies.[5]

One of the members of the committee was E. L. DeGolyer who was at the forefront in the application of explosion seismology to exploration for petroleum. His participation in itself was indicative of Field's thinking.

> It is an interesting and highly important fact that, through the powerful urge of economics, the geological and geophysical staffs of a number of oil companies have not only invented and applied geophysical and structural apparatus and techniques which have proved intensely practical as aids in the discovery of oil-pools but, what is still more important to our Committee, have already developed methods and machinery for the exploration of subsurface structures. Several of these methods, with only slight mechanical changes, may be made particularly effective in procuring much-needed data relative to the structure of ocean basins and the drowned margins of the continents.[6]

A program of research was outlined. It included applications of various geophysical techniques to solving "fundamental problems reasonably certain of solution." The problems included,

> (1) Investigation of the drowned continental shelves . . . ; (2) the study of oceanic islands, principally in the Pacific, but including Bermuda, especially in relation to the thickness of coral reefs; (3) the study of enclosed oceanic deeps similar to the Gulf of Mexico . . . ; (4) the thickness and stability of great delta deposits; (5) the investigation of structural island arcs.[7]

With regard to the continental shelves, "it is best to work from the known to the unknown," from the shore out to sea.

The committeemen did not propose to go to sea. In fact, they had almost finished their work. All that remained was to locate some promising young chaps and inform them of their destinies to be elected to the National Academy and win the Penrose Medal.

William Maurice Ewing (1906-1974) was born in Lockney, a small town in the Texas panhandle. He was one of the 10 children of an intellectually stimulating couple whom fate cast in unsuitable roles in farming and small enterprise.[8] Nonetheless, his interest in scholarship was belated, and it appeared that he would not be admitted to a first-class university. Fortunately, his high school mathematics teacher wrote convincingly of his promise, and at age 16 he won a scholarship to Rice Institute, the finest scientific college in the south.

For the next five years he poured out his enormous energy in Houston. He received a B.A. in 1926 and an M.A. in 1927, worked nights and in the summers to support himself, plunged into college activities, and wrote his first scientific paper "Dewbows by moonlight." In later life he never deviated from the work habits he established during that period—or even earlier on the farm. He simply worked night and day. His charm, however, kept him from appearing a drudge. As a fellow student, Avarilla Hildenbrand, recalled, "When he came striding down the street working his trombone slide in and out, my heart stood still." They were married in 1928.

The most important thing that Ewing did at Rice was learn how to be a research scientist. As one physicist memorializing another, Teddy Bullard wrote,

> He learnt not only the subject but the attitude of mind. All his life he preferred simple arguments; his theory was set out in detail, well understood and carefully explained, his instruments were ingenious and often made by himself. . . .[9]

Ewing was fortunate in that physics was then undergoing the revolution of quantum mechanics, and papers destined to be classics appeared in rapid succession. He was doubly fortunate to have H. A. Wilson as a physics professor. Wilson was from the Cavendish Laboratory at Cambridge and held seminars with tea in the Oxbridge style. To these he attracted a stream of international visitors whom the young physics student "would otherwise have thought not mortal."[10] It was exactly what Ewing needed; the one thing a student is least apt to be exposed to is the illustration of style. He later said, "I think I am the most grateful person who ever went to Rice." Wilson did one last service for Ewing. He showed that the exalted professor was human by advising him that "he had no aptitude for experimental work and should stick to theoretical physics."[11]

Ewing became an Instructor of Physics at the University of Pittsburgh in 1929 and transferred to Lehigh a year later at the same level. He received his doctorate from Rice in 1931 with a thesis titled "Calculation of ray paths from seismic travel-time curves." He was promoted to Assistant Professor in 1936, at age 30, after seven years as an Instructor. It is a simple indicator of the shift of supply and demand that two decades later the rank of Instructor had virtually vanished in science departments at great universities. Scientists were hired as Assistant Professors directly after receiving a doctorate. In contrast, the rank of Instructor always existed in the humanities and arts. Sadly, it is reappearing in the sciences. The new Assistant Professor at Lehigh had already established himself as a scientist and teacher by publishing 14 papers. Almost all were collaborative, so his lifelong style of research was formed very early. He was to be a sun among planets, or sometimes a double star, but always in constant mutual attraction with colleagues.

The event that shaped Ewing's future occurred on a slightly snowy day in November 1934 when two gentlemen wearing derby hats and coats with fur collars called on the young instructor in his basement office. Field and Bowie had found their man. "I remember it like yesterday," said Ewing 40 years later.[12] His own account of the meeting, however, is not as revealing as Bullard's reconstruction of it based on his own meeting with the same pair in 1937.

Field would have been persuasive, persistent, talkative and irrepressible, while Bowie lent an air of solidarity and charm; together they were irresistible, particularly when they offered funds and ships.[13]

The first fruits of this recruitment occurred early the next summer when Ewing and two students, A. P. Crary and H. M. Rutherford, put to sea on the Coast and Geodetic Survey Ship *Oceanographer*. The Congress, however, did not appropriate funds to the C and GS for scientific research, so the written orders for the ship specified that Ewing's work was permitted only if it did not interfere with the funded bathymetric surveying. The captain understood that he was to interpret his orders liberally. Not for nothing was Bowie the Chief of the Division of Geodesy of the Coast and Geodetic Survey. Unfortunately, the informed captain was injured at the last moment and could not sail. His executive officer, like a normal nautical bureaucrat, took the orders as written. Fortunately, the ship anchored at night (!) and then Ewing could test equipment, although he could get no geological information. From then on Ewing would have wanted absolute control over his own ship.

Pending the availability of such a ship, Ewing shot a seismic profile across the coastal plain of New Jersey so it would be "known" before he proceeded to the "unknown." The indefatigable and irresistible Dick Field persuaded Henry Bigelow, Director of the Woods Hole Oceanographic Institution, to offer the Research Vessel *Atlantis* to Ewing for two weeks in October 1935 (Figs. 25 and 26). Ewing, Crary, and Rutherford then made the first seismic profiles in the open sea although in the shallow water of the continental shelf. They shot four refraction lines off Cape Henry and three more on a line toward Woods Hole. At each station R/V *Atlantis* was anchored and a seismometer lowered to the bottom 100-200 m below. Then the ship's boat sailed out as far as 11 km, stopping occasionally to lower an explosive charge to the sea floor and fire it by wire. Ewing discovered sedimentary rock 3,800 m thick above "basement" rock of uncertain lithology. This was the first discovery of an offshore basin of the type from which much of the world's oil is now produced. Ewing was accustomed to working for oil companies, and he had the report of Field's committee to certify that the companies would value offshore exploration. He approached an oil industry executive in 1936 with a request for support. "He was told that there was no shortage of oil and that the company was not the least interested in looking for it at sea."[14] The "powerful urge of economics" works both ways.

In 1937 Ewing and students published their results in a paper with a characteristic title, "Geophysical investigations in the emerged and submerged Atlantic coastal plain. Part I: Methods and results."[15] The number was the key to his thinking. He had written home at the beginning of the marine work,

> This is by far the most important project with which I have been connected. It is so arranged that I see no possibility of anyone stealing the credit from me.[16]

Thus he staked his claim with a number that implied that more would follow and that other investigators should steer clear. He followed the same procedure time and again. In 1959, for example, there appeared "The floors of the oceans. I. The North Atlantic." It was never followed by Part II because of the falling out between Ewing and Heezen.[17] In general, however, Ewing delivered papers and thereby paid the charge for staking a claim. The publications on the Atlantic coastal plain reached part VII by 1954.

Ewing gained widespread recognition by 1937, but then he had a sobering experience. His coauthor for four consecutive papers in 1930-1931 was L. Don Leet, a very young Assistant Professor of Geophysics at Harvard. In 1937, Leet wrote a response to Ewing's paper and was highly critical of its methods, interpretation of results, scholarship, and acknowledgment of previous work. Of 12 land and four sea stations on the Cape Henry line only the outermost pair made any addition to knowledge, and those two were not worth much.

> The one-shot, two-point "profiles" obtained at each of these stations are described as "incomplete" and "decidedly inadequate," respectively. That puts the situation mildly, to say the least, for so far as depth to the basement is concerned they supply no evidence, negative or positive, in any true sense of the word.[18]

The stations on the Woods Hole line were no better.

Perhaps this criticism was influential in Ewing's choice of research topics in the next few years because only his explosion seismology on land was published. He nonetheless continued work at sea. He used a submarine for gravity observations in collaboration with Harry Hess and later J. L. Worzel. He also developed and tested a new instrument, a "deep-sea" camera, which showed that sandy bottom at 150 m was rippled by currents. Geologists had thought that sand ripples form only in very shallow water.

Ewing also practiced seismology at sea and experimented with techniques that could be used in the deep water of the main ocean basins. First he tried a rig with instruments in watertight containers strung on part of a steel cable, with explosives attached farther along. Teddy Bullard was with him in 1937 and found the operation hazardous and difficult at best. The technique was abandoned, and in 1939 and 1940 a "balloon" approach was tried. The balloons, like those devised by August Piccard for the bathyscaphe, were filled with gasoline. The equipment and explosives were overballasted to drag the balloons to the bottom for the seismic experiment. The weights had soluble releases, and the balloons brought the equipment back to the surface. This also was hazardous. Nonetheless, Ewing made measurements at two deep stations at 2,600 m and at 4,800 m. With only one shot and one geophone at each station he was unable to obtain any indication of basement under the sediment, but it was a beginning. The astonishingly simple technical solution for his problems would not come for several years. Seismic refraction studies take advantage of the fact that the velocity of pressure waves increases with depth in the earth. By moving a source (explosion) away from a receiver (hydrophone) it is possible to determine the velocity structure and thickness of the different layers of the earth. The problem Ewing faced was how to get the explosives and hydrophones down through the water to the "earth." The solution was a matter of definition. If the ocean is considered to be just one more layer of the earth, the problem vanishes. The hydrophones and explosions could be at the surface of the sea. Certainly, after all that wartime research, the velocity structure of the sea was known.

In 1939 Ewing had spent five years following the course laid out for him by Field and Bowie. In all that time he had had only 45 days of shared time at sea. It was no way to do marine geology.

In 1940 Ewing and his associates Allyn Vine and John (Joe) Worzel (Fig. 35)[19] from Lehigh moved to the Woods Hole Oceanographic Institution (WHOI) where they spent the war working for the Navy.[20] The Director of WHOI, Columbus Iselin, observed that they worked night and day, seven days a week, and the place was never the same again. In 1941 the Ewings were divorced, and in 1944 Maurice married Margaret Kidder. That same year he was invited to join the geology department at Columbia as an Associate Professor—a great leap for an Assistant Professor at Lehigh. He moved there in 1946 and became Professor in 1947. In 1948, the widow of Thomas Lamont offered Columbia his estate at Torrey Cliff on the Hudson along with $250,000 to modify it. The University then offered it to Ewing as the Lamont Geological Observatory (LGO). Running it—and a ship or two—would take an unprecedented amount of money and support. Before following American geology to sea after the war, we should consider the source of that support and the career of the future leader of another oceanographic laboratory.

Roger Randall Dugan Revelle (1909–; Figs. 27, 29, and 41) was born in Seattle but soon moved to California. He was three years younger than Maurice Ewing, and they would both be of an age to lead the new wave. He was early recognized as a genius as a consequence of a widely publicized study by L. M. Terman of the IQs of California children.[21] He received a B.A. in geology from Pomona in 1929, and from 1931 to 1936 he was a research assistant at Scripps Institution of Oceanography (SIO). In 1931 he married a classmate at Pomona, Ellen Virginia Clark, the grandniece of Ellen Browning Scripps who mainly endowed SIO. In 1936 Roger received his doctorate from the University of California with a thesis on the bottom samples collected on the last cruise of the R/V *Carnegie*, which blew up in Apia harbor in 1929.[22] He spent 1936-1937 at the Geophysical Institute at Bergen and returned to Scripps to continue research on a grant of $1,200 from the Geological Society of America, which was also supporting Ewing.

From February to April of 1939, Roger was the leader of Scripps's first deep-sea expedition—on the R/V *E. W. Scripps*, which had been the palatial yacht of the movie actor Lewis Stone[23] and which was renamed after the newspaper magnate who also contributed to endowing SIO. On the expedition were Professor F. P. Shepard, who had been on the Field committee in 1933, and two students, K. O. Emery and R. S. Dietz—just 22 years from hypothesizing sea-floor spreading. The expedition crisscrossed the Gulf of California and took 25,000 soundings as well as a 17-foot core, which was the longest yet recovered. It was good training for what was to come, but unlike Ewing's efforts in the Atlantic it included no geophysics.

The war and its aftermath kept Roger out of the academic life for almost seven years—longer, I believe, than any other prominent figure in this book. He was not, however, in the fighting Navy like Hess; he was always close to research, even though he wasn't doing what counts himself. He went on active duty in July 1941 as a Lieutenant (jg) in the Naval Reserve and was posted to the Navy Radio and Sound Laboratory about 10 miles from his home.[24] There he worked as a staff officer concerned with radar and harbor defense. In December 1942 he was ordered to Washington as a technical aide to the Hydrographer and "Oceanographic Consultant to the Bureau of Ships and the Commander-in-Chief U.S. Fleet."[25] He was chosen to promote and organize oceanographic research and apply it to submarine warfare, amphibious operations, and air-sea rescue. It is hard to see how the Navy could have made a wiser choice.

In December 1945, Commander Revelle was detailed as head of the Section on Oceanography of Joint Task Force 1 for Operation Crossroads, the atom bomb test at Bikini. Planning was already underway, and the surveys and tests were carried out in March-August 1946. To oceanography the operations at Bikini had some of the characteristics that the Manhattan Project had for postwar physics. A large fraction of youthful American oceanographers got to know each other on a high-priority, highly classified, exciting project. Among them were W. S. von Arx, E. C. La Fond, M. C. Sargent, T. S. Austin, C. A. Barnes, D. L. Bumpers, G. C. Ewing, John Lyman, and Walter Munk, and the geologists K. O. Emery, H. S. Ladd, and J. I. Tracey. Many of them found roles as leaders in federal oceanography after obtaining doctorates at Scripps after the war. At the head of it all was Roger Revelle.

The war was over, but Roger stayed on active duty. In November 1946, he became Head of the Geophysics Branch of the just-formed Office of Naval Research. This was a critical move for the future of oceanography just as the establishment of ONR was critical for American science. The staff of brilliant young Naval reserve officers included James Wakelin, E. R. Piore, and Roger. They were charged with creating an organization that would nurture basic research with an unprecedented amount of federal money. Normally the result would have been the establishment of a Sizable Bureaucracy. This bureaucracy inevitably would have shifted responsibility for the selection of recipients of grants to some outside consultants, and would have spent most of its time considering the details of small grants. In fact, this sequence of events was shortly to occur with the establishment of the National Science Foundation in 1950.[26] The men the Navy picked would have none of it. Roger created a small group who identified the best scientists and the most promising problems and gave money in large bundles where it would do the most good. It was as though Dick Field's committee had been given the keys to the U.S. Treasury. All was sunny, but clouds would eventually form. The system was conceived by geniuses to be run by geniuses. However, Herman Wouk generalized in *The Caine Mutiny* that the Navy is a system conceived by geniuses to be run by idiots. The supply of Revelles, Lills, and Maxwells might not last as long as ONR.

Roger himself was soon back in the field. In the spring of 1947 he organized and led a resurvey of Bikini to determine the effects of the nuclear tests on the biota of the atoll. Somehow, Roger being Roger, the program included drilling several holes deep into the coral cap in a test of Darwin's theory of the origin of atolls.

THE EXPANSION OF OCEANOGRAPHY

In December 1947, Roger finally terminated his active duty, but he continued as a civilian in the same job until 15 March 1948. By that time the great expansion of oceanography was underway. Before the war there were three oceanographic laboratories in the United States: Scripps, Woods Hole, and the University of Washington. They had a total budget of less than $250,000 and with it supported three ships. In 1948 the Navy poured about $600,000 into oceanographic laboratories, which was a sizable expansion even after allowing for inflation. Up to 1958 it spent a total of $46 million on academic research in oceanography. The number of laboratories multiplied, and the Navy spent about $300 million for ships, facilities, and equipment. The Navy disbursements for three laboratories of the most interest here are shown in Table 1. By the end of the decade, NSF money was becoming abundant, and Scripps also received significant funding from the State of California. The first decade of postwar expansion, to 1958, was only the beginning. In the next seven years the Federal support for SIO and WHOI would triple to more than $10 million per year.[27] The total for all academic oceanography from ONR and NSF would reach $25 million per year—just 100 times what it had been in 1941.

TABLE 1. Federal dollars (in thousands) for selected laboratories for 11 years

Lab	1948	1950	1952	1954	1956	1958	Total
WHOI	300	550	1,100	1,020	1,420	1,300	10,600
LGO	35	410	420	360	1,040	520	4,600
SIO	200	305	1,010	450	2,040	1,040	9,900

Note. WHOI = Woods Hole Oceanographic Institution. LGO = Lamont Geological Observatory. SIO = Scripps Institution of Oceanography.

In 1948 no one knew that this would happen. Even then, when funding had only doubled, Columbus Iselin, Director of Woods Hole, wrote

The effects of this great outpouring of money on oceanography are by no means all healthy. In the first place nobody knows how long it will last.[28]

It seemed an outpouring, but it was a tide at the ebb, and just at that moment, in 1948, Maurice Ewing was offered Lamont Geological Observatory and Roger Revelle was offered Scripps Institution of Oceanography. Roger was home in La Jolla for the holidays, and he talked until late at night on 5 January with Harald Sverdrup, the Director of Scripps. Sverdrup was ready to retire. The next morning Roger sent him a letter summarizing his position relative to their conversation.[29] He wanted very much to return to Scripps, although he liked his position in Washington and considered it a necessary and important job. Sverdrup had said and Roger agreed that if he returned he would "inevitably have to take a major share of the responsibility for the research and development work." He could not assume responsibility "without adequate authority," and he did not mean administration by academic committee. He proposed a merger of Scripps and the Marine Physical Laboratory, which was a war-generated arm of the University of California in San Diego. If Carl Eckart, the Director of MPL, would consent to be Director of the merged organization, Roger would be happy to be his deputy. He had only one reservation about Eckart:

> he is perhaps somewhat overcautious and too conservative to seize and exploit the many opportunities which should arise during the next few years to develop and expand the science of Oceanography and the Scripps Institution.[30]

Roger knew an ebb tide when he saw it. Sverdrup departed for Norway. Roger resigned his job in Washington on 15 March 1948, and the Regents of the University of California appointed him Associate Director of Scripps effective 22 March. The older hands at Scripps muttered, but soon he was Acting Director and then Director from 1951 to 1964. With an incredible sense of timing he was to leave Scripps just as the flood tide of Federal money slowed, never to rise so rapidly again.

Maurice Ewing had spent the war and the years immediately afterward far from the center of power in Washington. Unlike Revelle, however, he had experienced the result of trying to work in borrowed laboratories and on borrowed ships. Moreover, he was already 40, bursting with ideas and energy, and could see the future dangling before him. The first contract for research written by the Office of Naval Research in 1946 was to Maurice Ewing. More soon followed, and there was no end in sight. Others had vision as well. While Ewing was at Columbia, MIT offered him a seaside laboratory, a professorship, and support for his group of 20 students and assistants. He declined. Columbia offered him its Hudson Institute with its own grounds, but the institute did classified research and he declined. When the Lamont property was offered he jumped. He was advised to use the accompanying funds as endowment because he would have to be self-supporting. Instead he confidently spent his capital in two years and rose thereafter with the Federal tide.

Revelle and Ewing had a priceless opportunity before them and the courage and will to exploit it. So the expansion began. An expansion is an administrative dream, quite unlike normal equilibrium or the dread contraction. New people can be hired; promotions are frequent; new buildings require festive groundbreakings and dedications; ships can be purchased and outfitted; and rules can be bent or ignored. Best of all, mistakes can be remedied and deadwood in the staff can be promoted to harmless officialdom. Next best, the leeches of higher administration are slow (naturally) to learn how to bleed off overhead from expanding funds. The bold utilization of expanding funds cures all. No wonder the next dozen years were so happy at Lamont and Scripps.

The oceanographic fleets grew; the Navy was contracting and happy to give ships away. Scripps acquired *Crest* in 1947, *Horizon* in 1948, *Baird* in 1951, *Stranger* in 1955, *Argo* and *Hugh M. Smith* in 1959, *Alexander Agassiz* in

1962, and so on. Lamont acquired *Vema* in 1953 and *Conrad* in 1962. I spent a lot of time on R/V *Horizon* and developed a certain affection for her, so I wrote a memorial tribute after we sold her in 1969.[31] She was a fat, slow, dirty, uncomfortable, but happy little ship, 143 feet long and displacing about 900 tons. She sailed 610,522 miles all over the world on 267 oceanographic cruises for 4,207 days at sea. Few ships since the days of Nelson have seen the continuous service of the U.S. oceanographic fleet (Figs. 3 and 4). Certainly, modern navies do not utilize ships with any comparable frequency except in time of war, and wars do not last 20 years. The U.S. Navy looks upon six months at sea per year as optimum ship utilization. Ewing once tried to bill the Federal government for 13 months of ship time per ship per year with some argument based on utilization. Teddy Bullard asked Ewing where he kept his ships, meaning where were his marine facilities and docks. ''I keep my ships at sea,'' was the reply.[32]

THE NEW STYLE OF EXPLORATION

What Ewing and Revelle developed, independently and in their own styles, were instruments for continual geological exploration. No comparable instruments had ever existed in geology. Their nearest equivalent was the ''great engine of research,'' as G. K. Gilbert described the U.S. Geological Survey in the nineteenth century.[33] The Survey, however, left the field to the winter snows and, in Gilbert's time, worked mainly in the vast openness of the mountainous western states. There Clarence King, John Wesley Powell, C. D. Walcott, Gilbert, and their colleagues electrified geology with their discoveries and became famous. They became members of the fledgling National Academy of Sciences, hobnobbed with Cabinet members at the Cosmos Club, won the medals, and presided over the learned societies. They remain the heroes of a later age, certainly mine, but if Gilbert had been born in 1890, someone else would have explained the laccoliths of the Henry Mountains, and someone else would have solved the puzzles of Lake Bonneville. The opportunity to explore the vast, arid, exposed outcrops of the west was unique, and scientists, very few of whom were arguably better than those before or after, toiled and reaped rich rewards. Whatever their abilities, continental geologists could never again have such luck. Consider, however, the opportunities for discovery and fame for those, regardless of their merits, who would be lucky enough to be first to explore an area a hundred times greater than the arid west—moreover, a region no one had ever seen. Consider what opportunities would open for those backed by an engine for research that grew mightier every year and could operate every day, every month, year after year.

The little oceanographic ships were not usable in winter gales, and before

World War II they, too, stayed in port and scientists worked up their observations. After the war it was realized, rather by chance, that it was never winter in both hemispheres. Thus it was that oceanographers, like ski bums, shifted from the northern to the southern hemisphere to continue their work in accordance with the changing seasons. Thus it was that oceanographers spent so many Christmases in the southern seas, and that it gradually dawned on them that there would never be any time in port for contemplation.

As the ships multiplied, so did the results. By 1956 Lamont, using Navy submarines, had tripled the number of gravity observations at sea. Scripps had taken none. There were no heat-flow observations in the oceans in 1946, but by 1963 Scripps had taken about 300 and Lamont had taken none. There were perhaps 100 cores of sediment from deep-ocean basins in 1948. By 1956 Lamont had taken 1,195.[34] Ewing was obsessed with cores and Lamont always led the world, but by 1962 Scripps had about 1,000. There had been no seismic stations in the deep sea, and by 1965 there were hundreds. Underway data, essentially continuous observations, had multiplied even more. The number of deep-sea soundings had increased by about 10^8 and the number of plotted soundings by 10^5. Nothing comparable to shipboard magnetic profiles had ever been known, and Lamont and Scripps had towed magnetometers for hundreds of thousands of kilometers. By the time plate tectonics was developed in 1967 the ocean basins were better known geophysically than the continents. Had they not been, there would have been no plate tectonics then. Even in 1964 I was only half jesting when I wrote of a "digression from the familiar ocean basins to the mysterious continents. . . ."[35]

With the ships constantly in motion like sorcerers' apprentices, the sorcerers had to devise some way to avoid drowning in data. In the 1950s, some scientists barely kept their noses above water and published only many years after they collected data. Some thought their older data were obsolete before they could be published. Some felt responsible to drop their analyses of data and help keep the ships moving. Some, it seemed, returned to sea to escape a feeling of guilt for not working up the data from previous cruises. The soundings were relatively simple to process, but Bob Fisher and I once spent three months plotting and reducing the soundings from a single expedition. When either of us was at sea it was a matter of pride to have all the navigation adjusted and the soundings plotted before we stepped ashore, but we could not always be on each of six Scripps ships. I kept hiring more technicians to plot data and sending students out to do research on marine topography. Still the problem grew. In the early 1960s Bruce Heezen and I attended several meetings at the International Hydrographic Bureau at Monaco. The hydrographers of the major navies of the world assembled annually to coordinate their deep-sea sounding programs. For some reason the United States sent the two of us as well as the hydrographer of the Navy. By the end of the first meeting Bruce and I realized that Lamont and Scripps were collecting more and better soundings than any navy in the world.

The U.S. Navy was paying for all the people that Bruce and I kept hiring, and the Navy wanted our data. By the middle 1960s, computers and automatic data-processing equipment became available, and a solution for our problems could be bought. Brackett Hersey, the Woods Hole geophysicist, had accepted a major position in Navy Research, and he solved the problem for me with a rich infusion of dollars. Starting late in 1966 I bought digitizers, key punches, and computer-driven plotters, and hired more people to run them. Thereafter, I never had to increase the data-processing group. The Navy treated Lamont equally generously, and soon we could exchange data with them by computer tape. The only remaining problem was that none of the hydrographers of the world could yet process our tapes. The systems we developed for onshore computer processing of soundings were ideally suited for handling magnetometer records. Thus the necessary systems came on stream just in the nick of time for the plate tectonics revolution. By now, the second generation of computers is going on shipboard, little processing is done ashore, and the students shake their heads at our first computerized system and say, "How quaint." Had the systems not existed when they did, however, the proofs of sea-floor spreading and the rigidity of plates would have been much less overwhelming.

THE "SEA-MEN"

The fleet prospered and the ships kept moving; the equipment steadily grew more efficient and the data poured in. The problem that arose was stated by Richard Hakluyt in 1599:

> that ships are to litle purpose without skilfull Sea-men . . . and since no kind of men of any profession in the common wealth passe their years in so great and continuall hazard of life; and since of so many, so few grow to gray heires. . . .[36]

Henry Stetson, my Professor of Marine Geology at Harvard, died at sea on an oceanographic expedition west of Peru. Bruce Heezen died in a research submersible in the Atlantic. In 1961 a man was killed on *Vema* when it was standard procedure to throw a 0.2 kg explosive charge overboard every two minutes. In 1954, Maurice Ewing, his brother John, and two mates were washed overboard by a freak wave in a gale with mountainous seas north of Bermuda. A skillful captain (a "Sea-man") saved the two scientists and one mate. Maurice was left with a slight limp and permanent although minor effects of internal injuries. In 1964, on R/V *Baird* in the endless gales south of Easter Island, I was cautiously moving along a lifeline to inspect a damaged instrument on the stern. A seaman (not a skillful ''Sea-man'') cut the restraining ropes on a deep-sea camera, the ship gave a monstrous roll, and the camera slid athwartship and smashed into my back, compressing one vertebra by a centimeter. My arms draped over the lifeline kept me from flying overboard unconscious. I walked off the ship 10 days later, but three vertebrae fused themselves permanently.[37] And so on. The oceanographers of the period worked constantly with explosives and very heavy equipment on the open decks of small ships in great seas. Many went out again and again. Few people seemed to have kept records, but they still have memories. Frank Press, then a student with Ewing, went to sea for seven straight summers. It was probably good toughening for the future Science Advisor to President Carter and President of the National Academy of Sciences. I have approximate records that indicate that I went to sea on 25 major expeditions between 1949 and 1984 and led most of them (Fig. 5). Joe Worzel was chief scientist on at least 20 expeditions from 1936 to 1965 and participated in many more. He or Bob Fisher (see Figs. 5 and 6), who has been active at sea from 1951 to the present, may hold the record for scientific exploration, anytime, anywhere.

The fact that the work was hard, cold, and dangerous does not mean it was not enjoyable. Most people, especially those immune to motion sickness, loved it at sea. Perhaps they loved it too much because there was a price to pay in obvious things like divorces and in less obvious things as well. Gradually the field oceanographers who worked the mighty engine found the price too high or alternatives too attractive. After the plate tectonics revolution, the ships generally were run by a new generation of men and the first generation of women at sea.

Chapter 4

FINDING SMOOTHER PEBBLES

I do not know what I may appear to the world; but to myself I seem to have been only a boy playng on the seashore, and diverting myself in now and then finding a smoother pebble or a prettier shell than ordinary whilst the great ocean of truth lay all undiscovered before me.

—Sir Isaac Newton (1642-1727)

When we put to sea on the Midpac expedition in July of 1950 we expected certain results even though our observations would be in unexplored territory (Fig. 5). Seldom have expectations been so far from reality. Even now, 33 years later, Russ Raitt's face glows when he recalls our constant astonishment at what the instruments were showing. So numerous were the discoveries that our second great geophysical expedition, Capricorn, in 1952-1953 was an anticlimax—at least in retrospect. On our first ventures we found most of the kinds of phenomena that could be found with the instruments then available. On Capricorn we merely began the lengthy search for second-order effects, geographical discoveries, and regional correlations between the major variables.

EXPECTATIONS

What did we expect to find? To begin with, we believed the Pacific was very old. There was no conflict on this issue as there was for the Atlantic. Even if Wegener was right, the Pacific basin was ancient and unaffected by continental drift. The possibility that the rather circular basin was the scar of the moon's separation was not to be ignored, but if so, it was an ancient scar.

Most of the larger features of the bathymetry of the basin were known because of a century of sounding. The deep trenches around the margins had been discovered and even surveyed locally while selecting safe routes for submarine cables. Likewise, broad swells and swales were known to exist; one, called the Albatross Plateau, was soon to become the East Pacific Rise and the proving ground for sea-floor spreading. The existence of submarine volcanoes was also established, but they were relatively small features and thus their distribution was less certain. Even so, most of the seamount clusters or submarine mountain ranges were already known in 1950.

What was hardly known at all was the small-scale relief of the deep-sea floor, the physiography, or the equivalent of a very detailed topographic map or an aerial photo. Most of the old soundings were spaced 10-20 miles apart and thus could not detect small features. The recording echo sounders available to the U.S. Navy in World War II could measure depth much more frequently, and with them Harry Hess made graphic profiles of the relatively shallow tops of guyots. The instruments were capable of sounding much greater depths, but they were not intended to do so and were very rarely used for the purpose. Moreover, whenever the ship's power varied, the instruments were subject to timing variations that produced spurious bottom relief.

Some echograms had already been collected in the Pacific by the Swedish Deep-Sea expedition of 1947-1948. This remarkable effort led by Hans Pettersson went around the world, sensibly in the tropics, studying aspects of physical, chemical, biological, and geological oceanography.[1] With regard to the last, it collected more than 200 piston cores, each about 10 m long. The corer was a spectacular advance on previous technology, and the analyses of the cores were on an unprecedented scale. The geophysical program was also without precedent, including as it did measurements of heat flow, seismic reflection, and continuous echo sounding. Waloddi Weibull identified reflectors by explosive seismology at more than 50 deep-sea stations. The thickness of what was assumed to be sediment ranged up to 3,475 m in the North Atlantic but was no more than a few hundred meters in the Pacific. Pettersson himself attempted to measure heat flow in the deep sea but was frustrated by faulty instruments. The echograms could discriminate variations of only 1 m at a depth of 9,000 m. The results of the echo sounding were not published until 1954,[2] but word of preliminary results reached us in San Diego long before then. In some places the deep-sea floor was essentially flat except for narrow, steep-sided valleys. The plains were interpreted as evidence of thick sediment and the valleys as graben caused by faulting. In fact, the Swedish expedition had discovered the first deep-sea channels caused by turbidity currents. The remarkable technical advances and the scale of this expedition put Sweden in the forefront in postwar oceanography. No one else was so organized and ready for a great scientific effort at sea. In a sense, however, this expedition was the last great successor of the Challenger expedition a century earlier. A large ship took a group of well-prepared scientists to sea for a long but finite time. In contrast, the Midpac expedition was one of the first of the new style with almost interchangeable scientists, rotating on and off small ships that ran until they rusted away.

Although we had few direct observations of the detailed relief of the deep-sea floor, we had powerful reasons to have certain expectations. First, there was the argument of Philip Kuenen regarding the volume of deep-sea sediment.[3] In essence, the continents are billions of years old; they are eroded at rates of at least 10 m/10^6 yrs or 10,000 m/10^9 yrs; the erosion products are transferred to the deep sea; the oceans have twice the area of continents; and thus the primeval rock of the sea floor is buried under at least 5 km of sediment. Inasmuch as the deep ocean was thought to be motionless, this sediment was not eroded into channels or piled up in bars or drifts. Nor was the sedimentary surface disturbed by faulting or folding. Any ancient tectonic relief was long since buried, and no new deformation was occurring. This was known because almost the entire Pacific basin was free of earthquakes. Indeed, this was later to be the reason for calling it a tectonic plate. A few exceptions existed. The Hawaiian region had some quakes, but they were clearly associated with active vulcanism rather than faulting of the deep-sea floor. There were also rather small earthquakes in deep water in the southeastern Gulf of Alaska and on the East Pacific Rise in the southeastern Pacific. Much larger earthquakes occurred in the western Pacific along the line of island arcs and trenches. Beno Gutenberg, 20 years older than when he first advocated continental drift and recently elected to the National Academy, managed to keep the Pacific free of earthquakes by defining the basin carefully.[4] Geologists had long thought the western Pacific island arcs to be ''continental'' in character because of the presence of andesite and because of the trenches and the earthquakes. Thus despite deep water for a thousand miles to the west, the Marianas trench was taken for geological purposes as the edge of the Pacific basin. The volcanic rocks of Easter Island, rising from the East Pacific Rise, were suspiciously continental also. This fact plus relatively shallow water in addition to earthquakes and certain characteristics of the earthquakes caused Gutenberg and Richter to identify the Rise as a continental strip within the southeastern Pacific.[5] The quakes in the Gulf of Alaska were also in a ''continental area.'' Gutenberg defined the Pacific basin west of this anomalous area to correspond almost exactly with what is now the Pacific plate.

Given 5 km of sediment and no earthquakes, we expected the sea floor to be smooth and gently sloping except where volcanoes broke through the pelagic oozes. Kuenen had diagrammed another type of topography to be expected. The load of a volcano would surely squeeze the ooze to the sides and the mountain would be surrounded by a moat.

As it happened, volcanic islands and seamounts were the focus of Midpac and, less so, of Capricorn. Hess, unwittingly confirming Darwin,[6] had discovered guyots and reasoned that they are drowned Precambrian islands, perhaps depressed by the enormous thickness of pelagic sediment that had since accumulated. Ed Hamilton and I, mere academic striplings, and Bob Dietz thus expected to dredge cobbles rounded by river and wave, pelagic fossils, and perhaps even fossil reef corals. Bob Dietz, an established scientist and in any event more of a plunger, speculated that the guyots were Cretaceous islands. He reasoned that most of the mountain building around the margins of the Pacific was Cretaceous and that if something happened at the edges it probably happened in the middle, too. This was early Dietz but a good vintage. The reasoning was wild, somewhat plausible, and only incidentally wrong, and the conclusion was correct. He was thirty-six.

So little was known about the gravity, magnetic anomalies, heat flow, and crustal structure of the earth in 1950 that it might be supposed that we had no particular expectations regarding them. Instruments and techniques for measuring these properties, however, were under intensive development, and much thought had been given to possible results. The competition was keen because it was evident to the spiritual descendants of Field and Bowie that a great opportunity was at hand. Consider the thickness and layering of the crust. Vening Meinesz had already demonstrated that the depth of isostatic compensation is about the same at sea as on land, namely about 100 km. Thus in terms of rigidity the thickness of the "crust" was known. At the time, however, there was confusion between that "crust" and the "crust" defined by seismologists as the layer above the Mohorovičić discontinuity (Moho).

Maurice Ewing and colleagues had set out on their solitary quest to study the oceanic crust in 1935, but they had not succeeded in reaching the Moho before the war. Immediately after the war, as they moved to Columbia and Lamont and sought ship time, other scientists entered the field. Weibull from Goteborg, Hill from Cambridge, and Raitt from Scripps were all in competition by 1949. Twelve years before, Ewing had thought no one could "steal the credit" from him, but it appeared that he might have to share it. He adopted standard tactics to defend his position. The table of contents of a 1949 issue of the *Bulletin of the Geological Society of America* indicates that pages 1303 and 1304 are devoted to a Short Note by Ewing et al. on abyssal seismic refraction.[7] All that exists is a single short paragraph, and the second page is blank. Apparently Ewing obtained a commitment of two pages, before he went to sea in February 1949, with the expectation of bypassing the usual lengthy reviews. All he could produce was a paragraph that could have been sent by cable from the first port after the seismic station. They had found only two layers. The upper one appeared to be sediment 1.7 km thick, but no velocity was determined and one had to be assumed to calculate the thickness. Below the sediment was material with a velocity of 7.58 km/sec, which was identified as basalt below the Moho. There was no granite at this deep Atlantic station nor, indeed, any crust at all. The Short Note attempted to emphasize Ewing's priority by referring to a paper in press. When this was published in 1950 it had much to say about technique but little to add about the meager results. In fact, the interpretation was even less definite than before. These pioneering results were grossly misleading. The thick sediment obscured a layer of volcanic rock, and the high-velocity layer was atypical at best. The crustal structure of the deep sea was wide open for discoveries on Midpac.

The study of heat flow was in about the same stage as the study of the oceanic

crust. Perhaps the opportunity for a spectacular advance was even greater because of the likelihood that crustal tectonics is driven by internal heat. The
probabilities of expansion or contraction or mantle convection might be con-
strained by measurements of heat flow. At the very least it should be possible
to determine something about the composition of the oceanic crust—or the
mantle if there was no crust, as Ewing's only deep-sea observation indicated.
There were few observations of heat flow on land, but they gave a rather uniform value. The granites and kindred rocks of the continents have a much
higher content of heat-generating radioactive minerals than the basalts of
oceanic islands. Assuming that the islands are a representative sample of the
rocky basement of the basin, the oceanic flow would be far less than the continental.

Hans Pettersson had attempted to measure heat flow in 1947-1948 on *Albatross* but without success, and the scientists at Lamont achieved little more in
the decade of the 1950s. As a consequence, the first profile of heat flow across
the Atlantic was collected by Scripps in 1962, and by that time the operation
was routine for us.[8] It was becoming so for Cambridge University. This success was the first benefit of the intimate connection between Scripps and Cambridge, which has continued through the years. What kept the two together was
Sir Edward Crisp Bullard (Figs. 40 and 41), who was universally known as
"Teddy" until 1953 when he was knighted and sometimes known as "Sir
Teddy."[9]

Teddy Bullard was born in 1907 in Norwich, England. His father was the
affluent brewer of Bullard's Ale, and Teddy always had a respect and taste for
business. He went to Clare College, Cambridge, received first-class honors in
physics, and became a graduate student at the Cavendish Laboratory. His thesis was on the scattering of slow electrons in gases, and Lord Rutherford advised him that he had no future in physics.[10] He spent the years from 1932 to
1940 as a Demonstrator in the Department of Geodesy and Geophysics at Cambridge—at what is now the Bullard Laboratory. He measured gravity in the African rifts and heat flow in Africa and England; most important, in 1936 he en-
countered Dick Field who pushed him into seismic refraction at sea. Teddy
spent the war as a highly unconventional and successful civilian scientist in the
Royal Navy and was ever after full of anecdotes about startling admirals. He
returned to Cambridge in 1945 and became Chairman of the tiny Department
of Geophysics. He transferred to the University of Toronto in 1948 as Head of
the Physics Department. There, perhaps because he lacked fresh field data, he
began to work on the dynamo theory of the earth's magnetic field. He returned
to England in 1950 to become Director of the National Physical Laboratory, an
organization comparable to the U.S. Bureau of Standards.

Teddy spent the summer in La Jolla in 1949, which was my first summer
there as well. He worked then at the development of a temperature-gradient

probe to to on the Midpac expedition.[11] He and a graduate student, Art Maxwell, designed and built a watertight instrument-container attached to a long spear. Teddy knew about O-rings from his naval work, and they were used for watertight seals. He never had one fail. He could obtain no help from the busy Scripps machine shop, so he and Art did the lathe work. They tested penetration of the spear from a "sort of gallow" in a local bay and then went to sea. Almost everything worked, but never the whole. Teddy went off to the National Physical Laboratory where the machine shops did his bidding. Art Maxwell, J. M. Snodgrass, and John Isaacs designed and built a different model with thermistors and an amplifier. That was the one that was ready in time for the Midpac expedition.

THE MIDPAC EXPEDITION

At last we went to sea. What we found soon began to be known from the messages Roger Revelle radioed to Scripps and from the press conferences that followed the return of each expedition in those days. Thereafter, we talked at professional meetings and communicated in the usual ways with the other scientists who were studying the same problems. All that was preliminary to the formal statement of results, which generally took two or three years. Delays are to be expected in the normal course of science because of recalculating and confirming measurements and editorial delays. They were compounded in the early 1950s in marine geology because new expeditions were at sea before previous results were published and the earlier work became outdated and might never appear. Consequently, the in-group or invisible college of oceanographers was far ahead of the scientists who merely read about the results. This led to some unfortunate misunderstandings about what was known and what was still a matter of speculation. Perhaps the misunderstandings were inevitable because the individual talks and eventually publications were directed largely toward specialists who might be interested in refraction seismology or sea-floor morphology but not often both. Thus the significance of the whole range of shipboard observations was scarcely appreciated except by the participants. This point cannot be too strongly emphasized. The involuntary confinement of different kinds of scientists for months on small ships helped to forge geophysics into the instrument that could demonstrate plate tectonics when needed.

In the early 1950s the only oceanographic tool that gave results in real time was the echo sounder. Thus the first puzzling result of Midpac was the discovery that the deep-sea floor was hilly. It was only a few days after leaving port, however, that Russ Raitt first measured the oceanic crustal structure; it seems

more logical to begin with his results because they colored all shipboard thinking about the echograms.

Russell W. Raitt (1907–; Fig. 28) was born in Philadelphia. He obtained all
his degrees at Caltech including a Ph.D. in 1935. Then he had a career in oil
exploration. He says that "doodlebugging was getting boring" when he heard
about the University of California Division of War Research in San Diego. The
physics of sound transmission in sea water is the same as for rock layers on
land, so he changed jobs. After the war he combined the old and the new and
began to measure crustal structure at sea. The region off Southern California is
ideal for the purpose because it is a "borderland," intermediate in character
between shallow continental shelf and deep ocean. He had shot enough stations
to present a paper on the structure of the borderland at the Geological Society
of American meeting in El Paso in 1949. Thus all was ready when *Horizon*
stopped on Midpac station #1 in water 4,176 m deep about 300 miles southwest of La Jolla. The accompanying, nameless, Navy ship *EPCE(R) 857* began to drop explosives, the oscillographs recorded the first wiggles, and Raitt's
revolution had begun.

The sediment was not 5,000 m thick; it was only 260 m. The crust above the
Moho was not 30 km thick but only 7.4 km. Moreover, the crust consisted of
only three layers that Russ with typical clarity called the First, Second, and
Third layers. The first was clearly sediment because we cored the top, and the
seismic-wave velocity was typical for sediment. The second, at station M1,
was 0.9 km thick and had a velocity of 5.9 ± 0.2 km/sec. After Midpac, Russ
gave a talk about his results at a geological meeting in Los Angeles. In the
question period, an impatient geologist asked what kind of rocks he was talking
about, basalts or whatever. Russ is sometimes painfully deliberate but he an-
swered directly, "These are rocks that transmit seismic waves at about 5.9 and
7.0 km/sec." He brought down the house. The rest of us, mere mortals, tended
to speculate just a bit.

Russ published his results at last in the *Bulletin of the Geological Society of
America*, which had the slowest editorial procedure ever known, in December
1956.[12] He shot 16 stations on Midpac and 26 on Capricorn, and the extraordinary fact was that he always recorded three layers with relatively constant
properties at the 38 stations in deep water. In part, this was an artifact in that
the actual observations are interpreted as though they are in layers. The kinds
of analyses one can now do with computers indicate a more complex structure
but not much more so. It says a lot for the quality of Russ's old data that they
are still utilized for the most sophisticated computer modeling. The layers as
he found them were quite simple (Figs. 15 D, 16 H), and that is how all of us,
including Hess, visualized them through the discovery of sea-floor spreading
and plate tectonics. The sediment of the first layer was 0.17-1.14 km thick and
commonly was about 0.2-0.3 km or about 5% of Kuenen's estimate. The sec-

ond layer was the most variable, ranging in thickness from 0.35 to 5.4 km, and in velocity from 4.3 to 6.3 km/sec. The third layer was variable in thickness, 3.3-9.6 km, but almost half the stations were between 4 and 5 km thick. The velocity was remarkably constant, ranging only from 6.4 to 7.1 km/sec and with a standard deviation of 0.2 km/sec and a mean of 6.8 km/sec.

What were these layers that shortly would be known to be characteristic of two-thirds of the earth's surface? Russ spelled out how little could be firmly established by his measurements. The first layer is unconsolidated sediment. The second layer, by its variability, probably consists of various materials. Among those with appropriate velocities are limestones, "sial" or granite, schist, chert, and other consolidated sediments. The velocity is not characteristic of basalt as measured in a carefully selected, unflawed, specimen in the laboratory. In the course of these expeditions, however, Russ had determined that in bulk the velocity measured on the shelf around basaltic volcanic islands is only about 4 km/sec. Thus the layer might be basalt. Russ speculated that the second layer is volcanic in the western Pacific, which is peppered with large extinct volcanoes. The eastern Pacific with velocities in the 6 km/sec range, comparable to granite, he thought might have a layer of sial about 1 km thick. Shades of Wegener; echoes of Gutenberg (whom Russ heard lecture at Caltech).

The explosion seismologists at Lamont had not been idle in the interval between the Midpac observations in 1950 and their publication in 1956. Brackett Hersey and others shot a station in deep water in 1949, but the results like Ewing's in the same year were ambiguous. It was not until Officer and others shot numerous stations in 1951 that the crustal section in the northwestern Atlantic began to be revealed.[13] They measured some sections with sediment (1.69 km/sec) above volcanic rock (4.31 km/sec) above 2.49 km of crust (6.64 km/sec) over mantle. The stations that reached the mantle, however, were very few, and they were all in a relatively small area. Meanwhile, Russ Raitt, almost by himself, had measured many more complete crustal sections and showed that they were essentially uniform in an enormous area. Maurice Ewing rightfully deserves credit for taking explosion seismology into the margins of the sea. Russ Raitt has hardly received the credit due him for establishing some of the fundamental characteristics of the deep oceanic crust.

The rocky oceanic crust was not buried under kilometers of sediment, so the reason for expecting it to be smooth had vanished. This was fortunate because the echograms that began to form a loose web across the Pacific showed that the sea floor is hilly (Fig. 7). Dietz, Ed Hamilton, and I worked on the Midpac soundings together,[14] and Bob Fisher and I studied those from Capricorn. I was, however, the only person interested in geomorphology who went on both expeditions—plus three others in between. Thus it transpired that although I was employed at the Navy Electronics Laboratory, I became the curator, perhaps residual legatee is more accurate, of all the Scripps deep-sea soundings that were not special surveys. All the scientists—physical oceanographers, biologists, seismologists—kept the echo sounders going, and I compiled the data and rewarded them with fresh-drawn maps for their next cruises. I don't know just when I became aware of it, but in the course of this ongoing acquisition it became obvious that almost all of the sea floor was an endless expanse of hills. The most common landform on the face of the earth was the previously unsuspected abyssal hill. Largely on the basis of Midpac and Capricorn echograms, I concluded in 1956 that 90% of the Pacific sea floor is a hilly terrain and that the remaining 10% is smooth only because hills have been buried by sediment or fluid lava flows.[15] With more data the percentage of hills dropped to 80%-85%, but the conclusions about smoothing held.[16]

By 1960 or so, the hills were known to be 50-1000 m high and 1-10 km wide from countless profiles on echograms. A typical hill is 200-300 m high and 8 km wide. Length was harder to establish, but a few were 30-40 km long. Considering the thinness of the sediment, the hills could not be sedimentary features. Thus it was established that some process had produced a relatively uniform landform on the entire surface of the relatively uniform oceanic crust discovered by Russ Raitt. I reasoned that the process was volcanic or intrusive. I had moved to Scripps in 1956. My colleague there, Gus Arrhenius (Fig. 27), grandson of Svant Arrhenius of Arrhenius' Law, had pointed out to me that a rising intrusion of dense lava might be expected to spread out as a sill or laccolith between the second layer and the overlying, low-density sediment. Inasmuch as the hills, so far as we knew, were covered with sediment, this seemed a very attractive idea. Little did we know that the crest of the East Pacific Rise is glossy fresh volcanic rock and that the hills and the oceanic crust are produced at spreading centers.

Roger Revelle, the Acting Director of Scripps and organizer and leader of Midpac and Capricorn, took a particular interest in the heat-flow program. When he and Art Maxwell undertook to study heat flow on Midpac, there were few reliable measurements on land because of the heating attendant upon drilling in rock, the flow of ground water, and near-surface climatic fluctuations. The best values ranged from 0.5 to 3 μcal/cm^2 sec (or *Heat Flow Units*), and most were between 0.8 and 1.4 HFU. The mean was 1.2 HFU. The first six measurements, scattered across the Pacific basin on Midpac, by Revelle and Maxwell ranged from 0.9 to 1.3 HFU.[17] Even the 0.9 HFU value was suspect for operational reasons, and all of the remainder were within 10% of the mean continental value.

Revelle and Maxwell sought explanations for their astonishing results. Although the temperature of the deep-sea floor is close to freezing and remarkably uniform it could have varied in prehistoric time. If it was several degrees warmer 500-2,000 years ago, half the high heat flow could be accounted for—

but there was no other reason to suppose such a warming had occurred. Otherwise, all the heat was coming from the crustal rocks and mantle below. Raitt had established the depth to the Moho, so if the crust was basalt, only 15% of the heat originated by radioactive decay within it. Another 20% could possibly be attributed to the cooling of the earth—still a viable hypothesis 40 years after Wegener sought to discredit it. The remaining 65% (or 85%) came from the mantle below with Moho. However, for any reasonable distribution of radioactive minerals it was impossible to remove the heat by conduction before melting would occur below 130 km. Thus from six scattered observations with a 3 m spear plunged into frigid sediment, it appeared from Midpac that the hot mantle, deep in the earth, was convecting under the ocean basins. Paradoxically, similar convection could not be occurring under continents or the heat flow on land would be twice as great.

Teddy Bullard, who was then directing the National Physical Laboratory at Teddington, commented on these results. The effect of a change in bottom-water temperature was dismissed as improbable. Radioactivity might possibly be distributed in the oceanic mantle without convective overturn. It was only ''pure speculation'' that a convection current rose under the Pacific ''at some not too remote time'' because there was no other evidence in favor of the idea.

The evidence proved Teddy wrong, although the nature of the evidence might not have made it compelling to a physicist. Holmes and du Toit, for example, proposed that mantle convection is the cause of continental drift. Vening Meinesz offered mantle convection as the cause of oceanic trenches and the associated gravity anomalies. Hess, too, had done his early fieldwork in a submarine measuring gravity over trenches. Knowing of the geology of adjacent islands he was able to make the cogent point that the trenches and gravity anomalies had persisted for 50 million years (Ma).[18] Only mantle convection appeared to explain the persistence of intense crustal deformation for so long a period. Not very long before Midpac, David L. Griggs introduced another concept into the timing of mantle convection.[19] If surface cooling could not dissipate the heat brought up by convection, an overturn would occur, the temperature gradient would vanish, and convection would cease. The overturn cycle lasted 100-200 Ma. Hess, who was already attracted to convection currents in 1951, did not fail to use them in the years that followed to explain the origin of the oceanic crust, the subsidence of guyots, and finally the spreading of the sea floor. Hamilton noted that a cooling, contracting mantle convection cell could account for the regional subsidence of guyots. I appealed to convection to explain fracture zones and the evolution of mid-ocean ridges. The sudden popularity of convection derived from several factors, but I think a one-two punch on Midpac was critical. First, there was the heat flow and then Hamilton's discovery that the multitudinous guyots of the Midpac Mountains are about 100 Ma old.[20] Somehow it seemed that the two phenomena were linked by Dave

Griggs's experiments and that convection was not merely the pure speculation of Teddy's conservative comment. Certainly, the evidence for continental drift did not enter into our discussion.

The guyots that Hess discovered when he was 38 were never long from his thoughts thereafter.[21] Many of our conversations and letters over the years were concerned with the clues that guyots gave about the development of mid-ocean ridges. Thus it was a pity that Hess was not present when the mid-Pacific guyots began to be sampled from *Horizon* on 1 September 1950. More fortunate people included the established men, Dietz, Emery, and Revelle, but the man of the hour was a 35-year-old graduate student from Stanford, retired Marine Corps colonel Edwin Lee Hamilton (1914–; Fig. 29). Ed had attended an obscure school called North Texas Agricultural College and then transferred to Texas A and M to graduate in civil engineering. He entered a career as a regular officer in the Corps that was ended in 1943 by a Japanese mortar shell in the dense jungle of Bougainville. He came to visit Dietz at our section on Marine Geology in the Navy Electronics Laboratory late in 1949 and expressed interest in a thesis based on fieldwork on our ships. Dietz agreed to assign all rights and responsibilities related to the as yet unsurveyed and unsampled—and, at that time, unnamed—Mid-Pacific Mountains to Ed for his thesis. Ed went up to the University of Southern California to talk to Kenneth Orris Emery (Fig. 29) who had surveyed some guyots among the atolls of the Marshall Islands near where we would be working. As it turned out, K.O. had also attended North Texas Agricultural College, so the school has a distinguished record for producing pioneering marine geologists.

We lost the first dredge when the cable parted—a not infrequent event. Deep-sea dredging for bedrock is like fishing for tuna with trout line. We recovered a few small core samples from the top of Horizon guyot, but they were just enough to tease. Ed's Memoir was finally published in 1956, and it summarizes the record of disappointingly tiny samples, smashed core barrels, and the long struggle to obtain datable material instead of chips and manganese, basalt, and unfossiliferous limestone. The fourth guyot sampled we named after Harry Hess, and on the twenty-second effort at sampling, we pulled calcareous rock from the dredge. It resembled modern beach rock. Later, after almost a year of preparation, on the ninth day of dredging, on the thirty-second try, Ed Hamilton was suddenly confident that he had a doctoral thesis. The dredge was full of Middle Cretaceous reef corals. When the magnitude of the discovery was appreciated, a senior scientist approached Ed about a joint paper, but Dietz, as his advisor, held firm against it. A commitment was a commitment; a graduate student must be allowed to do independent work. Ed's professors at Stanford later told him he could have submitted his results on the back of an envelope and gotten a degree.

Hess was right about guyots being drowned ancient islands, but Dietz was right about when they drowned. The geological history of deep-ocean basins was at last leaving the realm of pure speculation. At the time the Cretaceous fossils were unexpectedly young, but by a curious chance the first datable fossils dredged were also the oldest recovered during the next 15 years. As the number of samples slowly increased, this fact was to loom increasingly large. Where were the older rocks in this primeval Pacific basin?

CAPRICORN EXPEDITION

In most respects the Capricorn expedition of 1952-1953 (Fig. 5) was merely more of the same, but in the South Pacific.[22] Revelle was leader. Art Maxwell, who was out for the whole five months, made seven more determinations of heat flow—mean value 1.3 HFU. Russ Raitt measured the structure of the oceanic crust at 24 more deep stations; the thicknesses were almost the same as on Midpac.[23] The velocities in the upper mantle, however, had a larger range, from 7.4 to 8.8 km/sec, and it appeared that regional variations existed. A decade later, Hess and I each proposed a cause for these variations. They were real. Many other prominent oceanographers, such as John Isaacs and Walter Munk, were on this expedition, but several more obscure (at that time) scientists were to become more important in this account. One of these was a graduate student R. P. von Herzen who would find the first organized pattern of heat flow associated with a mid-ocean ridge on our Downwind expedition in 1958. Another was Arthur D. Raff (Fig. 31), an engineer who had been improving the characteristics of Raitt's hydrophones. He sailed only from San Diego to Kwajalein and thus missed the entire scientific cruise under Revelle, but a few years later he would be helping to map the magnetic stripes off California. Meanwhile, the future discoverer of the stripes, Ronald Mason, joined the expedition at Suva, and from there to San Diego he collected our first magnetic profiles. This work, in itself, was mainly notable in proving what could be done and in setting the stage for the *Pioneer* survey to which we shall return. Before that, however, I want to describe the early work on another type of tectonic feature that was explored in the course of these expeditions—the great fracture zones. Although they were only discovered on the last days of Midpac, by the first leg of Capricorn two years later we had learned enough about them to inspire Ed Hamilton with confidence in my maps. A few days out of San Diego, with no star fixes, he told the captain of *Horizon* that the ship was way off course. The depth showed him he was south of the Murray fracture zone, a feature on no published map, and the ship was supposed to be in the deeper water to the north. The captain found that his aiming point had been misplotted to the south by exactly one degree of latitude.

Chapter 5

DISCOVERY OF FRACTURE ZONES AND SEA-FLOOR PROVINCES

After I received my Ph.D. at Harvard,[1] my wife wanted to return to California, and my friend Edwin Conger Buffington said that the U.S. Navy Electronics Laboratory in San Diego was a good place to work. Thus I joined him in the Sea-Floor Studies Section, headed by Robert Sinclair Dietz (Figs. 29, 36, and 37), which was a subdivision of the Oceanographic Branch under Eugene LaFond. This was a very fortunate association for me. In most government laboratories, particularly those run by the military, the spirit is bureaucratic; action, if there is action, is in response to directives, and initiative is suspect. Indeed, many parts of NEL seemed to have those very characteristics. Dietz, with the cheerful support of LaFond, assumed the Navy had no way of knowing enough about marine geology to direct our activities to its own advantage. We should work hard at basic research with the hope that what we discovered would be useful. Management growled, but I noticed that we were promoted faster than most.

I had hardly arrived, in June 1949, when everyone else in the section departed for a cruise to the Beaufort and Chukchi Seas of Alaska. That left me on my own, and I began to study the soundings that covered large sheets of paper in vast stacks of locked filing cabinets. The soundings were classified but at the lowest level—"Restricted"; in any event, as a Naval Reserve officer and Naval employee I had clearances.[2] The files contained every known sounding ever collected in the Pacific Ocean all compiled at 4 inches per degree of longitude. Most of them had never been seen by any scientist, and as far as I know I was the first person to study all of them. I had taken courses in geomorphology and spent three and a half years as a photo interpreter during the war, so I was quite prepared to try to make geological interpretations of the soundings. As it turned out, it was not only my job but a priceless opportunity even if no one realized it at the time. Much later, Harry Hess was to point out to me that whereas land geomorphologists merely study the surface of the continents, I was contouring the surface of the mantle. Later still, the extent of my good luck was brought home to me when Allan Cox told me he used my early work to illustrate to his students that even the most unpromising career choices may lead to something.

The sounding compilations were a mess. The bureaucrats of the Hydrographic Office had recorded not only every correct sounding but every conceivable form of erroneous sound.[3] Navy ships had collected sonic soundings as a routine part of keeping watch with no quality control. These suspect, but possibly correct, numbers were then compiled—again with no quality control. Digits were dropped and transposed with careless abandon. Then one track after another was superimposed in the main traffic lanes while in little-traveled regions they often were misplotted by even amounts, commonly 1° or 10° of longitude. The plots almost always had the minutes plotted correctly; generally it was only in large matters that the bureaucrats went wrong. The exceptions were at the equator, date line, and Greenwich longitude where the directions of latitude and longitude change. There all positions were suspect. Even now, the latest published bathymetric chart of the South Atlantic has a seamount plotted at the right distance but on the wrong side of the longitude of Greenwich.

I spent the summer devising internal checks for this confusing material, which we would never unscramble completely until we had collected our own crossing network of recorded echo soundings. Meanwhile, a hint of a remarkable feature began to appear on my charts. In a general way the sea floor slopes away from the central part of North America toward Hawaii. This was shown on published charts in 1950, so it was not remarkable that my charts confirmed it. It did set me thinking, however, that this slope must have a cause and that the northeastern Pacific must be peculiar in some way. Elsewhere in the Pacific the floor sloped down toward marginal trenches. What was even more remarkable was that west of northern California it appeared that the slope was offset. The soundings were sparse, but some kind of unknown boundary seemed to follow the fortieth parallel.

Being new in the field I consulted the literature. Henry Stetson, my thesis advisor at Harvard, had introduced me to the pioneering work of Francis Parker Shepard and K. O. Emery off the California coast, and here I was only a few miles from Shepard in La Jolla. Dietz, who had been Shepard's student, introduced me before he left for the Arctic, and all the publications I needed were convenient. They were few, which is one of the best reasons for entering a new field if it isn't too unpromising.

H. W. Murray was a cartographer with the U.S. Coast and Geodetic Survey, and in addition to his assigned charting he studied the new sounding lines across the Gulf of Alaska. The C and GS, like the Navy, had an enlightened policy of spacing sounding lines to obtain uniform coverage of the sea floor whenever time and budget permitted. The difference was that the C and GS ships were staffed by professional cartographers and they collected reliable soundings. The only time I met Murray he told me that his scientific publications were all written on his own time and, he felt, despite the disapproval of his supervisors. Not long after, and apparently overworked, he committed suicide. By 1939, Murray had discovered that the continental slope off Cape

Mendocino, California, is offset along a steep, north-facing scarp that trends east-west for 70 miles.[4] It is the only such offset in North America. Shepard and Emery soon used C and GS soundings to map the same area and noted that "sporadic" soundings suggested that the scarp continued at least 400 miles to the west.[5] The C and GS continued to extend its network of spaced sounding lines, and the existence of a scarp at 200 and 300 miles offshore was established although its continuity was conjectural.

I discussed this work with Dietz when he returned from the Arctic seas. We then began a collaboration that was to result in coauthored papers on the origin of the shelf break; the marine geology of the Gulf of Alaska; the Mendocino escarpment; and the Hawaiian deep, swell, and arch, as well as a three-author paper with Ed Hamilton. Our practice was to identify a subject for a paper in the course of ongoing discussions. The whole marine geology group was jammed into a few tiny rooms with a leaky roof, so communication was constant. When we had a subject, Dietz and I sometimes flipped a coin to decide who would be first author, with the understanding that whoever it was had to write the first draft of the manuscript. And so we produced the five papers. I confess that I sometimes have to consult my bibliography to see which of us wrote the Gulf of Alaska and the Hawaiian papers. On the Mendocino paper I certainly identified the target, did all the fieldwork on two expeditions, and wrote the first draft and drew the figures, but what determined who would do these things I do not recall. In any event, the ideas in all four papers were the consequence of a thorough amalgamation of our discussions.

Roger Revelle began to generate the Midpac expedition about this time and soon drew Dietz into the planning. No one had any plans for *Horizon* on the return leg from the Marshalls to San Diego. We managed to insert a few days of ship time, at a princely $300 per day, into the schedule to allow me to sail far north of the direct route and reconnoiter the suspected scarp.

DISCOVERY OF THE ESCARPMENT

Almost all the scientists flew home from Kwajalein in September of 1950, and Dietz, among others, never went to sea again on a Scripps ship. With a thin scientific group consisting of Jeff Frautschy, Lou Garrison, and myself we headed toward Midway Island on the long run toward the Mendocino anomaly. The logistics required that I survey from west to east, and thus I had to guess the location of the scarp about 900 nautical miles west of the coast and 600 miles west of the nearest certain crossing by the Coast and Geodetic Survey. About 1 October, in the usual appalling weather of the roaring forties, the time had come, and trailing persistent gooney birds I headed the ship more to the north. The echo sounder soon recorded the first of what turned out to be one of the standard profiles of a fracture zone. Starting in very deep water, we crossed a low asymmetrical ridge tilted south, a narrow trough, a steep scarp, 6,000 feet high, a ridge tilted north, and a low swale, and ended up in a relatively shallow region. It would be 30 years before enough was understood about heat flow and flexural rigidity of the crust to explain the origin of all these features in detail. Meanwhile, I made three more crossings—all more or less the same even though they were 100-200 miles apart. We reached San Diego on 28 October, and the discovery of a new kind of sea-floor "mountain range" was announced in the newpapers.

Dietz and I submitted an abstract, and soon I delivered a talk on the "Mendocino Submarine Escarpment."[6] I sent copies of a manuscript, requesting comments, to my former professors of structural geology, J. P. Buwalda at Caltech and Marland Billings at Harvard, and to A. J. Eardley, the master of the tectonics of the west coast, who was at the University of Utah. Dietz sent a copy to Hess. Billings found our arguments for a strike-slip origin to be unconvincing. Hess complimented Dietz on picking one of the two "most interesting outstanding problems" in the Pacific. Hess was "partial to a strike-slip hypothesis." Hess asked permission to quote liberally from this and the previous manuscript Dietz had sent him "citing you and your coauthors as the source." The other manuscript was on the Gulf of Alaska, and I was the first author of both, and thus to Hess doubly anonymous. The Matthew effect was at work.[7]

On 1 June 1951 we sent our paper to the *Journal of Geology*. On 3 June, Warren Wooster informed me he and John Isaacs were organizing an expedition, "Northern Holiday," in the Gulf of Alaska and asked if I would like to resurvey the escarpment. *Horizon* sailed on 28 July and reached Kodiak on 26 August.[8] I had an uneasy feeling that this sort of thing might go on year after year, so I used our new soundings to prepare a map while still at sea. I also revised the manuscript to include the new data. We resubmitted the paper in time to have it accepted only a month after I arrived in port. The editor, Francis Pettijohn, expressed "thanks for giving us the opportunity to publish this interesting paper on the extraordinary Mendocino escarpment." I record this chronology and some of the responses as examples of the style and novelty of the early days of the exploration. Today it would take longer to get an NSF grant than it took then to complete two expeditions. Today the Office of Naval Research will not even fund geological exploration with the kinds of instruments we were using—nor would we use them.

The second survey had not been easy. The echo sounder would not record in deep water, and graduate student Harris B. Stewart, Jr., and I (Figs. 30 and 36), assisted by the scientists standing watch, had to determine the depth at 5-minute intervals by the pings and flashing lights of the standby system. Typically, Stew, tied upright in front of the sounder, would call out soundings and I would record and plot them. I contoured a tentative map in real time in order to determine when to turn. We once did this for 32 straight hours. By these means we obtained 18 more crossings with an average spacing of 50 miles but no more echograms. The scarp was at least a mile high and remarkably straight for 1,200 miles along the fortieth parallel. At least it appeared straight on a normal Mercator-projection chart. Actually, following a line of latitude as it did, it was a curve on the face of the earth. We did not think in such terms at the time, but this very curve, then unique, would become the epitome of the transform fault along a small circle in Jason Morgan's plate tectonics 17 years later (Fig. 24 C). Meanwhile, for an unknown coauthor it was a lucky haul: the most prominent fault scarp on earth.

Dietz and I reasoned that the most important fact we had uncovered was not the scarp, spectacular as it was, but the existence of regions with different depth that were separated by the scarp. The regions are so large they must be isostatically compensated and the difference in depth (half a mile) must reflect a difference in crustal density. We equated the Moho with the depth of isostatic compensation, according to the custom of the time, and estimated that the shallow simatic crust north of the scarp was 10 km thick or twice Russ Raitt's central Pacific average value from Midpac. The crust south of the scarp was normal, judging by its depth, so the escarpment was equivalent to the continental slope—a sharp, steep boundary between different types or thicknesses of crust. Unfortunately, the origins of the continental slope itself was conjectural. It might be a normal, vertical-slip fault, or a transcurrent, horizontal-slip fault, or not a fault at all. The transverse profiles from the four Midpac echograms showed the typical asymmetrical ridges and graben of normal faulting. The long straight trace on the chart was typical of large-scale transcurrent faulting, but the sounding lines were too far apart to prove that, in detail, the scarp did not resemble a normal fault. We favored strike slip on the grounds that the San Andreas fault would be a complementary shear. This was a completely erroneous analysis, but it seemed reasonable at the time.

DISCOVERY OF THE GREAT FRACTURE ZONES

Our Mendocino paper was published promptly, as journals go, in May 1952, but by then I had discovered a whole system of comparable features and the long-awaited printed word was an anticlimax. I had left the Northern Holiday expedition at Kodiak, but Harris Stewart stayed on in charge of the geological program. The echo sounder was repaired. *Horizon* sailed south and crossed the Mendocino trend 200 miles farther west than when I was aboard. The scarp was there, but it was 60 miles south of the trend. After, 1,200 miles due west, the trend had changed. Stew continued sounding wherever Warren Wooster's study of physical oceanography took the ship as it zigzagged toward La Jolla.

Sometime after Stew's return I had his new data, the first echograms in a million square miles, plotted on the old sounding sheets. The geological information that could be extracted from our echograms was incomparably greater than from the old soundings. We had a continuous profile of reliable data that revealed fault scarps and troughs, volcanoes, guyots, flat plains, and so on. Stew had found a remarkable trough and made two crossings of a scarp about 700 m high. The older Navy soundings in the vicinity of these features were particularly vexing in two ways. First, they tended to vary erratically, and second, a ship that ran a continuous line of soundings for a thousand miles would just have a gap near the tectonic features. I remembered the trouble we had had determining the depth on the steep slopes of the Mendocino structure. I drew the boundaries of the vexing soundings and obtained a narrow band connecting Stew's three faults and trending almost parallel to the outer 200 miles of the Mendocino escarpment. Through the purest of chances the floor of the Pacific where we were exploring tends to have very low rolling hills as opposed to the Atlantic, which is all mountains except where they are buried by sediment. Consequently, the band of rough topography stood out in the Pacific.

I then knew of two ways to find what I would call "fracture zones"; I could look for narrow bands of linear ridges, troughs, and volcanoes, or for regional changes in depth. So guided, I found another zone to the south. It was dominantly a trough. Thinking to use alliteration as a mnemonic aid, I began to call them the Mendocino escarpment, the Murray escarpment, and the Maury trough from north to south. "Murray" was named after the nearby deep, which had been named after Sir John of the *Challenger* rather than after Harold Murray of the Coast and Geodetic Survey. "Maury" was the nineteenth-century American oceanographer. When first discovered, the two southern submarine features were totally unassociated with dry, named land, so I needed marine names. The redoubtable Warren Wooster was preparing to lead another expedition, which he was prevented from calling "Southern Holiday." The more stolidy named Shellback expedition was to leave on 17 May 1952.[9] Warren agreed to specially monitor the echo sounder when crossing the expected position of the Maury trough and let me know what happened. In due course he arrived at about 18° north latitude and radioed back that the trough was as predicted.

Shirley Fisher,[10] then Bob's wife, was plotting soundings for me, and she recalls very well the excitement of the next few days as we worked at a feverish pace and Warren headed farther south. The three features were close to parallel, which brought to mind the remarkable work recently published by F. A. Vening Meinesz (Fig. 33). The famous Dutch geophysicist had run out of unprocessed data during the German occupation of Holland during World War II. Like Wegener in the previous war, he turned to speculation on the deformation of the earth.[11] He reasoned that on a global scale the crust must deform plastically. Such deformation in laboratory experiments, whether in compression or tension, occurs within the otherwise undeformed solid by formation of narrow bands called Hartmann or Lüder's lines. The bands form two complementary sets that intersect at an angle of 110°. Vening Meinesz, with time on his hands, compiled a global map of tectonic lineations and sought a global cause. He found two that would produce the pattern: a change in the axis of rotation or a change in flattening attendant upon a decrease in the rate of rotation. I was in the process of discovering a set of long tectonic lines where he had no data so that I could test his predictions. Moreover, the Lüder's lines tended to be spaced evenly, so I had a basis for predicting not only the direction but the distance to the next undiscovered fracture zone to the south before Warren got there.

I radioed him to please monitor his sounder carefully at a certain position, and not many days later he radioed back that he had found another scarp as predicted. Jubilation! Sometime after this I decided that the new zone really should be named after Clipperton Island, which lay on it. Sticking to alliteration I renamed the Maury zone after Clarion Island, which I had realized was on it.

Warren was, however, still at sea, and I had not seen his soundings, so I mentioned only three fracture zones when I wrote to Vening Meinesz in June of 1952. He responded that he was "deeply stirred," which was to be expected because he was in the same position as Wegener receiving confirmation of the shift of longitude in Greenland. There is nothing like a confirmation of a prediction to cheer a scientist. He closed with "hearty congratulations." Unfortunately, neither prediction held up for very long. I wrote to Vening Meinesz again in October after Warren's return. There were four zones in the set, and the southernmost did not agree at all with his prediction. His mechanisms might still work, but I was attracted to an origin by the drag of a giant convection pattern under the Pacific. I had to take the heat flow into account. Moreover, his deformation had to be very ancient, and I had found that "near shore" the fracture zones are seismically active. A more significant statement would have been that they were active on the crest of the East Pacific Rise, but no one knew the rise extended into the northeastern Pacific at the time. I still had no clues about the slope of the sea floor from California to Hawaii. I apologized for my sketchy letter but said I was about to go to sea for three months.

I was not on the first leg of Capricorn because little could be done. The ships had to meet a tight deadline in connection with the test of the first hydrogen bomb "device" in the Marshall Islands. The track from San Diego to Kwajalein, however, was near the Murray fracture zone, and I was allotted surveying time of 12 hours or so on each ship in excess of that required on the most direct route. What was the most direct route? I obtained a great circle sailing chart, a special projection on which a straight line corresponds to a straight line on the spherical surface of the earth. Latitudes and longitudes look very odd. I drew a straight line between San Diego and Kwajalein and calculated the time the ships would need at cruising speed. Then I plotted the sparse data on the Murray fracture zone—they, too, fell on a straight line. The Clarion fracture zone was plotted; for 1,700 miles it deviated from a great circle by no more than 15 miles. The Clipperton zone followed a great circle for at least 3,300 miles. There was nothing known on earth that was remotely like these fracture zones. No two great circles can be parallel, but these were nearly so, and so was the western end of the Mendocino fracture zone once it deviated from the fortieth parallel. This totally unexpected discovery ultimately turned out to be grossly misleading with regard to the origin of the fracture zones. The great circles were actually successive arcs of small circles swinging around successive Euler poles.[12] Jason Morgan would have to ignore the apparent significance of these remarkable great circles in order to synthesize plate tectonics. All that, however, was far ahead. The fracture zones were not yet surveyed.

The Capricorn expedition began and *Horizon* and *Baird* each zigzagged 10 times at low angles within the Murray fracture zone and demonstrated for the first time the continuity of individual ridges and troughs for hundreds of miles. So much for so little time. It was strange that the first fracture zone followed a line of latitude and the second a great circle along a required course. This really was nautical topography.

I discovered the Marquesas fracture zone on Capricorn, so I knew that such zones extended to the southern hemisphere and were roughly parallel across an area of almost 10% of the earth's surface. By the time the expedition reached port we had two more crossings of the Clipperton zone, a total of six echograms in 3,300 miles, about one every 500 miles. When the first Lamont ship later entered the eastern Pacific it crossed over this zone. I think it was Chuck Drake who said it was pretty obvious and asked me what took me so long to find it.

I returned home on 21 February 1953 and submitted a manuscript to the Geological Society of America in October. They published it 23 months later. The GSA and I both sent manuscript copies to Harry Hess to review. I pointed out to him, a fellow officer in the Naval Reserve, that he would doubtless want to see large detailed bathymetric charts instead of my small diagrams and that I had them, but for "reasons of security" I was not attempting to publish them. When Genral Eisenhower became President he curtailed the obsessive military security, which he knew was counterproductive. He abolished the classification "Restricted," which should have made the soundings open information. However, the Navy exercised its option and instead reclassified the soundings upward to "Confidential," so I could not publish them at all. Fortunately, the secretive Navy bureaucrats had no policy regarding, or probably knowledge of, echograms or physiographic diagrams. Thus they did not limit publication of a track chart and an echogram for the track, or a plot of such echograms on a chart. It was this fact, and not desire, that drove marine geologists to substitute physiographic diagrams for contoured bathymetric charts and ultimately earned two gold medals for Bruce Heezen and one for Marie Tharp.

Hess was generous in his report to the GSA. "The results of this study certainly must mean wholesale revision of generally accepted ideas on large-scale tectonics. . . ." I had made a "valiant effort" to explain the phenomena, but Hess doubted that my preferred explanation would long survive. He also observed that fracture zones could not be found unless one was looking for them and speculated [as a reviewer] about several possible locations. In his covering letter to me on 24 January 1954 he was again highly complimentary, and my joy at this recognition from the master was only slightly muted by the fact that it was addressed to Howard S. Menard.

In my paper with the grandiose title "Deformation of the northeastern Pacific Basin and the west coast of North America" in September 1955, I described the regional bathymetry and the topography of the fracture zones to the extent that security permitted. I showed that the region is divided into physiographic provinces with different depths, numbers of volcanoes, and turbidite deposits, and that the boundaries between the provinces are fracture zones. I also showed that structures with abnormal westerly trends lie along the continental extensions of the enormously long, straight fracture zones.

Those were the observations, but what on earth did they signify? I was 32 years old, only three years past my doctorate, and I had discovered a whole new class of major geological structures. They resembled nothing known on land. Indeed, they seemed to display a bizarre combination of the classical features of both horizontal-slip and vertical-slip faults. Moreover, they were individually enormously long, and collectively they spanned 5% of the area of the earth. They had to be caused by some global or very large-scale phenomenon. What could it be? None of the existing hypothesis of global tectonics had predicted the existence of fracture zones. The mere fact of their existence, however, was hardly enough cause to propose a new hypothesis, so I attempted to explain the fracture zones in terms of the old ones.

Before seeking an ultimate cause I attempted to explain the regional differences in depth and the spacing and parallelism of the fracture zones. As to depth, it was possible to eliminate the influence of sedimentation and volcanism because the sea floor consisted of an endless field of abyssal hills. It was the top of normal oceanic crust unburied by anything. My views on the relation between crustal structure and depth of the sea floor had changed since the Mendocino paper. Raitt in the Pacific, and Officer and Ewing in the Atlantic, had made many more measurements of the oceanic crust, and it was always the same thickness. Thus I attributed the differences in depth of the provinces between fracture zones to differences in density in the mantle below the Moho. I had taken the critical step of distinguishing the seismologists' "crust" from the gravimeterists' "crust." Joe Worzel, the principal investigator of marine gravity at Lamont, never accepted this distinction. I had no doubts because I was familiar with Russ Raitt's abundant observations. So what caused differences in density below the Moho? Presumably, it was something related to the ultimate cause.

I thought that the distinctive geography of the fracture zones was probably related to the physical properties of the crust rather than to the ultimate cause. The fracture zones formed a system of almost parallel, narrow bands of deformation that were widely spaced within undeformed crust. Moreover, the zones were somewhat elevated on the average, even though they included deep troughs. I assumed they were isostatically compensated, and if so they had a root. In other words, the deformed zones were thicker than the undeformed regions between them. I interpreted the fracture zones as almost ideal examples of Lüder's lines on a gigantic scale. They were equivalent to the thickened, straight, narrow bands of plastic deformation that occur in an otherwise undeformed metal sheet when it is compressed in a laboratory (Fig. 8 A and B).

But what caused the compression? To answer this question I had to consider the available ideas on global tectonics in the early 1950s. These ideas had been summarized and discussed in 1951 by Beno Gutenberg and at a colloquium of geophysicists in 1950, so it is easy to reconstruct the thinking at that time. Gutenberg began with a generalization that set the tone of the paper:

> Books and papers dealing with hypotheses on the development of the earth's crust are as the sands of the sea.[13]

He then went on to show that the hypotheses of crustal contraction, subcrustal convection, polar wandering, and continental drift of some sort were all viable at that time. Evidence that could not be refuted seemed to support each of them, and the importance of each was only a matter of opinion. Global tectonics was in a sorry state.

The Colloquium on Plastic Flow and Deformation within the Earth was held in Hershey, Pennsylvania, on 12-14 September 1950. The list of invited, supported participants gives a fair sample of the major figures in global tectonics at that time. Included, among the people who appear elsewhere in this book were Benioff, Birch, Bucher, Griggs, Gutenberg, Hess, and Rubey from the United States; Bijlaard, Umbgrove, and Vening Meinesz from or formerly from the Netherlands; and Bullard from England. There was a thorough discussion of plastic deformation in the crust and its relation to geosynclines, trenches, and other major geological structures. The deformation was thought to be a result of compression and "the cause was sought in convection currents below the crust."[14]

I tried to sift these papers, like the sands of the sea, for clues regarding the origin of the fracture zones. Not surprisingly, considering what their origin turned out to be, I was not very successful. I fell back on the method of multiple working hypotheses in which every known possibility is considered equally. The hope is that a scientist thereby will avoid premature commitment to a favored hypothesis. My own rather lengthy experience with this method leads me to believe that it is the only way to reason about a scientific problem. A scientist who comes to the point of writing about such reasoning, however, would do well to conclude that the available data are inadequate to solve the problem. Such an approach would have reduced the volume of geological literature by half or more in the 1940s and 1950s.[15]

I believed that I could eliminate contraction as the cause of the fracture zones. That left either a double migration of the pole of rotation or Pacific-wide mantle convection to explain the deformation. I favored the latter for several reasons. It certainly was simpler. Moreover, Harvey Brooks had proposed in 1941 that such convection should exist because of mantle cooling in the central Pacific.[16] The pattern of convection that he predicted was in the right place and in the right direction to produce the fracture zones by plastic deformation. The geographical correspondence was remarkable even though it turned out to be meaningless a decade later. Meanwhile, it had a powerful influence on my exploration of the Pacific. I kept trying to establish whether the fracture zones ended where Brooks predicted that horizontal convective flow changed to a divergence. Hess and I were to correspond on this question.

Granting the existence of a large mantle convection cell, I believed that small surficial eddies might be generated and confined between the roots of the fracture zones. Thus the regions between would have different thermal histories. I correlated an area of greater heating with the shallow region between the Murray and Clarion fracture zones. This was also an area with extraordinarily abundant volcanoes, which seemed to be a corroborating indication of heating. In contrast, the region between the Murray and Mendocino fracture zones was deep and volcanoes were very few. This appeared to have a cooler mantle. The correlation however, was not perfect. The region north of the Mendocino zone was the most shallow of all, and volcanoes were present but not abundant.

I distinctly remember thinking some time late in the evolution of these ideas that somewhere in discoveries of this scale there must be a Penrose Medal waiting. Geologists, unlike molecular biologists, do not think of Nobel prizes. I was right. Tuzo Wilson received it 15 years later for his insight in converting fracture zones into transform faults. Meanwhile, there were other rewards. I attended my first international meeting, the International Geological Congress in Mexico City, in 1956, to talk about marine geology. I learned of a symposium on the tectonics of western Central America and wandered in to find a debate on the continental extensions of the Clarion and Clipperton fracture zones. The Chairman, J. V. Harrison, a famous geologist from Oxford, asked that each commentor or questioner from the floor identify himself. I had a comment to make after a talk, arose, and identified myself. There was a pause, and the Chairman said, ''If I may say so sir, it is a pleasure to see you in the flesh.'' I realized that I might never experience such a moment again.

Fortunately, I did not know that the Matthew effect was in operation. When Dietz and I were collaborating, I had just received my doctorate and published only two obscure papers, whereas Dietz had published 14 papers, mostly in major journals, and was an established geologist. Although we had early hints, we did not realize it when the Matthew effect then took over. Even much later I did not realize what had happened because I still had not heard of the Matthew effect. At that time Tuzo Wilson as Chairman of the award committee for the Bucher Medal proposed to the other commitee members, including me, that the medal be given to Dietz. Everyone was enthusiastic, time was short, and we were not all on the same continent, so we all approved the Chairman's draft of a citation as soon as it was received. The citation acknowledged his contribution in discovering the Pacific fracture zones. I would have preferred one of his many individual contributions, but under the circumstances it seemed acceptable. The important thing was to give him some of the praise he deserved. A few years later a friend informed me that my biography was in the new version of the *Encyclopedia Britannica*, so I hastened to read it and see who else was in the books. There was Dietz, whose principal contribution it seemed was the discovery of the Pacific fracture zones. He was as surprised as I, since neither of us had been interviewed. At last I began to wonder and, upon inquiry, discovered that Tuzo Wilson and Teddy Bullard both thought that Dietz had discovered all the Pacific fracture zones. This despite the fact that by the time of my inquiry Teddy and I had been shipmates on three expeditions and I had spent two sabbatical years with him in Cambridge. Once Dietz and I were side by side on a stage being honored by a corporation because we had been among its founders 30 years before. Among Dietz's many achievements mentioned by the master of ceremonies was the discovery of the Mendocino escarpment. We both burst into laughter. All this because he was the beneficiary of the Matthew effect long ago.

GLOBAL DISTRIBUTION OF FRACTURE ZONES

During the next few years by far the most important development regarding fracture zones was the discovery of the offsets of magnetic stripes that will be recounted in the next chapter. It is useful at this point however, to describe the continuing discovery of fracture zones because the timing influenced the evolution of tectonic ideas. From the time of their discovery it was realized that such zones might be global phenomena that would be found in all oceans, but through the fifties they were known for certain only in the Pacific (Fig. 8 C). There the number known slowly increased as I acquired more data. In short, there was no early refutation of my hypothesis that they were produced by a sub-Pacific convection current. By 1959, I had added the Easter fracture zone and a conjectural Eltanin zone. They extended through the whole eastern Pacific. George Shumway believed that he had found another fracture zone running west from the Galapagos Islands. The echograms never looked quite right to me and there was no regional change in depth, so I always wondered what it was. Two decades later it turned out to be a spreading center.

By October 1955, Harry Hess not only had my name straight but was writing to me as "Dear Bill" to thank me for the data I kept sending him. Bruce Heezen and I constantly exchanged data, but Hess had none so I gladly gave him data and exchanged ideas. I had proposed that the Clarion and Clipperton fracture zones extended across Central America and formed the northern and southern boundaries of the Caribbean. The assumption was that the Pacific convection merely flowed under central America. Hess proposed that the Clarion zone continued as the Barracuda fracture zone as far east as the middle Atlantic.[17] He was quite proud of this zone because he had, unknowingly, discovered it himself in the late 1930s on the U.S. submarine *Barracuda*. He had collected a sounding line across an asymmetrical ridge with a trough at the base—a typical fracture zone—and he remembered it. The next time he visited the Navy Hydrographic Office in 1955 he inspected the sounding compilation sheet for the area to see if the feature could be traced. A fine bureaucratic hand had annotated the compilation sheet beside his personal trough, "ignore, obviously erroneous."

Hess also proposed that the Romanche Trench was a fracture zone, perhaps an extension of the Clipperton zone. This "trench" was a deep east-west trough. Physical oceanographers had shown it was continuous across the Mid-Atlantic Ridge by comparing the water masses on each side. There had to be a deep sill. Hess found that the western end of the Romanche Trench connected to a great scarp leading to St. Paul's Rock. It is one of the few oceanic islands composed of peridotite and serpentine, which were just the crust and mantle materials that Hess expected to be exposed in a major fault.

Bruce Heezen was looking for fracture zones in the Atlantic just as I was seeking median rifts in the Pacific, but he had no luck. His basic problems were three. First, the Atlantic sea floor is mountainous, so fracture zones do not stand out as they do in the Pacific. Second, if he did find a trough or asymmetric ridge on a single echogram, he tended to think it was a manifestation of the median rift. Third, everyone tends to cross the Atlantic on east-west lines, which are more or less parallel to the fracture zones. Bruce told me that Ewing would not allow him ship time to survey suspected fracture zones. Thus it was not until Lamont ships began to expand into the southern hemisphere that sounding lines of the right quality were acquired in the right places. Heezen put together a few sounding lines acquired across the equatorial Atlantic from 1956 to 1960 and was able to identify a whole sequence of fracture zones from 7° to 10°N latitude (Fig. 9 H). The troughs were partially filled by turbidity currents, so it was easy to trace the fracture zones by the existence and slope of the flat sedimentary plains. He talked about this work in 1961, and in 1964 he published a description of the Atlantic fracture zones.[18] Later he described one that he had found in the Indian Ocean. Heezen and colleagues did not credit Hess with the discovery of the Romanche fracture zone nine years earlier.

I had already demonstrated a connection between fracture zones and the East Pacific Rise. It is noteworthy, however, that in the period 1958-1960, when tectonic syntheses were evolving rapidly, there were no certain fracture zones offsetting the mid-ocean ridge in the Atlantic where continental drift and sea-floor spreading were postulated. Ewing and Heezen thought there were sinuosities in the ridge but no offsets. Hess simplified his model of rising convection to eliminate both sinuosities and offsets and believed that fracture zones were unrelated. I knew about offsets but did not associate them with the Atlantic. It remained for a detached Tuzo Wilson to balance the data from both oceans.

Chapter 6

DISCOVERY AND USE OF MYSTERIOUS MAGNETIC ANOMALIES

In the middle of the 1950s, two marine geophysicists from Scripps were to discover an incredible pattern of magnetic anomalies in the eastern Pacific. A decade later such patterns would have been found in all the ocean basins and recognized as symmetrical around mid-ocean ridge crests:

> The straightness, symmetry, extent and ubiquity of these patterns is without parallel in geology and their discovery was completely unexpected.[1]

So it was. After his moderate success in towing a magnetometer on Capricorn, Ron Mason shuttled back and forth between London and La Jolla for a few years pursuing his magnetic studies. The early 1950s were a time of growing ferment in paleomagnetism and he was in the midst of it.

Perhaps the greatest interest was focused on the possibility that the earth's magnetic field has reversed frequently during geological time. Magnetic minerals in modern and historical lava flows are oriented parallel to the magnetic field when the lavas cool. By 1950 it was known that this orientation may be preserved for geological periods. In 1906, however, Bernard Brunhes had discovered that some rocks in France are magnetized in a direction opposite to the present magnetic field. Other reversely magnetized volcanic and sedimentary rocks were subsequently found around the world. In 1951 Alexandre Roche studied a series of Tertiary volcanic rocks in France and found that, although in all other respects identical, about half were normally and half reversely magnetized. In the same year Jan Hospers made essentially identical discoveries in Iceland.[2] Both Hospers and Roche considered the possibility that the reversely magnetized minerals had become so by self-reversal while cooling in a normal field. Both rejected this possibility and concluded that the field had reversed frequently during Tertiary time.

So far so good; perhaps the reversal of the field was on the point of universal acceptance, but in that year it was also found that a lava from Haruna volcano in Japan became magnetized reversely when it cooled.[3] "A number of mechanisms that would explain this bizarre behavior"[4] were soon suggested, but the subject of field reversals became clouded until cleared by meticulous and extensive studies in the following decade.

MASON AND MAGNETIC REVERSALS

Meanwhile, it was only reasonable that Ron Mason should investigate the possibility of detecting field reversals in the magnetizable minerals in red clay. The rates of sedimentation then being measured by Ed Goldberg indicated that red clay cores about 30 feet long would require 5-10 million years of deposition. Judging by the Tertiary lavas in France and Iceland, that should be enough time to capture several reversals. So it was that Mason and Art Raff found themselves with Bob Fisher and Jeff Frautschy on R/V *Baird* in 1954. Three piston cores, 29 feet long, and many shorter cores were collected in red clay west of the Revellagigedo Islands off Mexico. The coring expedition, appropriately named Yo-yo, featured corers with compasses that were immobilized when the corer was pulled out, thereby establishing the declination.[5]

Mason and Raff sampled the cores at closely spaced intervals, and Mason took the samples to London. There he determined their magnetic orientation on Blackett's sensitive magnetometer. Reversals were detected, but they did not occur at the same level in all the cores. The pattern of the vertical sequence of reversals in the cores was hauntingly similar to the horizontal variations in intensity of the field recorded by the towed magnetometer on Capricorn. The exciting possibility that the vertical and horizontal variations had the same cause was entertained. One of the great virtues of oceanographic laboratories and ships in the early days was the intensity of cross-fertilization. The Scripps micropaleontologists were learning that piston corers often failed to core all the layers of sediment. The tops of cores were commonly missed. It was for this reason that Scripps substituted a small gravity corer for the trip weight utilized by the Swedish inventors of the piston corer. Mason and Raff began to realize that unidentifiable gaps in their cores might degrade any vertical correlation of reversals. They tried to correlate the whole sequence of reversals with depths rather than assuming that the top of the core was necessarily the top of the abyssal sediment. They became convinced that they had a correlation from core to core and thus proof of field reversals. Mason circulated a manuscript to that effect among his peers. They told him he would look like a fool if he published; he did not publish. Still, if there was one subject on Ron Mason's mind in the next few years it was surely the reversal of the earth's magnetic field.

THE PIONEER SURVEY

During this same period in the early 1950s the Navy gradually expanded its submarine detection capabilities, and our Sea-Floor Studies Section did its bit to help. Dietz had departed for Japan on a Fulbright fellowship for 1952-1953 and did not return for long before joining the Office of Naval Research scientific liaison group in London for several years. That left me as head of a section that included Ed Hamilton, George Shumway, and Bob Dill. Among the problems we could address was the chance that the noise of a submarine would be concealed from a detector by an intervening seamount. Our echograms were showing one unknown mountain after another, and this was a sensible cause for alarm. We had become bored with them on our interminable cruises, and I had long since calculated the probability of finding a new one while standing a four-hour scientific watch. I figured out a way to tell the height of a seamount without passing over the top by assuming all the seamounts were simple cones with constant slopes. That enabled me to predict that several undiscovered banks existed off Baja California. Our San Diego fishermen were informed, and they soon discovered one and landed from it 10% of the United States tuna catch for the year.[6] J.F.T. Saur, one of Gene LaFond's physical oceanographers, and I felt encouraged to publish a report relating seamount number and size distribution to long-range sound transmission. In it I calculated that there probably were 10,000 undersea volcanoes with more than 1,000 m of relief in the Pacific. Naturally the report was classified ''Secret,'' so the details of the calculation never appeared in the scientific literature. I published the end numbers, however, with a sketchy outline of how they were derived. I mention this as an example of how greatly the operational needs of the Navy during this period influenced the collection of data that provided the basis for the forthcoming geologic revolution. Hess later used these numbers as an indicator that the sea floor was new.

No more serendipitous relation can be imagined than the needs of the Navy and the discovery of magnetic stripes off the west coast of North America. The Navy wanted to deploy a new very-long-range listening system on the California sea floor in the mid-1950s. It was essential to know the location and size of every seamount that might affect the performance of the system. We were consulted for background advice because of our research on fracture zones and physiographic provinces in the area. I was called back to Washington to talk to King Cooper and associates in the Bureau of Ships, and we discussed the optimum spacing and orientation of the sounding lines. Earlier we had discussed a more delicate matter, namely, the relative incompetence of the Hydrographic Office. I was glad to learn that the impoverished but competent Coast and Geodetic Survey had contracted for the work using its ship *Pioneer*. As I prepared to leave, King asked if there was anything else that the ship could do underway that would not interfere with running straight sounding lines. I remembered the Capricorn expedition. Ron Mason was still at Scripps working on his red clay cores; I suggested that they let us tow a magnetometer.

Mason took on the job. The survey would be precisely located electronically by LORAN-C, an advanced system developed by the C and GS, and the line spacing would be only 5 miles. No comparable survey existed anywhere else in the world, and the magnetics would be unclassified. I suppose that if the results of the survey had been suspected, they would have been classified. Fortunately, no one conceived of an enemy ballistic-missile submarine approaching Oregon and the captain saying ''All right men, we've reached the Jaramillo event, prepare to fire.'' In fact, no one in the Office of Naval Research or any of its advisors could see any point in towing a magnetometer on the *Pioneer* survey. The U.S. Geological Survey said it would be a waste of money, and, ONR refused to fund the magnetic survey. Thus it was that one of the most significant geophysical surveys ever made was wholly financed by the minuscule discretionary funds of the Director of Scripps Institution of Oceanography.[7]

Unfortunately, the magnetometer was not ready in time for the first cruise of the lengthy survey, so in the southernmost area only topography was mapped. The soundings were classified, but I had clearance and regularly obtained copies. These surveys give an example of the misadventures of classification. The C and GS plotted soundings every half hour on shipboard and sent to the Navy in Washington a contoured chart based on those soundings. That was the final product as far as the Navy was concerned. Meanwhile, I had received sheets with soundings plotted every few minutes, and I made the most detailed charts of the deep sea then in existence. The scale was 10 times larger than our normal plots. For years I also tried to obtain the echograms that should have been collected with the recently invented Precision Depth Recorder. I never could get them, and since the C and GS knew nothing about physiography, I have a suspicion that the records were lost or discarded.[8] In any event, we had extremely detailed, but classified, charts to compare with the magnetics.

MAGNETIC LINEATIONS

Mason, assisted by Arthur Raff (Fig. 31), began his survey. Magnetic data poured in, and Mason began the analysis that was to continue for years as his colleagues grew increasingly impatient. Bob Fisher and I were in adjacent offices, and quite frequently we had to walk past the small area where Ron worked. His labors appeared to be vexingly tedious. He had to plot thousands of magnetic observations along the ship's track just as I was doing with the soundings. Having plotted, however, I merely contoured and was finished because I had a known datum, namely, sea level. Mason had no datum. The raw data showed remarkable lineations, so the trend of something magnetic was known. In order to determine its intensity, however, it was necessary to subtract the magnetic field of the earth. This was not known in sufficient detail for the work, so Ron had to calculate by hand a moving average of his own values in an area and subtract it from the individual observations. It was endless. Ironically, the procedure ultimately obscured the most vital fact: the anomalies were symmetrical north of the Mendocino escarpment. The raw profiles, as Vacquier would show in a few years, contained most of the important information.

The results of Mason's labors, as is now well known, were baffling. He hauled off his data to consult with L. L. Nettleton and N. C. Steenland in Houston. They were experienced at interpreting geophysical data. Mason then returned to Imperial College in London, and it fell to Raff to continue the grueling field program and forward the data to Mason for analysis. Raff was directed to do something else by his supervisor, but apparently for the first time and certainly most fortunately, he ignored his orders and kept on with the fieldwork. Art would have enjoyed working for Bob Dietz.

Time passed; by 1958 Roger Revelle, as Director of Scripps, was getting impatient that some results appear to justify his bankrolling the research. Victor Vacquier, who had invented the towed magnetometer as a geophysical tool for Gulf Oil and developed the magnetic airborne detector for submarines for the Navy, had been lured, if that is the word, to Scripps from the New Mexico School of Mines.[9] Roger proposed that Vic and I publish a note about the correlation between topography and magnetics discovered by the *Pioneer* with the hope that Mason would be needled into publishing his survey. We were uneasy, but the *Office of Naval Research Review* wanted to print something and it was paying for much of our work at Scripps, not to mention ONR was staffed by friends. Moreover, I could see that ONR could hardly refuse to declassify my bathymetric charts for its own public relations, and I didn't see how else I could get them declassified.[10] For all these reasons and still uneasy, Vic and I chanced to be the authors of the first publication on the results of the *Pioneer* survey.[11] We showed that the scale and trend of the abyssal hills and magnetic anomalies were the same in a small area near the Murray fracture zone.

Roger's stratagem worked. Ron published by the end of the year.[12] Even then he wrote only about the results of the very first cruise when the magnetometer was used. He concentrated on "the practical aspects of the survey" but speculated as well regarding causes. Ron demonstrated that the anomalies are relatively shallow and probably in the volcanic layer of the crust. The sides of the anomalies are steep or vertical. The magnetic anomalies show "little or no correlation with topography, as far as can be judged from the best available topographic maps." Can it be he was not at Scripps after I drew my classified maps? Mason also showed that the transverse profiles of the anomalies are higly distinctive and that the partern is offset 84 nautical miles right-laterally across the Murray fracture zone. The importance of Mason's painful work now emerges. If the sides of the anomalies are vertical, the 84-mile offset represents a horizontal component and the fault is strike slip. If the orientation of the anomalies could not be determined, they might be caused by almost horizontal lava flows. Thus the large apparent offset might be the result of, for example, a mile of vertical offset.

OFFSETS OF ANOMALIES

Vic Vacquier's interest was aroused, and a man who has had a carrier task force at his disposal as a research tool expects quick results. The *Pioneer* survey had discovered another fracture zone, with subdued relief, parallel to the Mendocino escarpment and about 80 miles south of it. The zone resembled a smaller version of the Mendocino fracture zone in every respect. Topographically it appeared to connect with a similar feature that I had discovered on the Midpac expedition about 200 miles to the west. Mason's preliminary map showed an offset of the mysterious magnetic stripes along the fracture zone, but unlike the Murray fracture zone the pattern could not be matched across the fault. Vacquier assumed that a match could be found if the stripes could be traced west of the *Pioneer* survey, which extended only 300 miles from shore.

Thus began the active mapping of magnetic anomalies as opposed to the passive program on *Pioneer*. Vic put to sea in December 1958 on *Horizon* and made two east-west lines south of the fracture zone and one to the north. That was all it took, a few days of skillfully planned research, to demonstrate an offset of 138 nautical miles.[13] Moreover the analog records from the magnetometer required no processing or reduction at all (Fig. 10). For the first time, the profiles of magnetic stripes could be perceived as containing a wealth of geological information. Nature here was being very kind, even tutorial. If the fracture zones had had any other orientation than east-west, it would have been necessary to apply a regional correction to the magnetic field as Mason had been doing. In fact, nature was kind in almost every way in laying out these fracture zones. They were conspicuous because the surrounding abyssal hills were so subdued. Like the magnetic stripes, they were right off one of the few deep-sea oceanographic laboratories in the world. They were even oriented so they could be surveyed by ships that were urgently needed for Naval research. As Einstein said, however, "Subtle is the Lord."[14] The fracture zones and magnetic anomalies did not appear to be associated with a mid-ocean ridge.

Vacquier noted in his stunning report that the magnetic stripes are hardly distorted by the great offset and that this implied an "unsuspected mobility of small blocks of oceanic crust." His blood was up. If a minor fracture zone like the Pioneer had an offset of 138 miles, what was the offset on the great Mendocino escarpment? Off he went again, on *Baird*, in August of 1959 to run more east-west profiles. But was the offset left or right lateral? He had only enough ship time to test one possibility. It would be another year before I realized that the bathymetry was offset exactly the same amount as the magnetic anomalies, so Vic had to guess. He guessed that the offset was right lateral, meaning that he should survey farther west to the south of the Mendocino in order to find a match for a stripe that would be offset to the right on the north side. This he did with customary dash but no success. The stripes continued farther west, and they continued to trend north-south, perpendicular to the fracture zones, but there was no match. The offset on the fault was enormously greater than any other, but how great and in what direction?

Vic was out again on *Baird* in April 1960 and speedily found an incredible offset of 640 nautical miles. The direction was left lateral, and when added to the Pioneer offset, the total was 780 nautical miles. The raw analog data from the towed, flux-gate magnetometer that Vic had invented two decades before overwhelmed any skepticism about the correlation of the offset magnetic stripes. Like most scientific discoverers, however, Vic himself sought for some way to confirm the correlation with statistical rigor. A few wiggly lines, even though almost perfectly matched, do not provide the same satisfaction as a proper correlation coefficient. Accordingly he consulted with a statistician, only to be told that a statistical approach was a waste of time. The wiggly lines matched perfectly enough. For this work and his pioneering application of the magnetometer to geologic problems, Vacquier received the Fleming Medal in 1973 from the American Geophysical Union. About that time he also received the Albatross Award from the American Miscellaneous Society for, as I recall, suggesting that the sea floor is cracked and moves about. This unusual award is an actual stuffed albatross, which the hapless recipient realizes is a rather large bird as he carries it home and then back again a year later for the next recipient.

PUBLICATIONS

The paper on the Mendocino offset appeared in August 1961, next to a paper by Mason and Raff on the southern half of the *Pioneer* survey and one by Raff and Mason on the northern half (Fig. 11).[15] This smorgasbord of magnetic maps and data creates some confusion about priorities in interpretation, especially if one also considers that, with Vacquier's permission, I had already analyzed some of his data in December 1960 in my East Pacific Rise paper, to which I shall return.[16] I rely on Vacquier's memory as well as my own in what follows.

The Mason and Raff paper noted that the magnetic stripes north of the Mendocino escarpment were more intense and closely spaced than those to the south. This was attributed to the fact that they were much more shallow, and thus the magnetometer at sea level could record more details. The stripes were attributed to slabs of magnetized rock under the positive anomalies. It should be noted here that the towed magnetometer measures only variations in intensity of the present magnetic field. What Vine and Matthews would recognize as reversals are merely stripes of less intensity (in this region) to a magnetometer, and so they were to geophysicists in 1961. Mason and Raff noted the presence of a pattern of anomalous stripes trending northeasterly near the California coast. They ventured no explanation for the pattern, and it remained a puzzle until Tanya Atwater and I apparently solved it in 1968.

The paper by Raff and Mason, three pages and a map, covered the area from the Mendocino scarp to 52°N. The magnetic stripes are offset in many places along lines trending northeasterly and northwesterly. Some seamounts in the area give strong magnetic anomalies, but others do not. It was "not at all clear" why not. The crustal rocks that contain the magnetic anomalies appear to dip under the edge of the continent. Positive anomalies are associated with linear ridges 65% of the time. It is maddening to look at their map and see what is now so obvious but what was then concealed not only from the authors but from all of us who had access to the maps for several years (Fig. 12). How could we miss the symmetry of the magnetics around the ridge crests? I have speculated that everyone was numb by the time the *Pioneer* survey reached the Juan de Fuca Ridge.[17] Mason produced his maps in sequence as the survey expanded from south to north. There was no symmetry in the anomalies south of the Mendocino escarpment because the ridge crest was subducted. The section of the survey that included the escarpment terminated at 42° latitude and included the southern part of the Gorda Ridge, which was actively spreading. But the magnetics there are not symmetrical! Those on the west flank are straight and trend parallel to the ridge crest. Those on the east flank trend at an angle, and moreover they are curved. The trend is similar to that of the anomalies off central California and thus appeared to be related in some way to the edge of the continent. On the basis of the survey from 32° to 42°N, there was no symmetry. Meanwhile, the anomalies, whatever their cause, provided evidence for spectacular, unprecedented displacements of the earth's crust. By 1961, perhaps in frustration but more likely with hopes of glory, interests had shifted from understanding the anomalies to using them. Then, the magnetics of the Juan de Fuca Ridge were mapped, and they are the ones that are now so obviously symmetrical. It would have helped, perhaps, if the topography had been symmetrical, but it is not. The west flank of the ridge is plastered with volcanoes, and the right flank is buried by turbidity currents. Moreover, although I knew that the Juan de Fuca and Gorda features were ridges, my highly detailed bathymetric maps were still classified "Confidential."

Vacquier, Raff, and R. E. Warren, authors of the third paper in this 1961 issue, focused on the offsets and speculated more as a consequence. Perhaps the San Andreas is not a conjugate shear of the great fracture zones as I had proposed. If not it is younger because it is still active. The offset and rate of slip on the San Andreas suggest an age of about 100 million years (that magic age again). Therefore the fracture zones were at least that old, and considering their great offsets the magnetic stripes were probably at least 200 million years old. To quote from this paper:

> The displacements along strike-slip faults in the ocean floor are thus of the same magnitude as the distances through which continents have been presumed to drift. . . .

> continents are pictured in the literature as rigid plates drifting without distortion through the viscous ocean floor. Now we have evidence for the rigidity of the upper part of the oceanic crust that carries the magnetic pattern, which is just as good as the evidence for the rigidity of continents exemplified by the fit between Africa and South America.[18]

Eight months earlier I had stated, "The crustal blocks between fracture zones have areas of a few million square kilometers, but they are hardly more distorted by displacement than small blocks of wood." It is apparent that somewhere in the group at Scripps the idea had germinated that the sea floor, particularly the Pacific sea floor, is rigid.

The only related study of magnetic anomalies in the Pacific prior to the discovery of their origin was a curious amalgam of hypothesis and experiment. Sometime after Vacquier measured the offsets I was intrigued by the possibility that there was a close correlation between the length and offset of great strike-slip faults that were then called wrench or transcurrent faults.[19] The Mendocino fracture zone was the longest such fault known and it had the greatest offset, so the basis for this line of reasoning is obvious. I talked about this with colleagues, but my correlation was weakened by the great length but small offset of the Murray fault. All the great wrench faults of the continents, the Garlock, Bocono, Dead Sea, Great Glen, Alpine, and Caribbean faults, agreed with the correlation. The magnetic offsets of the Cape Blanco, Pioneer, and Mendocino fracture zones also agreed, as did the less certain topographic offsets on the Molokai and Clarion fracture zones. What was wrong on the Murray?

Solely on the grounds that nature ought to agree with my graph, it occurred to me that maybe the offset was small on the Murray because we were looking at one end. Perhaps the offset was greater in the middle of the fracture zone farther to the west. I convinced Art Raff that he should go to sea again, and he soon ran three long magnetic profiles parallel to the Murray trend and found a 350-mile offset—right on the graphed correlation.[20] I couldn't understand the reason for the correlation except the trivial one that a fault has to be at least as long as the offset. Dave Griggs convinced me, however, that I should publish my data with the hope that someone could explain it. In retrospect, the correlation was meaningless but nonetheless fruitful. Raff had discovered a distinctive characteristic of fracture zones: the offset can change abruptly. Moreover, he identified as the "disturbed zone" a region south of the Murray fracture zone where the magnetics are jumbled:

> it bears on the question of crustal material apparently gained or lost—for in terms of east-west miles of oceanic crust, the disturbed zone is 300 miles wider on the south side of the fault than on the north side. If this is truly an area of dilation and more recently formed crust, one would expect to find some topographic evidence.[21]

Such evidence already existed. The disturbed zone was the only region in the northeastern Pacific that did not slope toward the west, and the topography was unusually mountainous. We had long since surveyed routes for submarine telephone cables across it. The jumbled magnetics, however, were entirely new. Raff had discovered what would be recognized later as the first jumped or abandoned ridge.

Chapter 7

MEANWHILE DRIFTING ASHORE

We have been following marine discoveries in the early 1950s in the Pacific where the data had no obvious relation to continental drift except to show that the crust is rigid and highly mobile. We are now at the time when a discovery of the first magnitude was made in the Atlantic: the crests of mid-ocean ridges are tensional rifts. The interpretation of this discovery, however, being in the Atlantic, was linked to continental drift. Perceptions of this subject had changed since we last considered it at Wegener's death in 1930, so before turning to the rift it seems advisable to revert to developments ashore.

CONTINUING RESPONSES IN THE UNITED STATES

In the 1920s, the reaction to Wegener's theory in the United States had been intense, and generally neutral or negative. In the 1930s, perhaps because of Wegener's death, perhaps because of ennui, the controversy generally subsided. American geologists were divided among believers and nonbelievers; they found practically nothing new to discuss, nor did they have much reason to be other than urbane. Most of the participants in the discussion in this decade had been through it all before. For example, in 1933, A. P. Coleman wrote that the evidence of Permian glaciation failed completely to support the concept of continental drift.[1] He was, however, 81, and he seems to have forgotten, or at least he does not mention, that he wrote much the same paper with an identical title in 1924. That paper began as follows:

> This is a time of unrest when the solidest foundations are shaken, so that it is not surprising that even the continents themselves begin to drift about in eccentric ways.[2]

It was a hard style to match, but the second paper was amiable in its refutation of drift.

We last saw William Bowie in November of 1934 joining Dick Field in a siren song to pull Maurice Ewing into the sea. We have no indication from the report of the Field Committee regarding the attitude of the members concerning continental drift. However, we need not speculate regarding Bowie's thoughts because he published them in 1935. Like Coleman he was even-tempered, but unlike Coleman he favored continental drift. He discussed older ideas on tectonics, but they conflict with the "laws of physics and mechanics and therefore seem to be untenable." Only the Darwin-Fisher hypothesis did not abide his question. Sir George Darwin had proposed that the moon had separated from the earth and retreated to its present distance because of tidal drag. Osmond Fisher elaborated this hypothesis by asserting that the continents would drift toward the Pacific basin—the great, deep scar left in the solid crust by removing the moon. This occurred more than 1.6 billion years ago. As of 1935, that was the oldest known age of the continents. Moreover,

> It is rather interesting to note that the two coasts of the Atlantic are so nearly alike that they have the appearance of the shores of a great river. Is it not possible that North and South America could have been torn away from the crustal material that forms Europe and Africa just before the [removal of the moon] occurred?[3]

So Bowie offered a new kind of continental drift, a drift that was dead, sterilized by time, and free of refutation by modern measurements of longitude. It was a kind of drift that was appropriate for the official leader in geodesy in the United States at a time when the Danish geodesists were finding Greenland as firm as a rock. Perhaps it was a kind of drift particularly attractive to Americans. Certainly, Maurice Ewing and Arthur Meyerhoff would be attracted to something similar a few decades later. It did have the flaw, in Bowie's version, that it completely ignored, or reduced to mere coincidence, all the evidence of relatively young, matching, trans-Atlantic geology that Wegener found so compelling. Still, for a government official it was a bold espousal of a dubious cause. Compare a publication, in the same year, that came out with the permission of the Director of the U.S. Geological Survey.[4] Covered in the discussion are field observations of pre-Devonian rocks in Scotland and eastern North America—two regions with remarkably similar geological structures striking into the Atlantic. As it turned out, the wrong ideas of paleontologists about the foundering of a trans-Atlantic continent were within the scope of the discussion, but drift was not. Conservatism might have been expected; it had been a quarter of a century since a Geological Survey Director had been involved in what counts.

Another man who had been through it all before was Beno Gutenberg (Fig. 33).[5] In 1936 he restated his Fleisstheorie of 1927 in the light of recent seismology. Naturally, the theory was in agreement with the new observations. Gutenberg was a master of interpreting seismograms. Frank Press remarked two decades later upon his genius in identifying a low-velocity layer in the upper mantle when other seismologists saw no evidence of it. Nonetheless, something led him astray in the 1930s because he believed that the seismograms showed a layer of sial under the Atlantic and Indian oceans (but not the Pacific). He surmised that the layer was 20 km thick. Consequently, the Atlantic could not have opened as Wegener proposed because there would have been an enormous initial surplus of sial. A proto-Atlantic continent must once have existed between the continents now flanking the Atlantic. If the sial is now half the normal thickness, the original width of the proto-Atlantic continent must have been half the present width. I was to advance exactly the same arguments regarding thinning on the crest of the East Pacific Rise. Gutenberg's problem was that he considered his Fleisstheorie to be a mere improvement on Wegener's basic theory, but by invoking a proto-Atlantic continent he had weakened or invalidated much of Wegener's supporting evidence. The margins of continents should not be congruent if they could flow out to fill the Atlantic.

Gutenberg's version of continental drift was an interesting example of interpreting data in the light of a hypothesis, and there were many more in the 1930s.[6] Ross Gunn developed quantitatively the driving forces for drift arising from an initial asymmetry of the earth. G. W. Munro observed that the only possible energy source was heat. W. W. Watts explained drift in terms of alternating expansion and contraction of the earth, although Walter Bucher proposed the same alternation but no drift. R. W. Chaney found that the distribution of Tertiary forests gave no indication of drift, but in general the skeptics were mute. If it was not certain whether the earth was heating or cooling, contracting or expanding, whether continents were drifting or not, perhaps the only common-sense approach was to be mute—particularly with regard to educating the young. I took introductory and historical geology at Caltech in 1939, and no word of these matters was allowed to trouble our unfurrowed brows.

The circumstances were rather different for our professors at that time because the literature was full of reviews of a new book supporting continental drift, a book full of those firsthand field observations that were the core of geology but that were so lacking in Wegener's books, a book by the man who Reginald Daly called the "world's greatest field geologist."

NEW FIELD EVIDENCE

Alexander L. Du Toit (1878-1948) was a South African geologist. As a young man he had accompanied Daly on his Shaler Memorial expedition in 1920-1921, and Daly helped him obtain a Carnegie Institution grant to study the geology of Argentina, Brazil, and Uruguay. He published his results in 1927.[7] South American geology was strikingly similar to that of South Africa. The fieldwork so strongly supported continental drift that Wegener included Du Toit's map in the 1929 edition of *The Origin of Continents and Oceans*. By

1937, Du Toit had analyzed his extensive observations and at age 59 published *Our Wandering Continents*, which was to be reprinted in 1957.[8] In the preface he noted that most discussion of drift had focused on the northern hemisphere because it was more familiar to most geologists. But in this book,

> fuller attention is paid to the Southern Hemisphere, for which the evidence is, as it happens, clearer and less equivocal.

> So numerous then become the congruities, so remarkable the so-called "coincidences," and so close the agreement between prediction and observation, that whether the explanation here offered for the *causes* of such postulated Drift be valid or not, the author feels that a great and fundamental truth is embodied in this revolutionary Hypothesis.[9]

And so it was; Du Toit had discovered or confirmed most of the continental evidence that now so obviously supports continental drift. A late Paleozoic glacial center spread out upon what are now four widely separated continents (Fig. 1 E). A geosyncline formerly was continuous from South America across South Africa to Australia. The floral and faunal affinities between the continents were very close in Paleozoic time. The boundaries of geological provinces in the bordering continents match closely if the Atlantic is closed and the Cape belt of fold mountains continues in Argentina. Du Toit did not strengthen his case by offering the curious proposal that continents that move toward the poles are compressed and folded because the meridians converge on a globe.

The many reviewers were genial. S. J. Shand considered the book timely and stimulating; W. A. Rice was fair and neutral; R. T. Chamberlin was reserved about the conclusions but found them stimulating; and Chester Longwell thought that many geologists were not as hostile to the hypothesis as Du Toit believed. Nonetheless, the reviewers were not convinced by the geological evidence from the southern hemisphere. It seems hard to believe that this evidence, uncontaminated by a theory of origin, would not have prevailed had Wegener's integrated concept of continental drift not already been rejected. As it was, the war approached; the data were in another country; and besides, the controversy was dead.

SHARPENING CRITICISM

The mobility and global compass of the second World War did not bring science to a halt in the western sanctuaries, nor did time moderate the prose used by the critics of the great poet. In the 1940s the controversy was reheated. Thus George Gaylord Simpson expressed his views in the *American Journal of Science* in 1943:

The fact that almost all paleontologists say that paleontological data oppose the various theories of continental drift should, perhaps, obviate further discussion of this point and would do so were it not that the adherents of these theories all agree that paleontological data do support them. It must be almost unique in scientific history for a group of students admittedly without competence in a given field thus to reject the all but unanimous verdict of those who do have such competence.[10]

Simpson had been elected to the National Academy three years before at the extraordinarily young age, for a paleontologist, of 39. He would receive the Penrose Medal in 1952.

Du Toit ceded Simpson a Pyrrhic victory in the same journal in 1944 by quoting himself from 1927:

> geological evidence almost entirely must decide the probability of this hypothesis for those arguments based on zoö-distribution are incompetent to do so.[11]

So much for paleontologists.

Much of that issue was devoted to comments on continental drift. Perhaps Vening Meinesz was not the only person to run out of unprocessed data during the war. In any event, Bailey Willis, then 87, made his notorious reference to German fairy tales in that issue and said that the "hypothesis should, in my judgment, be placed in the discard."[12] That year he was honored with the Penrose Medal. Alfred Wegener would have been 87 in 1967; had he lived as long as Willis he would have seen the confirmation of sea-floor spreading and plate tectonics.

A NEW CHAMPION

Another lifelong champion of continental drift emerged in Arthur Holmes (1890-1965), who was born in England near Newcastle-upon-Tyne. His academic career began as a physics major at Imperial College in London where R. J. Strutt was then making some of the first determinations of geological age based on radioactivity. Holmes undertook to use the uranium-lead method for this purpose and in time shifted his research to geology. By 1928 his interests had broadened to embrace tectonics, and echoing Daly in 1923, he thought that there was a far stronger case for continental drift than either Taylor or Wegener had put forward.[13] Again following Daly, he set forth his support for continental drift in the Lowell lectures in Boston in 1932. He proposed mantle convection as the driving mechanism for continental drift and offered an elaboration of Wegener's ideas on the origin of the Mid-Atlantic Ridge.[14] Convection currents rise under continents because of their high content of radioactive minerals. The currents diverge, split the continent in two, and carry the fragments apart (Fig. 19 A and B). Above the divergence zone, however, a stretched fragment of continent remains, and being thin, it is the submerged mid-ocean ridge. A new ocean is created, and it is floored with sima that is exposed as continents drift apart. Holmes, like Daly, was a petrologist, and he conceived of an important addition to the concept of continental drift. Zones exist where convection currents converge and plunge into the mantle. The crust above such zones is in compression, and Holmes reasoned that the common rock basalt is converted to a high-pressure phase called "eclogite." This conversion is observed in compressed areas of some mountain ranges. Being dense, the eclogite sinks along with the plunging convection currents. Holmes, at a blow, had conceived of what now may reasonably be taken as precursors of sea-floor spreading, continuity arguments respecting the surface area of the earth, and the plunging lithosphere of tectogenes.

In 1944, as the controversy boiled in America, Arthur Holmes published the first edition of *Principles of Physical Geology*, which became the standard textbook in English outside the United States. The young British geologists and geophysicists who would invent so much of paleomagnetism and plate tectonics were brought up on it. Naturally, Holmes taught about continental drift. In addition to presenting the evidence assembled by Wegener and Du Toit, he generated a new version of the evolution of drifting that is driven by convection currents (Fig. 19 C and D). In the initial stage, continental blocks lie on a basaltic layer, and the accumulation of heat causes mantle convection cells to rise and spread under the block. The continent and basalt above any region of diverging mantle are put in tension, and above any region of convergence they are in compression. Then the continent and basalt layer split and drift apart. Between them a new ocean basin is generated with an island, such as Iceland, or a swell in the middle. The large area of new crust is accommodated on an earth of constant size by compressing an equal area of the basalt layer into eclogite at mantle convergence zones. The dense eclogite then plunges with the descending limb of the convection. The continental block, in contrast, is buoyant and is merely thickened at the convergence zone. As it will develop, the initial stage is almost exactly what I proposed is now happening on the East Pacific Rise, yet I do not believe I ever read or saw a reference to Holmes on this subject. It will also turn out that the final stage is very similar to sea-floor spreading as proposed by Hess, who merely cited Holmes as an advocate of mantle convection.[15] Textbooks are not normal scientific literature. However, for those brought up with Holmes, the images were striking:

> To sum up: during large-scale convective circulation the basaltic layer becomes a kind of endless traveling belt on the top of which a continent can be

carried along, until it comes to rest (relative to the belt) when its advancing front reaches the place where the belt turns downwards and disappears into the earth.[16]

The present volume and other histories of the evolution of continental drift and plate tectonics inevitably leave the reader with the impression that a great controversy was constantly under debate. For decades, nothing was farther from the truth. Chester Longwell set forth the facts.

I associate with numerous men in the science and find them, on the whole, a rather open-minded group, not given to arrogant judgments on the many abstruse problems of the earth, and emphatically not uniform in their thinking on these problems. They find much in their geologic work to occupy their attention aside from the conflicting hypotheses on the history of continents.[17]

Most people at most times were doing other things. Scientific medals were awarded, people were elected to the Royal Society, the journals were filled with papers, geological meetings were held several times a year, and continental drift was rarely mentioned. Even the principals in this account spent only a tiny fraction of their time in writing, and presumably thinking, about continental drift. Consider, for example, Arthur Holmes and Harold Jeffreys who, for present purposes, are respectively for and against drift. Jeffreys' collected works fill several volumes, but having long since disposed of the possibility of drift he had little reason to waste his time on it. In 1945, just after Holmes's book appeared, Jeffreys was elected to Foreign Member of the U.S. National Academy of Sciences and the following year became the Plumian Professor at Cambridge. Medals cascaded upon him; he was knighted, and to this day he remains skeptical about drift. Scientists are judged on the quality of the best of their work just like winners of the Pulitzer Prize in literature or the Academy Awards in motion pictures.

Arthur Holmes is an example of a great scientist who was appreciated by his contemporaries—not, like Jeffreys, despite the fact that he was wrong about continental drift but, like Daly, despite the fact that he was right about it. In 1956 the Geological Society of America gave him the Penrose Medal. He was 66 and had just retired from an extraordinary career: elected member of the Royal Society in 1942, recipient of the Murchison and Wollaston medals from the Geological Society of London, Foreign Honorary Member of the American Academy of Arts and Sciences, and Foreign Member of the academies of science of France, the Netherlands, and Sweden. The Penrose Medal was presented in style with a Citation by Hollis Hedberg, who would later win the same medal, and a Response, in the absence of the ill Holmes, by Kingsley Dunham, future Director of the Geological Survey of Great Britain. Holmes was extolled at length for his work on radioactive dating, the age of the earth, and isotopic tracing in geology. The only references to his work on continental drift were the rather oblique:

He has contributed to clarification of thinking with regard to granitization and subcrustal convection flow and he is presently deep in the study of the origin of the peculiar potassic ultrabasic lavas of Uganda.[18]

(You see I have not even had time to mention his contributions to sedimentary-rock petrology as well as to ore deposits, tectonics, and many other branches of geology.)[19]

Was Holmes offended or did he feel slighted at the omission of his pioneering work on continental drift and sea-floor spreading? If so, he certainly was not so rude as to say so, but it seems unlikely that he had any such feelings. Certainly, Hedberg gave him every opportunity to express them in private if they existed. He asked, on a thoughtful visit to Holmes in Scotland before the presentation, what work gave Holmes the greatest satisfaction. Holmes replied, (1) The attempt in 1932-1934 to establish a time scale based on the decay of radioactive potassium, an attempt that was unsuccessful; and (2) Estimation of the age of the earth. It appears that Hedberg wrote his citation just as Holmes himself would have done—omitting nothing of importance.

PALEOMAGNETISM

About 1877 the leading American geologist, G. K. Gilbert, wrote,

Of late years the most important contributions have come from the physicists, and in their scales have been weighed the old theories of the geologists.[20]

Perhaps he was thinking of William Thomson, Lord Kelvin to be, who had been estimating the age of the earth by physical means and had it down to 50 million years by 1876—thus vexing a mere naturalist, Charles Darwin, who believed the geologists' estimates of billions of years (American billions, not British). Kelvin was wrong by two orders of magnitude and the geologists about right.

In the early 1950s the physicists again rallied round to help solve geological problems, this time by the use of rock magnetism. What finally occurred was summarized by Teddy Bullard in 1975:

The clarity which was finally achieved in the interpretation of paleomagnetism should not obscure the complexity and difficulty of the route by which it was attained. To establish the facts of continental movement and field reversal in the face of doubts raised by the existence of many unstable rocks, the existence of self-reversing rocks and the complexity of the relations between movements of the continents and the pole is a major achievement.[21]

Now it is, it most certainly is, but what about in the 1950s? Did the new evidence of paleomagnetism help in the groping toward a new paradigm of global tectonics, or did it merely add a degree of confusion? The record indicates that at least as late as 1960 the evidence of paleomagetism was no more than suggestive, and acceptance of the conclusions was a matter of faith.

The prophet of the faith, in my experience, was Keith Runcorn, a student of Patrick Blackett. He had the energy for it, constantly exercising and swimming daily whenever he visited La Jolla, and he was a rugby international; he keeps racquets and sports shoes in file drawers in his office in Newcastle. Stanley Keith Runcorn (1922–) was born at Southport, Lancashire, and attended local schools. He went on to Gonville and Caius College in Cambridge University in 1941, and received a B.A. and the Johns Winbolt Prize in 1944. From 1943 to 1946 he was an Experimental Officer working on radar. After the war he became a Lecturer in Physics at the University of Manchester and received a Ph.D. there in 1949. He returned to Cambridge as Assistant Director of Research in Geophysics and fellow of his college during the late 1950s. It was during this period that he held visiting appointments at the University of California, Los Angeles, the Dominion Observatory, Ottawa, Caltech, and the Jet Propulsion Laboratory in Pasadena. He became familiar with many American geologists in the course of fieldwork in the western states. Despite an education as a physicist he became enough of an insider in geology to publish in the sedate *Bulletin of the Geological Society of America*. He went on to be Professor of Physics and Head of the School of Physics at the University of Durham from 1956 to 1963, and then assumed the same position at the University of Newcastle-upon-Tyne.

At that time, there was a general ignorance of paleomagnetism among geologists, but with increasing knowledge ignorance was soon to flower into disbelief. A few pioneers had established decades before that some historical lava flows record the direction of the earth's magnetic field. Other rock magnets were identified, and in 1906 Brunhes had discovered rocks that are magnetized in a direction that is the reverse of the present field. It appeared by 1930 that paleomagetism might help map any drifting of the poles of rotation or of continents if they had occurred—provided, first, if ancient rock magnets were stable recordings of the earth's field when the sediments were deposited or the lava cooled; and provided, second, the magnetic field was always a dipole superimposed on the pole of rotation.

Runcorn dealt with the status of these assumptions in 1960 at one of the long series of fruitful symposia he organized at Newcastle-upon-Tyne. He is nothing if not candid regarding the theoretical underpinnings for the dipole assumption.

It is natural to assume that the mean field is dipolar as well as axial, as this is the simplest type of field. The theory of the geomagnetic field does not provide a simple argument for this assumption which would be acceptable to the skeptic.[22]

As to the second assumption, John Graham had established by ingenious field tests that rock magnets can be stable for geological periods. He believed, however, that it was unwarranted to draw conclusions about the time that the rocks acquired the magnetism "in the absence of exact knowledge of the way the rocks became magnetized." Runcorn gave the status as of 1960.

However unpalatable it may be, it is not at present possible to decide this question, and the development of this subject, and other scientific disciplines, would not have made progress if this kind of philosophy had guided the early work. It has been much more fruitful—and I believe much more in the tradition of scientific inquiry—to obtain accurate observations and ask what sense can be made of them, i.e., to compare the observation with a simple theory.[23]

Runcorn was to win the Vetlesen Prize and other awards for being right, but meanwhile what were the observations and theory that a skeptic had to digest even if he were willing to accept the dubious dipole and the possibly permanent rock magnetism? In the early 1950s it seemed that papers on theoretical and experimental ways to generate self-reversals would fill the journals. The reversal of the earth's magnetic field was consequently in question. Then in the mid-fifties Patrick Blackett showed that Paleozoic and Mesozoic rocks record pole positions far from the present ones. Teddy Bullard was not convinced by this evidence at the time.[24] Soon after, Runcorn and his chaps decided that it was more a matter of polar wandering than drift. Next (Fig. 13) it was polar wandering and drift.[25] After that, it was thought that parts of continents rotate relative to the rest. Walter Munk and Gordon J. F. MacDonald summarized what I believe was the general reaction to these developments.

It is usually a bad omen for any method if the degrees of freedom required to interpret measurements grow at the same rate as the number of independent determinations.[26]

Their book, *The Rotation of the Earth*, received the Monograph Prize of the American Academy of Arts and Sciences for 1959. It was dedicated to Sir Harold Jeffreys.

In that same year (1960) Blackett, Clegg, and Stubbs published a global summary of paleomagnetic results[27] that convinced Bullard of the reality of continental drift, and Harry Hess wrote,

> One may quibble over the details, but the general picture on paleomagnetism is sufficiently compelling that it is much more reasonable to accept it than disregard it.[28]

In 1960, Bruce Heezen agreed:

> The method is beset by technical difficulties but in general it appears to work.[29]

THE HOBART SYMPOSIUM

Apparently the significance of paleomagnetism was yet to be wholly convincing in 1960-1961, when the first syntheses of the new marine exploration appeared. It was, however, enthusiastically accepted by the converted who promptly launched the series of symposia on continental drift, which apparently will never end. The first of these was held in 1956 in, as we might expect, the southern hemisphere, almost as far south as an English-speaking geologist can live, in Hobart, Tasmania.[30] There S. Warren (Sam) Carey assembled a cast of believers and one man left over from the symposium of mainly disbelievers in 1928—Chester Longwell, who could be counted on to be gloriously rational as before.

Longwell's conclusion was that the theory of continental drift was not proved but not disproved either.[31] All around him the converted, scenting blood from paleomagnetism and marine geology, began to pursue the details of drift. Lester King from the University of Natal reassembled the predrift Arctic and North Atlantic by utilizing rigid, curved continental shapes on a globe.[32] E. (Ted) Irving discussed the evidence of rock magnetism. Faunal evidence was analyzed in several papers despite the concession by Du Toit, in 1944, that it was not pertinent. Experts will be experts.

One of the two longest papers was a reprise by Lester King on *The Origin and Significance of the Great Sub-Oceanic Ridges.*[33] A 41-page paper, it contains not a reference to Heezen, or to Hess's papers on mid-ocean ridges, or to mine on Pacific tectonics, so it offers an independent view apart from the American-English invisible college. The Lester King who was so well informed on continental geology simply did not have enough information to draw significant conclusions about marine tectonics. His paper shows very well how inconclusive speculation on continental drift might have been without the marine data that were just being acquired.

By far the longest paper, 179 pages, was by the convener, and Carey shows just how much of sea-floor spreading and plate tectonics one man could visualize by 1956.[34] In retrospect, he conceived of several crucial elements of global tectonics, but could anyone at the time be sure what was gold and what was dross in his work? Carey begins by putting normal tectonics in perspective:

> the folds and faults which have absorbed the attention of geologists since the birth of structural geology are not the first order structures but are of the second and lower orders.[35]

He visualized an oceanic crust at least as strong as continental, and both so rigid that the most careful methods of reducing plotting errors were necessary to appreciate the congruence of Africa and South America (Fig. 1 F). He gives no sign of knowing Euler's theorem but goes so far as to express the opening of the North Atlantic as a constant angular displacement (Fig. 24 A) and identifies a pole. Thus he anticipated Bullard, Morgan, and plate tectonics. He visualized faults that cut through the whole crust and chasms, bounded in part by such faults, that open in continental crust and dilate to form ocean basins. Rifts, he thought, are produced by simple rifting, and he knew just where and how it occurred.

> The marginal coasts may long since have ceased activity, but there will be an active median line of faulting and seismicity following a ridge composed of two tilted rims with an intervening narrow trough on the site of the latest extension.[36]

Thus did he anticipate Heezen and Ewing. But how much was anticipation and how much was by word of mouth, newspaper, and preprint? He says, "The presence of the central trough has only just been discovered" but cites no reference. He shows a figure of "the mid-oceanic crack system (after Ewing)" but cites no reference. Apparently, some scientific newsflash or letter had come to hand, and he used it to confirm well-developed concepts.

Carey coined about a dozen terms to describe the great faults, triangular and rhomboid chasms, and deformed orogenic belts that were the main elements of his tectonics. "If any proposals are rejected, my terminology will go out with them, so it will not encumber our language."[37] The proposals generally have been accepted, but the terminology was superseded by the geometrical terms of plate tectonics. It is necessary to define a few of his terms, however, in order to understand his ideas. A *megashear* is a strike-slip fault that cuts through the whole crust. An *orocline* is a once straighter orogenic belt that has been bent by rotation. A *sphenochasm* is a triangular gap in continental crust created by rotation of one block with respect to the other. The gap is "occupied" by oceanic crust. A *rhombochasm* is a similar gap but with parallel sides and is caused by dilation. Given all those definitions, he can write,

None of these structures can develop in isolation. An orocline implies a sphenochasm on one side of it, and this cannot go on diverging indefinitely. It must end against a megashear or perhaps another orocline. A megashear must either go right around the globe or begin and end at oroclines or rhombochasms. A single structure of this scale implies a chain of other structures to absorb the implied movement.[38]

Thus did he anticipate the ridge (rhombochasm), transform fault (megashear), and trench (orocline) of Tuzo Wilson and all the tectonic models involving crustal continuity and equilibrium that were proposed about 1960 by Heezen, Hess, and myself.

One of Carey's most advanced insights was the recognition of what he called *nemataths*, namely, "submarine ridges across Atlantic-type ocean floors." We call them aseismic ridges. "They join points which were originally closer, and mark the path of the separation movement." Carey's concept was not quite right. Nematath means "stretched thread," and that is how he describes them, as hot filaments drawn out of the mantle. He also clouds his meaning by referring to sial threads because the nemataths originate when continental splitting begins. If we make allowance for these two factors, his discussion is remarkable.

If I break a slab of toffee which is cold and brittle except for one warm spot, the slab will break cleanly except at the hot spot where a thread of toffee will be drawn out across the rift. The thread will be straight or curved according to the path of separation . . . in view of the density difference it will endure permanently as a submarine ridge on the ocean floor.[39]

I find this a bit confusing. The aseismic ridges were produced where a median rift happened to lie over a hot spot. So far so good, but somehow the blob of hot sial keeps feeding material into the lengthening ridge. Thus it appears that each ridge is growing from the wrong end. Perhaps, however, I am being obtuse in order to avoid adumbrationism, "the denigrating of new ideas by pretending to find them old."[40] It does appear to me that Carey discovered the implications of hot spots and aseismic ridges for tracking plate movement. If not, he came so close that it would require but a minor metamorphosis to reach modern concepts.

These spectacular conclusions were issued in a multilithed book by the Geology Department of the University of Tasmania in 1958. Following the custom of the time, many of us received complimentary copies through the mails. I have no record of when the book arrived.

That was the last symposium on classical continental drift for some time. The next symposium publication I received was from the Alberta Society of Petroleum Geologists in 1958, and it was confined to paleomagnetism and paleowinds for field evidence and ''causes'' and orocline tectonics for interpretation. Likewise, a symposium volume on drift, organized and edited by Keith Runcorn in 1962, is full of paleomagnetic and sea-floor evidence and theory, but after half a century of disputation the work of continental geologists and paleontologists no longer merited inclusion. It would not be until after seismologists, paleomagicians, and marine geologists had accepted plate tectonics that land geologists would rally to repel the new barbarians and quiet the ''sound almost imperceptible, so slight, so little different from silence itself'' that had turned to thunder.

Chapter 8

DISCOVERY OF MEDIAN RIFT

In 1956 the geological world was electrified by stories in newspapers and magazines that Lamont scientists had discovered a world-girdling rift 2 miles deep, 20 miles wide, and 40,000 miles long. This discovery was remarkably fruitful in that it immediately stimulated exploration of mid-ocean ridges and speculation on global tectonics. The history of the discovery may also prove fruitful in illuminating how science works. Hardly any of the factual material in the "discovery" was new, and hardly any of the speculation was correct. Nonetheless, the hypothesis of a world-girdling topographic rift could not have been more stimulating or valuable. The continuation of this history in later chapters also provides unusual glimpses of how scientists work and how pride, prejudice, stubbornness, and discord may intrude in the laboratory and ultimately lead to tragedy.

As Bruce Heezen was to emphasize in the 1960s, there is no significant difference between oceanic and continental rifts with regard to topography or structure. Thus the geological exploration of rifts began on the continents in the eighteenth century. The standard geological term for rift, used in technical papers in all languages, is the German word "graben." The name of a folk character in letters to *Geotimes* used to be "Dr. Grabenhorst." A graben is a downfaulted block, and a horst is an up-faulted block. Both structures are produced by tension.

GRADUAL DISCOVERY OF MID-OCEAN RIFT

Entirely by coincidence, the first submarine rift valleys were also discovered by German scientists pursuing economic objectives. Furthermore, although coincidental to this history, the stimulus for their fruitful research was an erroneous hypothesis. Gold was discovered in sea water by E. Sonstadt in 1872 who reported a concentration of 65 mg per metric ton of sea water. That was enough to be of economic interest. After World War I, Germany found itself with enormous war debts and without its former resource-rich colonies. Fritz Haber, a famous German chemist, proposed that gold extracted from the international seas might solve the problem.[1] The Meteor expedition went to sea between 1925 and 1927 to study the waters of the South Atlantic. Unfortunately for the basic objective, the improved methods of chemical analysis in the 1920s showed that the gold content of sea water is extremely low. Fortunately, the German government had an enlightened vision of the value of science, and the expedition made the most advanced and detailed oceanographic survey then in existence. The Germans developed an echo sounder and made the first transoceanic traverses with closely spaced although discrete soundings.[2] The 13 profiles across the center of the Mid-Atlantic Ridge were relatively evenly spaced from north to south, and most ran east-west. Of these profiles, six definitely show a deep valley in the center of the Ridge. By this I mean that the valley was far deeper than any place nearby on either side. The remaining seven profiles also show a valley in the center of the Ridge, but it is not conspicuously deeper than other valleys on the adjacent flanks. The report on these soundings by Theodor Stocks and Georg Wust in 1933 naturally became famous in oceanographic circles because little like it existed for over 20 years. It put no emphasis, however, on the central valley or on the possibility that it might be continuous.

Another section of the mid-ocean ridge system had been discovered, with a single crossing, by the Danish Dana expedition in 1928-1930. It was named the Carlsberg Ridge in honor of the brewery in Copenhagen, which had helped finance the expedition. The brewery continues to help science, and Carlsberg is known as the "academicians' beer" at the oceanographic laboratory in Copenhagen. The British John Murray expedition surveyed the northwestern Indian Ocean in the early 1930s. It made 22,000 miles of echo-sounding profiles, many across the Carlsberg Ridge, which was traced for 200 miles. John Wiseman and Seymour Sewell wrote,

> Throughout the greater part of its length from Socotra to the Chagos Archipelago this ridge appears to be double. Along the two crests numerous soundings between 2,291 and 3,059 metres have been made and the least water detected over them is 836 metres near the northern end, and 1,569 and 1,752 metres on the two ridges respectively in about latitude 1°30'N where there is a depth of over 3,383 metres in the enclosed gully.

> The Carlsberg Ridge seems to cease just to the south of Rodrigues but it is continued as a low ridge, with some 3,650 metres of water over it, far to the south. . . .[3]

The Carlsberg Ridge thus had a median valley where mapped in detail, and it connected to the Mid-Indian Ridge, which was not so mapped. It "reminds one of the similar Mid-Atlantic Ridge." Moreover,

> We would here like to call attention to the apparent similarity between the topography of the floor of the Arabian Sea and the region to the west of it that is characterized by the presence of the Great Rift Valley.[4]

Wiseman and Sewell proposed and attempted to trace a direct link between the African rifts "and the deep gully that runs along the length of the northern part of the Carlsberg Ridge." They noted that the marine valley and ridge formed a mirror image of the land rifts. The only difference was that the land features were offset, whereas the marine one was continuous. They conjectured that this was an artifice of the intensity of mapping and that similar offsets might exist at sea. Thus, if anyone could have perceived it, they predicted fracture zones, and their thinking was closer to reality than Ewing and Heezen would be with their continuous world-girdling rift in 1956.

The questions of the age and origins of the Carlsberg Ridge remained. The Ridge was already known to be seismically active, so it was still developing. By analogy with the African rifts, the development began in Tertiary time. As to origin, Wiseman and Sewell's findings were ambiguous. They made a close analogy with the structures of the rift valleys, but at the time the tensional origin of the rifts was in doubt. The young Teddy Bullard had mapped gravity in East Africa, and it is noteworthy that the head of the Cambridge group in the marine discovery phase of the 1950s was certainy interested in and familiar with continental rift valleys.

Wiseman continued to publish marine geology for decades and had an abiding interest in the median valley. He reasoned that the seismically active crest of the Mid-Atlantic Ridge probably had a similar valley and planned to survey it in 1940 but was prevented by the war.[5]

It is hardly surprising that American oceanographers returned to the Atlantic before the British, who had been battered and depleted by the war. The Americans, led by Maurice Ewing, quickly began to reap the benefits of new instruments, notably the recording echo sounder, developed during the war. The instruments were similar to the ones we used in the Pacific, bulky and subject to many errors, but they made continuous profiles across the Atlantic. The results of *Atlantis* cruise #150 in the summer of 1947 were published in October of 1949 by Ivan Tolstoy and Maurice Ewing.[6] For the first time in the Atlantic, the existence of physiographic provinces could be demonstrated. Of interest here was the identification of a central "Main Range" of the Mid-Atlantic Ridge. It was characterized by a series of parallel ranges and valleys trending parallel to the ridge axis. The Main Range had thin sediment, whereas the flanks, a succession of smooth shelves, had thick sediment.

This was an important observation that went unappreciated. Lamont would later expend much effort in the early 1960s in trying to establish whether sediment thickness increased with age—as implied by the then unproved hypothesis of sea-floor spreading. A local feature was also noteworthy:

Close to 31°N Lat. a deep east-west trench extends from about 41° to 43°W Long. and cuts deep into the Main Range. Its deepest point is at 2,800 fathoms. Crushed and metamorphosed ultrabasics were brought to the surface by dredging its flanks.[7]

Elsewhere in the paper it is clear that the trough is at 30°N, not 31°N. Tolstoy and Ewing had discovered the first fracture zone (Atlantis) in the Atlantic, although Bruce Heezen would later think it was part of his continuous median rift. The new information was useful but more echograms were accumulating on every cruise of *Atlantis*, so the writers were in no hurry to make a speculative synthesis.

> Thus the status of the problem of the origins of the Mid-Atlantic Ridge is somewhat confused. The present authors are not prepared to take any definite stand on this aspect of the problem.[8]

Tolstoy continued to study echograms from the North Atlantic and published again on the subject in May 1951.[9] The new information, however, did little to supplement or change the conclusions of the paper in 1949. Meanwhile, a new graduate student had begun to study Ewing's echograms; Bruce Heezen had arrived.

After the war, marine geology came into the hands of a new generation in Britain. Maurice Neville Hill (1919-1966) found and surveyed the median valley in the North Atlantic in September 1953 on HMS *Challenger*, the second oceanographic ship of that name. Maurice Hill was born in Cambridge.[10] His father was A. V. Hill, who would soon win the Nobel Prize in Medicine for his physiological research. His mother was Margaret Keynes Hill, and his uncles were Maynard Keynes the economist and Geoffrey Keynes the surgeon and biographer. Maurice Hill interrupted his undergraduate studies at Cambridge to accept a post as a civilian scientist in the Royal Navy in 1939. In 1944 he married Phillipa Pass. His war work on mine and submarine countermeasures ended in 1945, and he returned to Cambridge, a second-year undergraduate at age 26. As a graduate student he worked with Teddy Bullard, and he went on to become one of the pioneers of deep-sea geology and geophysics. He fathered five children; led a happy, harmonious, and very hospitable family life; charmed and unobtrusively supported his students; was munificently, imaginatively, and quietly charitable; was elected to the Royal Society in 1962; and received the Chree Medal of the Physical Society in 1963. Then, wrote Teddy Bullard,

> gradually, and at first almost imperceptibly, things started to go wrong. His overwork became obsessive, the high standards he set himself became a belief in his own inadequacy, in the last year or two of his life his scientific judgment and firmness of purpose on his expeditions faltered. . . . He believed that his brain was deteriorating and there is strong medical evidence that he was right.[11]

He committed suicide in January 1966, loved by everyone who knew him.

The 1953 survey at the start of Maurice Hill's career obtained nine echograms across the ridge crest and traced a conspicuous median valley for 70 miles. It had a maximum depth of 4,000 m between flanking ridges with a minimum depth of 1,260 m. The average relief was about a mile, the top width 12 miles, and the bottom width 5 miles. The survey included some seismic work, but it and further bathymetric surveying were ended by a storm. The Atlantic at 47°N is hardly an ideal site for a research program.

Undaunted, with his interest whetted by the plans frustrated by the storm, Hill returned to the survey area on RRS *Discovery*, also the second oceanographic ship of that name. The bathymetry was surveyed in greater detail, but attempts to obtain rocks by coring and dredging were unsuccessful. Hill returned again on *Discovery* in 1956 and carried out more seismic refraction, coring, and one of the first determinations of heat flow in the Atlantic; storms, however, prevented further bathymetric surveying.

The whole story is typical of field oceanography at the time; it resembles, for example, the mappings of the Mendocino escarpment over a two-year period. Hill made a major discovery and one that British geologists had anticipated. The ship and equipment, however, were marginal, so he could not quite finish the work. Back on shore he would of course describe what he had found but would want to finish the job before publishing. So he tried again, still was not satisfied, and went a third time. During 1956, he published an oblique reference to "a deep steep-sided valley" on the Mid-Atlantic Ridge while reviewing a new chart of the North Atlantic.[12] Otherwise, he never published even a description of the fieldwork until 1957,[13] or a map of the topography and geology of the median valley until 1960.[14] This delay in no way impaired the awareness of oceanographers regarding his work. There were few ships, few cruises, and few people, but many linkages in oceanography in the 1950s. We all knew that Maurice Hill had discovered a seismically active median valley in the North Atlantic.

Knowledge of the seismicity of mid-ocean ridges had been growing in parallel with the bathymetric surveys of the Carlsberg Ridge and Mid-Atlantic Ridge. In 1935, Heck called attention to the seismicity of the ridges in the Atlantic and Indian oceans as well as to the rifts in Africa.[15] In 1945, Beno Gutenberg and Charles Richter issued one of their periodic reports on global seismicity. They located many new earthquakes: "Most of these are close to the well-defined line of the Atlantic Ridge and its Arctic continuation."[16] They mapped the seismicity of the Carlsberg and Indian-Antarctic ridges, tracing the latter to south of New Zealand. Activity of the African rifts was very low.

Attention as again focused on the seismicity of mid-ocean ridges by J. P. Rothé at the meeting of the Royal Society for *A Discussion on the Floor of the Atlantic Ocean* in 1954. Teddy Bullard presided, Maurice Ewing participated in discussions, and Harry Hess presented a paper at the same meeting. Rothé, who was Director of the International Bureau of Seismology, had located many new earthquakes. His map showed a narrow, almost continuous, belt of epicenters from north of Iceland, through the Atlantic, past South Africa, through the mid-ocean ridge in the Indian Ocean to the Carlsberg Ridge, and on into the African rifts and the Red Sea. "It seems, therefore, that these two ridges are related structures."[17] Indeed, the "Discussion" was too limited. The mid-ocean ridges in the Atlantic and Indian oceans, to the extent they are defined by seismicity, are continuous for more than 30,000 km and wrap right around three sides of Africa.

In the same year (1954), Gutenberg and Richter published another global summary of seismicity.[18] It was to be the standard reference during the last half of the 1950s while the integrated hypotheses about the origin of mid-ocean ridges were taking form. (At Lamont, however, seismologists were active at compiling and refining data during this period, and their work would also influence their thinking.) The continuity of the mid-ocean seismic belt through the Arctic, Atlantic, and Indian oceans was confirmed. "North Atlantic shocks are very favorably placed for epicentral determinations" and "Numerous minor shocks could have been added to the catalogue. . . ."[19] Elsewhere most locations were less reliable. Nonetheless, the seismicity of the mid-ocean ridges was traced along the Indian-Antarctic Ridge into the South Pacific. A branch extended to Chile, but the main trend was along the crest of the East Pacific Rise and into the Gulf of California. There the marine trend became indistinguishable from the seismicity of the San Andreas fault and the Basin Ranges. Another band of earthquakes extended into the sea from the northern end of the San Andreas fault in Northern California. This northern region was known to be shallow, and Max Silverman and I had already surveyed the crest of the Juan de Fuca Ridge and discovered the median valley of the Gorda Ridge in 1952. We did not think of them, however, as mid-ocean ridges, and I do not believe I had mentioned our survey to Gutenberg. Despite all the evidence of seismicity, little that was new could be stated about the origin and structure of mid-ocean ridges:

> It is suggested that these ridges, originally produced by folding, are now being broken up by block faulting consequent on a redistribution of tectonic forces.[20]

EARLY WORK AT LAMONT

The oceanographers at Lamont had a golden opportunity in the 1950s to use the new instruments to explore and explain the Mid-Atlantic Ridge, which was right off their home port. Following Ivan Tolstoy's departure, Bruce Charles Heezen (1924-1977; Fig. 32) became Lamont's principal investigator of echograms and a collaborator with Ewing and the other older hands in marine geology. Bruce and I had remarkably coincidental lives. He was born in Iowa and grew up on a farm in Muscatine, which was my mother's girlhood home. He graduated, at 24, from the University of Iowa, majoring in geology. His first paper was in paleontology,[21] as was mine, and his coauthor was Walter Youngquist who, by the time the paper was published, was my officemate and fellow graduate student at Harvard. Bruce and I went on to work with the same kinds of data and simultaneously to discover about the same things in different oceans. We were both consultants, concerning our different oceans, for the American Telephone and Telegraph Company, Cable and Wireless Limited, Rand NcNally, Life Magazine, and Creative Playthings, Inc., which made maps. We traveled to several international meetings together and stayed in each other's homes. For one who was not at Lamont, I knew him very well.

Bruce was remarkably energetic, enthusiastic, and single-minded about science. Manik Talwani recalls,

> I still remember the relentless way in which Bruce cored site after site, jumping up and down gleefully every time the coring tube came up wrecked and the rest of us mournfully contemplating the work necessary to straighten out the bent tube to set up the rig for the next coring station a few hours away.[22]

Bruce had predicted the existence of sand layers under an abyssal plain, and such layers tend to wreck cores. Bruce concentrated his early efforts, as did I, on the question of the existence of turbidity currents in the deep sea, which is why he was interested in the sand layers and abyssal plains. In this work he, a graduate student, collaborated chiefly with Ewing and with an already noted marine geologist, Dave Ericson. The style at Lamont was far more toward multi-authored papers than at Scripps, where we collaborated at sea and helped each other on land but generally published separately. Counting abstracts as well as papers, Heezen published 14 times with Ericson mainly in the early fifties but tapering off as their interests diverged. He published 44 times with Ewing in one of the best known and most widely cited collaborations in modern geology. The interests of Ewing and Heezen in global tectonics and marine geology, however, never diverged. Thus it is a remarkable fact that, remaining in the same institution, they never published together from 1964 to 1977, when Bruce died. Allowing for printing delays, they ceased active collaboration about 1962. Clearly, something happened between them.

In the early 1950s, there was no inkling of what was to come. Lamont expanded, exciting problems abounded, and the means to solve them were on hand. The Navy upgraded the security classification of its soundings at that time, and Bruce turned to physiographic diagrams just as I did. He was fortunate in 1952 to establish a close relationship with Marie Tharp, which was to endure for the rest of his life. She plotted echograms for him, and they began to prepare the first of the physiographic diagrams that would make them both famous by the time they were published by the National Geographic Society. What then happened was later recounted by Bruce to William Wertenbaker:

> In three of the transatlantic profiles [Marie] noticed an unmistakable notch in the Mid-Atlantic Ridge, and she decided they were a continuous rift valley and told me. I discounted it as girl talk and didn't believe it for a year.[23]

What changed his mind about the rift was that, as a consequence of consulting for AT&T he had a plot of Atlantic earthquakes prepared about a year later. Earthquakes and quake-generated slumps were known to damage submarine cables. The quakes were in the vicinity of Marie's proposed rift (Fig. 14 A). Moreover, if the circles of error in locating the earthquakes were plotted instead of points for epicenters, all the circles overlapped the rift. Thus it was possible that they were actually confined to the rift in the North Atlantic. Bruce then reasoned that if the median rift has earthquakes, it follows that earthquakes on a mid-ocean ridge must be a rift. In Marie's words,

> Using earthquake epicenters where there were no soundings, plotting of the valley was continued about the globe.[24]

Marie remembers that Bruce became convinced that a world-girdling valley existed by mid-1953, which is in accord with his interview with Wertenbaker. This was a very remarkable conviction on two counts. First, he concluded from the distribution of earthquakes that a continuous mid-ocean ridge existed even though soundings were lacking or even suggested otherwise. It was a bold conclusion and, allowing for offsets on fracture zones, a correct one. Second, he concluded that a median valley existed along the crest of the ridge. This was even bolder, and on the scale he meant, it was wrong. On a smaller scale, however, that no one could then map, it was correct. The conclusion was bold; Bruce was bold; his behavior in the next few years thus was out of character.

Late in 1953, Heezen, Ewing, and Miller published a paper on a trans-Atlantic profile of topography and magnetic intensity.[25] It included data collected as late as June 1953. The topographic profile was one of the six that Heezen later published again as examples of a median rift and with which he said that anyone would have found the rift valley. Thus, given his spectacular discovery, Bruce apparently had a priceless opportunity to include it in the paper or at least add a note in proof. Nothing was said about a median valley on the profile—let alone around the earth.

Heezen published not a word about the median valley through 1954, 1955, and most of 1956—the entire period during which Maurice Hill was actually surveying the valley. Heezen, however, did not lack time, energy, or other data to publish eight times in 1954, and seven times in 1955, mostly in collaboration with Ewing and Ericson. Can it be that he thought the soon-to-be-famous rift was not much of a discovery? Was he waiting for Maurice Hill to publish as I waited for Russ Raitt? Was it all too speculative even for Bruce Heezen? Or was it so important that he was saving it for a special occasion, as Manik Talwani surmises?

EXTRAPOLATING TO A GLOBAL RIFT

The correspondence of ridges, rifts, and earthquakes finally emerged in 1956 in several places. In an abstract by Heezen alone and based on a talk delivered in the fall of 1955, there is the following passage:

> The provinces of the Mid-Atlantic Ridge established in the North Atlantic probably continue through the South Atlantic, Indian, South and East Pacific, and Arctic Oceans. This great mid-ocean median ridge is remarkably similar in topography and seismicity to African plateaus and rift valleys, suggesting a common origin and similar age.[26]

The similarity was also mentioned by Ewing and Heezen in another abstract, including the extension of an active rift zone into Africa.[27] The rift also appeared as a continuous feature on the first physiographic diagram of the North Atlantic displayed by Heezen and Tharp in 1956.[28] The first discussion of the world-girdling rift, however, was in a paper with Ewing as first author and Heezen as second:

> Oceanic ridges of the Mid-Atlantic Ridge type are believed to be continuous over great lengths. They apparently have the median rift zone as a characteristic feature throughout, and this rift zone is the locus of the shallow focus earthquakes. The ridges may be traced through poorly sounded areas by the aid of an epicenter map. It must be borne in mind that the rift zone may be the primary feature in this combination and the ridge simply a consequence of the rift."[29]

The last sentence was remarkably prescient, but what had they really contributed? They had only a trivial amount of new information on which to base a hypothesis. They credited "the magnificent studies" by Gutenberg and Richter for the epicenters and noted the median rift on profiles by Stocks, Hill, and Wiseman and Sewell. It is curious that they did not mention the six Lamont profiles in the North Atlantic. They then made a big jump to a world-girdling rift (Fig. 14 B) that "apparently existed."[30] Right or wrong, they did two important things. They provided a new and stimulating way of thinking about mid-ocean ridges, and they provided a target that unified global exploration during the International Geophysical Year.

The media quickly spread the news of the great discovery, and a map of the world-girdling rift valley appeared in the *New York Times* in February. Meanwhile, marine geologists who were not at Lamont awaited reprints of the Ewing and Heezen paper to find out what was happening. The book in which they published was not widely available. My reaction to these goings on appears in a letter to Heezen on 20 March 1957:

> I just received a copy of your "Some problems of Antarctic submarine geology" and I am very gratified to find a discussion of the much publicized world-girdling rift. I have been increasingly distressed to read one account after another in the press and magazines of this fabulous rift 2 miles deep, 20 miles wide, and 40,000 miles long. I have had to say to more than one reporter that I believed that you were misquoted because I knew no soundings existed to support such a continuous rift in the south Pacific and Indian Oceans. The constant repetition had begun to wear, however, and I was just reaching a point where I intended to write to you with the thought of issuing a contrary statement regarding the Pacific. After reading your paper this hardly seems necessary. I do not mean to imply that we have soundings to prove that the rift does not exist in the south Pacific; perhaps it does. On the other hand, north of N10° latitude in the eastern Pacific we can prove that it does not exist as you describe it in the press.

Bruce responded in a few days later, in a letter dated 29 March 1957:

> The rift is not always 20 miles wide by 2 miles deep as the papers stated . . . the general depth below its rim is usually only 500-1000 fathoms. . . . However, we have found no profiles across this belt which fail to show a topographic feature which could be interpreted as a rift. I would very much like to see your evidence that a rift does not exist north of 10°N in the Pacific. I believe that your published profiles can be interpreted to support its existence in this area. I realize, however, that you have a vast amount of unpublished data which I have not seen. If you can establish to our mutual satisfaction that you are right in this regard it might save me future embarrassment and the scientific literature further confusion.

In my letter I had called attention to the similarity between his rift and my fracture zones. He, however, had come to believe that any valley near the crest of a mid-ocean ridge was a median rift.

THE SEARCH FOR THE RIFT

All was was not yet certain; Ewing and Heezen recommended systematic exploration of the ridge and rift as part of the upcoming International Geophysical

Year. This recommendation was one of their most useful contibutions. As I
wrote to Bruce in my earlier letter,

> We shall be taking two ships into the area next winter as part of the IGY, and
> I plan to make every effort to confirm or deny the existence of the rift.

They predicted a rift on the East Pacific Rise and the existence of a ridge and
rift between Easter Island and southern Chile. The latter prediction was particularly bold because it was based solely on the epicenter distribution and defied
the few available soundings.

I shall shortly describe the Downwind expedition of 1957-1958 and its relation to concept of the global tectonics. For present purposes, we are concerned only with the search for the predicted rift and ridge. My memory is that
we were very excited by the Ewing-Heezen hypothesis and fully expected to
find the rift. During the April meeting of the American Geophysical Union, I
had invited Bruce to join me and see for himself. In a letter of 3 July 1957, I
repeated the offer:

> My invitation to join us on Downwind still stands. We leave November 1,
> go to Tahiti via the Tuamotus, thence to 45°S, across the East Pacific Rise to
> Valparaiso, arriving 23 December. . . . IGY budgets being what they are, I
> do not believe that I can cover your travel expenses but if you have travel
> funds and want to come, please let me know and I shall hold a place for you.

He could not make it.

I arrived home on Christmas day, and a few days later (6 January 1958) I
told him what had happened:

> I have the mournful honor of informing you that the median trough of recent
> fame does not exist along the crest of the East Pacific Rise between 48°S and
> 43°S. On our Downwind expedition, from which I have just returned, I man-
> aged to make seven crossings of the crest of the rise in that area with the
> express purpose of surveying the trough. I hesitate to dwell on the frustra-
> tions of staying up hour after hour on the original crossing waiting for a
> trough to appear so that I could survey it. In any event the crest is rather
> smooth except for the usual concentration of seamounts.

I went on to say that the issue of the small rise between Chile and Easter Island
was in doubt from the soundings we had taken on *Baird*, but that I had not yet
seen the soundings from our sister ship *Horizon*. In fact, the two ships together
confirmed the bold Ewing-Heezen prediction. There was a ridge, and it appeared to be rifted. I shortly informed them of their success when I passed
through Lamont in April.

Bruce was unconvinced about our data on the rift, so I invited him to see for
himself. We unrolled the echograms and compared them with the navigation

plots. All I could see were a few low abyssal hills and a few small circular volcanoes. Bruce was convinced that one of the typical shallow valleys between elongate hills was "the rift valley." I put a footnote, added in proof, to a paper I published in September 1958:

> Doctor Heezen has inspected all the Downwind echograms in this area with the writer, in a preliminary attempt to standardize topographic terminology in the Pacific and Atlantic basins. Heezen believes that the soundings may show a small median trough and certainly do not rule out the possibility of one.[31]

Heezen and Ewing responded in kind in 1961:

> From the equator to the mouth of the Gulf of California, the epicenter belt seems to follow a ridge similar to other mid-oceanic ridges, but Menard (1955), who has studied much unpublished and unavailable data, has come to a diametrically opposed view, that the area is dominated by east-west trending "fracture zones," and that no distinctive topographic feature is associated with the epicenter belt.[32]

All this was very friendly and cooperative.

By 1959 we had completed three IGY expeditions—Downwind, Dolphin, and Doldrums over the East Pacific Rise, and it appeared to be the smoothest region with thin sediment in the Pacific. Now, a quarter century later, it is known to have the lowest relief of any region of abyssal hills in the world. Not uncommonly, there is a median horst rather than a rift.

Wertenbaker reports Heezen's memory of these events a decade or more after they occurred:

> But, says Heezen "this idea of a rift valley was so revolting to geologists that some of them from Scripps went right out and, they said, proved us wrong."[33]

That was, however, the very different Bruce Heezen of the early 1970s. Wertenbaker himself continues:

> A worldwide rift was less than wanted. In their eagerness to dispose of a rift, however, scientists confirmed the existence of a worldwide ridge.[34]

At the Colloquium on Deep-Sea Topography and Geology at Nice in May 1958, Bruce further expounded his ideas. The seismic belt on mid-ocean ridges was associated with a median rift valley. The entire ridge system, including the African rifts, was a product of tension. Large transcurrent deformations seemed required to explain crosstrends at right angles to the mid-ocean ridges. Global expansion deserved serious investigation.[35]

Heezen and Tharp finally published their North Atlantic physiographic diagram in 1957; it was an immediate success and was reprinted in 1958. It showed a continuous median rift, but the east-west trough, mapped by Tolstoy and Ewing as cutting across the ridge crest, was very obscure indeed. Every deep apparently was part of the rift. This was spelled out in their remarkable book in 1959 in which they published the first account of regional abyssal geology based on modern techniques.[36] It was a text to accompany the physiographic diagram. One of the illustrations showed that the Tolstoy and Ewing trough had become part of a continuous, sinuous rift. It was some years before it was transformed into the Atlantis fracture zone. Thus Heezen's ideas regarding the interactions of rifts and fracture zones during this period are obscure.

Global reconnoitering of mid-ocean ridges was intense in the next few years, and the continuity of the ridge in what had appeared to be gaps was triumphantly confirmed. Moreover, except for the pesky Pacific and a bit of the southern Indian ocean, there was always a rift. The Lamont group viewed the ridge as a sinuous but continuous feature with many sharp bends (Fig. 14 A, B, and C). According to their maps the rift had to be opening in all directions rather than sideways or toward the flanks.[37] No wonder they believed that the rift was tensional. No wonder that Heezen believed that it was evidence of an expansion of the earth. The wonder is that Ewing did not believe it—unless, of course, he was less certain about what they might have found.

THE RIFT BETWEEN EWING AND HEEZEN

It was at this time of mutual success that the relationship between Ewing and Heezen began to come apart. The split would lead to virtual banishment in 1966. Bruce complained that Ewing insisted on putting his name on reports to which he had made no contribution despite the fact that he did the same thing himself. Bruce began to expand his interests from marine geology to seismology—a field in which Ewing had other, more highly qualified, collaborators. Bruce twice cited a paper on the seismicity of the Mid-Atlantic Ridge as in press.[38] He was to be first author, Ewing second. According to their complete bibliographers, it was never published.[39] Thereafter, Bruce began to publish alone more often, and Ewing published with other collaborators about marine geology. In a sense, the split that would widen was formalized in an unusual joint statement.

> The present authors have concluded that the rifted mid-oceanic ridge is dominated by extensional deformation (Heezen, Tharp, and Ewing, 1959). However, they each favor a different primary mechanism of the deformation and differ in their estimates of the amount of extension indicated. Ewing (Ewing and Ewing, 1959) favors a mechanism drawn by mantle convection

currents, while Heezen (1959) believes that the extension results primarily from the internal expansion of the earth.[40]

At that time Maurice Ewing was the founder and Director of Lamont Geological Observatory and, at the very least, was among the leading geophysicists and oceanographers in the world. After nine years as a graduate student, Bruce Heezen finally received a doctorate in 1957. I assume that like some Scripps students he was indifferent about obtaining one. He already had a job, money, and a supporting staff for research, and was leading expeditions. What did it matter whether he had a Ph.D.? Moreover, one might be reluctant to undergo an examination by professors whom one regarded as colleagues. In any event there was no implication of slow progress at that point. Nonetheless, Heezen did not become an Assistant Professor until 1960 at the age of 36. There was no doubt which of the two authors was expressing a majority report for Lamont. Bruce, however, persisted in advocating the hypothesis of an expanding earth, which will be discussed in a later chapter.

Chapter 9

BLUE WATER, GREEN ROCK: THE ADMIRAL'S MANTLE

Harry Hess and Alfred Wegener were similar in many ways: cheerful men, professors, married to professors' daughters. Each fought through a long war, not as children straight from school but as established young scientists breaking a career. Wegener was 34-38 years old in World War I and twice wounded on the western front. Hess was 35-40 in World War II and earned four battle stars in the western Pacific. Both had such broad and deep interests that each appeared to have had quite different careers as viewed by contemporary specialists with narrower but no deeper concerns. The major difference in the two men lies in their acceptance by their colleagues. Wegener was an outsider who never penetrated the inner circle of academicians, medalists, professors of endowed chairs, and presidents of professional societies. Hess was a member of the establishment all his life.

THE EARLY YEARS

Harry Hammond Hess (1906-1969) was born in New York City, where his father was a member of the N.Y. Stock Exchange; one of his grandfathers was a leader in construction work, and the other operated a distillery.[1] In 1923 Harry went from Asbury Park High School in New Jersey to Yale, where he initially majored in electrical engineering. Tiring of "drawing cross sections of spark plugs"[2] he transferred to geology and, he said, failed his first course in mineralogy. Adolph Knopf told him he had no future in the field—just as Lord Rutherford told Teddy Bullard that he had no future in physics. Hess found himself as one of two undergraduate geologists among a galaxy of professors—Alan Bateman, Carl Dunbar, Adolph Knopf, and Chester Longwell—and an extraordinary cluster of graduate students. Among them were W. H. Bradley, M. N. Bramlette, James Gilluly, D. F. Hewett, Thomas Nolan, and W. W. Rubey. Young Harry had fallen into a nest of future academicians who would dominate American geology just when his radical hypothesis of seafloor spreading would emerge.

Upon graduation, Hess spent two years running geological traverses at quarter-mile intervals in what was then Rhodesia. He developed "leg muscles, a philosophical attitude toward life, and a profound respect for fieldwork."[3] He was examining rocks in South Africa only a month before his death. Returning to America, Hess considered graduate school at Harvard, but he was alienated by "No Smoking" signs and continued his lifelong puffing at Princeton instead. There he came under the powerful influence of professors Arthur Buddington, A. H. Phillips, and Richard Field. Buddington led him into petrology, Phillips into mineralogy, and Field into marine geology and geophysics. It was Hess's genius to excel in each of these diverse subjects and finally to meld them into a theory of the evolution of ocean basins.

The farsighted Richard M. Field arranged for Hess to join F. A. Vening Meinesz in a gravity survey by submarine in the Caribbean in 1931. In the following years, Field obtained another U.S. submarine for Hess to continue this work and encouraged him to obtain a reserve commission as a Lieutenant (jg) in order to smooth operations with the Navy.

Hess's first paper, in 1932, was on the interpretation of gravity anomalies and soundings in the Caribbean. In 1933 he gave a talk on submarine canyons in the Bahamas and submitted a doctoral thesis on a field and laboratory study of an altered peridotite body at Schuyler, Virginia. For a decade thereafter he followed the normal career of a successful young scientist. He taught at Rutgers, shifted to the Geophysical Laboratory of the Carnegie Institution of Washington, and settled down for life as a professor at Princeton in 1934. In that same year he married Annette Burns, daughter of a professor of botany at the University of Vermont. During the decade, he published 20 papers in refereed journals and a host of abstracts. The papers spanned all four aspects of Hess's interests; 11 of them were concerned with serpentinization of ultrabasic rocks, and another nine with gravity anomalies and island arcs. By 1939 Hess had realized they were related problems:

> For a number of years the writer has been occupied in research in two widely different fields; one, the investigation of ultramafic rocks, their origin and alterations, and the other, the study of gravity anomalies in island arcs. It was most unexpected that these two divergent fields should come together and afford solutions for problems in each based on data obtained in the other field.[4]

He had achieved his first imaginative synthesis relating serpentinization to global tectonics.

THE WAR

As a reserve officer, Hess took the 7:42 train from Princeton on December 8, 1941 to report for active duty. He spent some time in scientific or operational research in and off New York but maneuvered himself onto the attack transport USS *Cape Johnson* first as navigator and then as captain. Despite the existence of Japanese submarines, and the detectability of 12 kHz echo-sounding pulses, he interpreted the fleet standing orders to maintain silence rather liberally.[5] The result was the discovery of deeply submerged, flat-topped, extinct volcanoes, which he called "guyots." Their existence was also recorded, but unnoticed or unremarked, on the deep-sea-sounding compilation sheets of the U.S. Hydrographic Office. Thus Hess had identified 160 of them when he gave his first talk on the subject in May 1946. Like Wegener in yet another way, he did not cease scientific research just because of a world war.

THE ADMINISTRATOR AND COMMITTEEMAN

The paper was a bombshell. Dick Field was vindicated again. The ocean deeps were beginning to disclose their mysteries. From that paper came the genesis of the Midpac expedition of 1950 and the foundation of much of my own career as well as the careers of others. Hess himself, aged 40 in 1946, began to be involved in those affairs that typify the mid-careers of scientists whose youthful research has brought them broad acclaim at home and global recognition among related specialists. This is the period when a real scientist is first exposed to the blandishments of recruiters, the flattery of professional office, and the appeal to responsibility for public service. From it, hardened by the fire, he returns to research or, instead softened, descends into the abyss of administration. Hess, for a geologist, was young when he reached this point. Perhaps as a consequence, he achieved an unusual accommodation whereby he simultaneously undertook both administration through National Academy advisory committees and research largely by imaginative reinterpretation of other people's data. To do geopoetry, all that is needed is a scratch pad. To win medals doing what is available for anyone to do, that is a triumph.

In 1944 Hess published the first of many reports on the activities of his advisory committees. At one time or another he was a direct advisor to the Navy, Coast and Geodetic Survey, Atomic Energy Commission, National Aviation and Space Agency, and Office of Science and Technology in the White House. He also advised government through the committees of the National Research Council. He was Chairman of NAS-NRC committees on radioactive waste disposal and the Mohole, Chairman of the Earth Sciences Division of the NRC, and Chairman of the Space Sciences Board during the critical years of the great plunge into space. In addition, he was President of the Geological Society of America, the Mineralogical Society, and two different sections of the American Geophysical Union: first Geodesy and later Tectonophysics. No wonder he had little time to visit a library to check references.

Hess's early achievements brought him election to the National Academy in 1952, which in part precipitated this "crushing load."[6] Activities in committees and professional societies, however, are not what counts, and it was not until 1966 that his geopoetry was confirmed and he won his first awards, the Penrose Medal and the Feltrinelli Prize, never before awarded to any scientist in the western hemisphere and worth $32,000.

THE OCEANIC CRUST

In later chapters we shall turn to the geopoetry and how it fared, but here we are concerned with the research that was a necessary precursor for the geopoetry and how it led to Project Mohole and what then happened. Hess wrote three papers about serpentinization and tectonics in 1954-1955 of the sort that a busy scientist produces when he has one theme but is invited to three meetings. The theme, however, was novel, and the papers give us snapshots of its evolution. The first was presented at *A Discussion of the Floor of the Atlantic Ocean*,[7] one of the timely symposia that so distinguish the Royal Society, and the one where Rothé spoke of mid-ocean earthquakes.

Hess's casual style is quickly revealed;

The ideas here presented are the product of a few weeks' thoughtful consideration, whereas a year would have been a more appropriate interval to do the data justice.[8]

The paper bears all the signs of being written on an airplane, and if so, we can be grateful that the propeller planes then in use took longer than jets to cross the Atlantic.

Hess states flatly that

the most momentous discovery since the war is that the Mohorovičić discontinuity rises from about 35 km under the continents to about 5 km below the sea floor.[9]

He cited no one, and in a sense he could not, because most of the data were collected by Russ Raitt on the Midpac and Capricorn expeditions—and they were unpublished. Here we see the enormous advantage of the members of an invisible college during a scientific revolution. A few insiders had about five years to digest the implications of a flood of observations before the outsiders ever saw the bare data. Judging solely by the results, it also appears that the most favorable position is to be an insider who is not distracted by actual collection of data.

The second most important "discovery" was the recognition that samples of the upper mantle are available at the earth's surface. They take the form of nodular inclusions or xenoliths in basaltic lava flows. Given the two new facts, Hess drew on his work on gravity to make the basic assumption that any large area of the crust is in isostatic balance. Thus above a "depth of compensation," perhaps 40 km, the mass of a unit area of a continent, deep-sea basin, or shallow mid-ocean ridge was the same. Wegener had made this same assumption in proposing continental drift, and it was well documented on land. Hess drew on the few marine observations of his old field instructor, Vening Meinesz, for a firm assumption that allowed him to interpret Raitt's and Ewing's sparse seismic data.

The compensation depth was far below the oceanic Moho, and therefore the density of the upper mantle varied from place to place. Why? Hess sought other data bearing on the question. How did oceanic ridges differ from basins? Hess, like everyone else at this time, made little distinction between what are now mid-ocean ridges, aseismic ridges, and continental and oceanic plateaus. Naturally, guyots attracted the attention of their discoverer. They occurred on the Mid-Pacific Mountain plateau, the Hawaiian Ridge, and possibly the Caroline Ridge; none were known on the Mid-Atlantic or Mid-Indian ridges. Thus they indicated that at least some ridges had once been shallower and had since subsided. Possibly the distribution indicated that different ridges were in different stages of development.

Among the other features of ridges:

Almost all of them are devoid of linear small-scale ridges parallel to their main axes. . . .[10]

I have no idea where Hess could have obtained data adequate to sustain such a conclusion. Certainly, as Heezen was shortly to demonstrate, it could not have been more erroneous. Here Hess illustrates the indescribable chaos of geological hypothesizing about ocean basins prior to the massive exploration of the fifties. On another subject, Hess drew on the recently published results of that exploration[11] to note that heat flow was higher on some ridges than in deep basins and that this, too, might be related to the evolution of the ridges.

For an explanation of these facts he went back to the phenomenon he had been studying for two decades, his thesis subject, the serpentinization of ultrabasic rocks. A famous paper by his old classmate at Yale, W. W. Rubey, had recently supported the concept that the oceans have grown gradually by outgassing of water from the mantle.[12] Experiments at the familiar Geophysical Laboratory had shown that olivine in the presence of water and at a temperature below 500°C gradually alters to serpentine exothermically with an increase in volume of about 20%. The temperature gradient measurements by Revelle and Maxwell indicated that the critical isotherm was far below the oceanic Moho.

Hess had identified a reaction that probably affected the probable rocks of the upper mantle. On a regional scale, the reaction was quantitatively adequate to elevate ridges above basins by the amounts observed. The reaction was exothermic and would produce the high heat flow observed on some ridges. It was reversible, and this could account for the large subsidence of guyots on some ridges. This masterly synthesis of marine geophysics and field and laboratory petrology indeed appeared to be a worthy output for a few weeks of consideration.

Hess described the first paper in 1955 as a modification of the one in 1954, but the changes are minor.[13] He went so far as to list the names of the marine seismologists who had made the most momentous discovery, but he still cited no references. He noted that serpentinized peridotite occurred at St. Paul's Rock and on abyssal fault scarps in the Atlantic as well as in xenoliths. One addition was a drawing that clarified his ideas on the influence of vertical migration of isotherms on the elevation and subsidence of ridges. A second addition was a table purporting to show "temperature for various gradients at various depths." From it he drew the conclusion that "Any reasonable estimate of the temperature gradient indicates that for a considerable depth below the Mohorovičić the temperature will be below 500°C (Table 1)."[14] This was typical of Hess in his geotectonic, as opposed to his mineralogic, mode. He cited neither references nor data. Nor did he do so in 1959 when he presented the table in graphic form for shallow depths. His "reasonable estimate" was based on gradients ranging from 20°-40°C/km. The only existing measurements were in the top few meters of sediment, but they indicated a minimum gradient of 21°C/km, an average of 60°C/km, and a maximum on the crest of the East Pacific Rise of 215°C/km.[15] Hess was misled but not actually wrong in his conclusion as long as he thought of static ocean basins and isotherms that moved only vertically. It was going to take a major reorientation in his thinking when he conceived of sea-floor spreading in which the crust is created where the temperature gradient is abnormally great and in which the crust drifts sideways through a thermal gradient.

The second paper in 1955 was more of the same. They were an electrifying series in 1954-1955, but in retrospect I am reminded of the advice I received from Kirk Bryan when I was a graduate student. "Keep writing the same paper until you are sure someone is getting your message." Hess's message was received although not necessarily believed. However, the composition of the upper mantle was widely discussed, and soon many petrologists were confident that they knew the nature of the mantle rocks. Unfortunately, the confident statements of the petrologists did not agree. The rest of us were baffled, and the stage was set for Project Mohole—an attempt to sample below the Moho by drilling.

It is now generally accepted that the oceanic crust and upper mantle have been thrust up from the deep sea and onto the continents in several places. Thus in the Blow Me Down complex at the Bay of Islands in Newfoundland, or in the Samail complex in Oman, the interior rocks of the earth can be sampled with a geology pick. What is found from top to bottom is the following:

pillow basalts		
sheeted dikes	layer 2—volcanics	
gabbro	layer 3—oceanic crust	
	- - - - Moho - - - - -	
ultramafics	mantle	mantle

This is more or less the sequence of rocks one would expect to result from sea-floor spreading. What is particularly interesting for this account is that one of the ways the sequence is identified as oceanic crust and mantle is by laboratory measurements of seismic velocities.[16] Thus the question put to Russ Raitt in Los Angeles two decades earlier has been inverted. Instead of asking seismologists to identify crustal rocks by petrographic names, petrologists identified crustal rocks by matching them with the seismologists' layers.

These interpretations were not known in the fifties for want of an accepted theory or the observations specifically made to test the theory. Hess thought that the second most important observation of the period was that xenoliths are samples of the mantle, but there was no consensus among petrologists on this matter. Consequently, it was not unreasonable that F. B. Estabrook of the Basic Research Branch of the U.S. Army should propose drilling a hole through the crust. He identified many desirable scientific objectives, but perhaps the general tone was set by the statement that "massive financial backing can with increasing ease be obtained for organized group attacks on basic problems."[17] Federal money, in short, was available for large projects in the earth sciences as well as for the little science that was familiar to geologists.

THE MOHOLE PROJECT

Prior to the initiation of the system of Federal grants for research in the fifties, academic geologists supported their research with grants of $100-$1,000 from professional societies or their universities. Government scientists in the U.S. Geological Survey worked with sustained support, but on an equally meager scale. The Office of Naval Research changed all that when it began to support research projects of individual scientists. Oceanographers, in those days, were a special case because they had to work in groups in order to use a ship efficiently, and ships had to stay out for months to work on the wide, deep sea. The ante in the oceanographic game was at least $100,000 per expedition. I was on a National Science Foundation advisory panel for three years around 1960, and it was quite striking how different kinds of earth scientists conceived of support for their research. Typically, a land geologist wanted $10,000 to run his jeep and analyze his samples, a land geophysicist or geochemist wanted $100,000 to build and utilize new instruments, and a marine geologist wanted $1,000,000 to develop new instruments and use them on a ship or two for a few months. The land geologists on the panel, although dazzled by the idea of $10,000 to run a jeep, were astounded by expenses elsewhere and appalled by the proportion of earth science research funds going to new disciplines. Still, they voted us the funds.

On the basis of professional interest, petrologists or seismologists or marine geologists might have followed up on Estabrook's idea of drilling to the mantle. The marine geologists, however, were the only "organized group" accustomed to "massive financial backing," and they acted. As far as anyone can remember, Estabrook's note in the most widely read scientific journal in October 1956 actually had nothing to do with the action in early 1957. Memories, however, were clouded, presumably because the important events only seemed so in retrospect. At the time, they were simply part of the ongoing activities of a group of interacting friends. Thus two early accounts by Hess and Willard Bascom differ about the time or even the sequence of early events.[18] It hardly matters. In the spring of 1957 an amorphous group calling itself the American Miscellaneous Society (AMSOC) assembled for a typical meeting, a sunny "wine breakfast," at Walter Munk's house in La Jolla. Walter lives in an academic enclave called "Scripps Estates," which was organized and subdivided by local scientists in the early 1950s. Among the characters in this history, Russ Raitt, George Shor, Jeff Frautschy, and I all lived within a block of Walter. AMSOC was distinguished by a total absence of organization: no bylaws, membership list, officers, or formal meetings. There was a lot of whimsy as indicated by the imaginary divisions including Etceterology and Triviology and the affiliation with the Society for Informing Animals of their Taxonomic Position. In short, it was a little in-group whose activities would have been good only for a few laughs, except for two things. First, among the amorphous nonmembers were Munk, Roger Revelle, John Isaacs, and Gordon Lill, and thus a very high level of competence, imagination, and practice in obtaining large research grants. Second, the American Miscellaneous Society, a militantly unorganized group, undertook to direct a multimillion-dollar attempt to drill to the mantle—the Mohole Project.

Either before (Hess) or after (Bascom) the sunny breakfast, which neither attended, there was a meeting of the NSF Earth Sciences review panel in Washington. Hess was Chairman, Munk a member. After reviewing the day's quota of submissions, Walter remarked that not a single one of them would result in a really major advance; he suggested the panel invent one and promptly proposed the Mohole. Harry had the imagination to take his colleague seriously, and in the few minutes remaining it was proposed to refer the project to AMSOC. By 1959 Hess felt called upon to state "This was not a joke."[19] By 1961 the name "American Miscellaneous Society" with its whimsical overtones had begun to be replaced by "AMSOC Committee" without any explanation of the derivation. Even the new chairman Hollis Hedberg, who gave the lucid Penrose citation for Arthur Holmes, came to accept the obfuscation.[20]

In 1964 the News and Comment section of *Science* contained a series of three articles titled "Mohole: The Project That Went Awry."[21] What could one have expected? If ever there were a doomed project, surely this was it. Imagine an amorphous cluster of academic scientists trying to organize and manage a multimillion-dollar project through meetings of a noncommittee. Imagine the cloistered National Academy of Sciences discarding its century-old advisory role and directly sponsoring these activities. The proper, the only reliable, way to conduct complex research and development projects was well known. Appoint a single project manager with complete authority, give him a tautly responsible support group, construct a PERT chart, and put the engineering, construction, and operation in the hands of a major corporation—preferably one in aerospace. How else could one hope to solve the problems of drilling from an unanchored ship in water 2½ miles deep amid the swells of the open sea? Ironically, reality defied reason. Under AMSOC the Mohole project was a spectacular success; in the hands of a giant corporation noted for turnkey operations it collapsed. Many of the people in this history became involved during the AMSOC years. In 1961 Ewing, Hess, Maxwell, Munk, and Revelle were on the Committee. Maxwell and I were on the Scientific Objectives and Measurements Panel, and Hess, Raitt, and Shor were on the Site Selection Panel. Much of what happened, however, can only be attributed to the skills and energy of Willard Bascom, who was appointed Executive Secretary and became Project Director. Bill Bascom had participated in the Capricorn expedition and was a familiar figure in oceanographic circles. By 1958, AMSOC had an NSF grant under the sponsorship of the NAS, and Bascom organized a technical staff that included several people who went on to leading careers in ocean engineering. The original study grant was for $15,000, and early estimates were that the Mohole would sample the mantle for about $5 million.

Phase one was the design and testing of an integrated, dynamically positioned, computer-controlled, drilling system in abyssal depths in the open sea. It was completed in three years for about $1.5 million—on time and on budget. Four holes were drilled, the deepest to 601 feet, in water 11,700 feet deep, despite 15-foot waves and 35-knot winds, between Guadalupe Island and Baja California. About 400-600 feet of Miocene-Recent sediment overlay basalt. The seismic second layer was sampled at last. *Life* magazine did a feature story.

The critical stage in developing a new business occurs when skill, energy, and luck have caused the endeavors of the founders to prosper and expand beyond their abilities to finance and manage personally. Bankers want a say in management, hired managers have not the spirit of entrepreneurs, and a winning long shot becomes a losing sure thing. So it was with Mohole in its second phase that began in mid-1961. The objectives and timing of the project came into question. Was it to drill a number of holes in the sedimentary layer or go straight for the Moho? With one or two ships? Would the phenomenally loose administrative structure survive the examination that accompanies federal expenditures in the $20-30 million range? Should the NSF bankers put a firmer hand on the tiller? Should the NAS sponsors withdraw to a more conventional role? A line item would be required in the NSF budget, and that meant specific approval by the Congress. Prudence and caution prevailed. The NAS eventually withdrew, and AMSOC retreated to an advisory role; Bascom's peerless group quickly dissolved, and the banker mentality took over.

A more conventional management was instituted, and giant corporations were invited to bid on a contract for further development and drilling. An elaborate evaluating procedure eventually succeeded in selecting a contractor. It was the bidder who was initially rated lowest, who tied in estimating the longest time for completion, who was the only bidder who asked for a fee in addition to expenses, and who was second highest in estimated costs. The winning bidder was Brown and Root, Inc., a highly regarded, multibillion-dollar construction company with exceptional experience in building and deploying offshore drilling platforms. It was based in Houston not far from the home district of Albert Thomas, the Democratic chairman of the House appropriations subcommittee with jurisdiction over the NSF budget.

While Brown and Root pondered and planned, the estimated cost escalated to $68-70 million in 1964 and about $75 million early in 1965. The Congress and the Bureau of the Budget both put restraints on the budgeting process, and the project appeared to be stalling as the cost of the Vietnam war continued its exponential growth. In August 1966 the House voted 108-59 not to restore the funds for Mohole that had already been cut from the NSF budget. The project was dead.

Not long before the vote it had been learned that the family of George R. Brown of Brown and Root had given $25,000 to the President's Club—a political organization that aided the incumbent Texan, Lyndon Johnson. Everyone denied any connection between the gift and the President's appeal for restoration of the Mohole funds. I believe that that denial was correct; politics was involved, but probably that was not the issue.

During the academic year 1965-1966 I was on a leave of absence to serve as a technical advisor in the Office of Science and Technology, which was the science staff of the White House. My boss was Donald Hornig, also a once and future academic, and his boss was the President.[22] The administration at that time was desperate for money because it was quietly bleeding it from every possible source of discretionary or unaudited funds as a transfusion for the hemorrhage in southeast Asia. The Congress was not unaware that something was up, and if there is anything calculated to upset the Congress it is any attempt to pare its constitutional authority over federal spending. In this atmosphere, I was summoned by my boss one day to accompany him to the hill and help explain why his boss was willing to put himself on the line in support of the Mohole project. We entered the spacious office of the chairman of one of the House subcommittees responsible for budget oversight. We talked about scientific objectives and schedules and escalating costs. When it was over and we climbed back into the White House limousine I confessed to bewilderment. What was going to happen to the earth scientists' first naive attempt to rival the space program? I was informed that the project probably would be killed. Some Congressmen conceived of the President's support for Mohole as a personal favor for a friend, and they would spurn it just to tweak his nose as a reminder of who controlled the money. If so, what began in the games of scientists ended in the games of politicians.

The game, however, was not over; all that happened was that the rules were changed. After a tweak, common sense reasserted itself; the Congress again manifested its abiding support for good science, and the Deep-Sea Drilling Program was launched on its gloriously successful career. Moreover, if we could not go to the mantle we could call it to us, like Glendower calling spirits from the vast deep. "But will they come . . . ?" said Hotspur.[23] They came for Harry Hammond Hess. We were attending the Seventeenth Symposium of the Colston Research Society held at the University of Bristol in April 1965. Walking back from lunch in town he suddenly asked me if I would like to see the earth's mantle. Naturally, I said "Yes," and he led me a few steps to a building made of coarsely crystalline, polished green rock. "They make the banks out of it around here," said Harry. He was always ahead of his time.

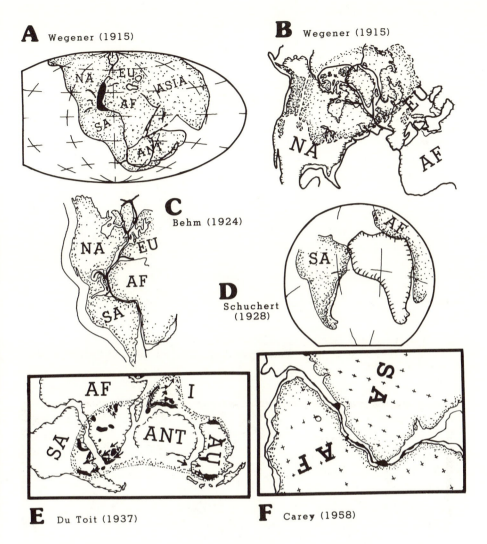

A Wegener (1915)

B Wegener (1915)

C Behm (1924)

D Schuchert (1928)

E Du Toit (1937)

F Carey (1958)

FIG. 1 The evolution of ideas on the fit of continents and glaciation as a confirmation of drift.
A. Wegener's continents were very rubbery, and thus the fact that he could fit them together was not impressive. B. Wegener thought that the Pleistocene ice caps (shaded) of Scandinavia and Canada were one and, consequently, that the North Atlantic had not yet opened a million years ago. This led him to the fatal error of predicting a rate of drift of Greenland that could be measured geodetically. C. By 1924 Wegener's followers were unrestrained in bending continents to emphasize the congruence of continental margins. The figure is from Schuchert (1928). D. The skeptic, Schuchert, made a plastic mold of the globe and, by an analogue of Euler's theorem, showed how rigid continents must move. He obtained a remarkable fit but was unimpressed. E. DuToit demonstrated that the Paleozoic glaciation of the southern continents could readily be explained by drift from a single central source. Arrows give direction of movement; black spots are glacial deposits; and shaded areas show the limit of glaciation. Peninsular India (I) was then a southern continent. F. Another analogue fit of rigid continents prepared by cutouts of a plastic cover of a globe. Carey found hardly any mismatch of the 2,000 m contour if the youthful Niger delta is ignored at the tip of Brazil.

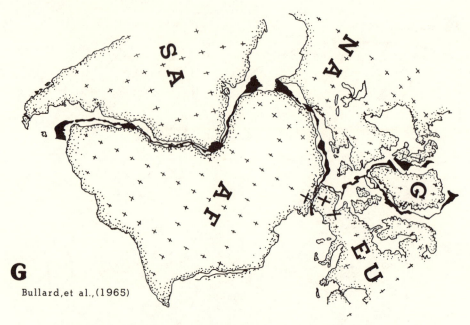

G

Bullard,et al.,(1965)

FIG. 2 G. The famous computerized fit, based on Euler's theorem, by Bullard et al. The fit on the 500-fathom contour was certainly no better than obtained by Carey a decade earlier, but more people found it convincing.

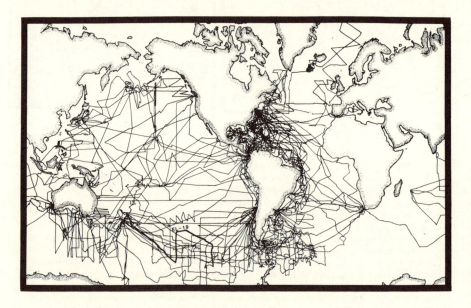

FIG. 3 Lamont ship tracks to 1968 when the data collected were used to demonstrate sea-floor spreading and the history of ocean basins. Ewing's calculatedly even spacing of the lines in most regions is evident. The famous Eltanin-19 track is emphasized. Obviously it was necessary to manipulate the data by computer to produce profiles perpendicular to the East Pacific Rise. This is basically a computer-generated illustration prepared through the kindness of Dennis Hayes.

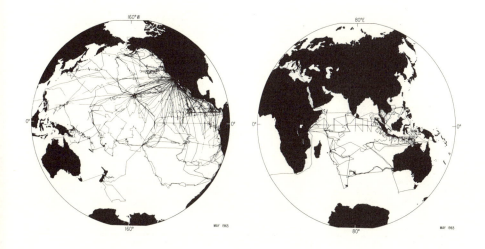

FIG. 4 (*Above*) Scripps tracks to 1965 showing the absence of any overall plan. Individual scientists used the ships for their own purposes, although the lines were spaced if possible.

FIG. 5 (*Right*) Fisher and Menard expeditions in the Pacific to 1968. Minor expeditions and a year at sea in small boats off California are omitted. Expeditions emphasized in the text are highlighted and major fueling ports identified.

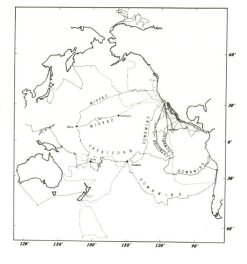

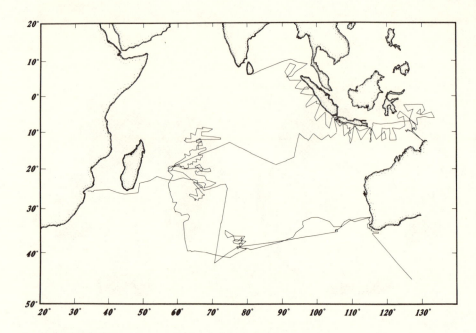

FIG. 6 Fisher expeditions in the Indian Ocean to 1968 showing the complex surveys typical of the Scripps geological expeditions.

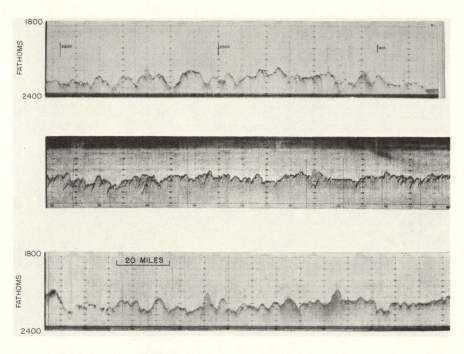

FIG. 7 The endless abyssal hills discovered on the Midpac and Capricorn expeditions in 1950-1952. It would be a decade before Dietz proposed that they were all created at spreading centers.

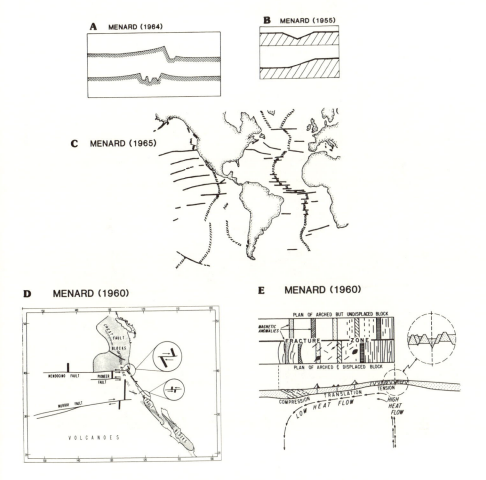

FIG. 8 Evolution of information and ideas on fracture zones. A. Profiles of fracture zone topography. B. Profiles of bands of plastic deformation produced in laboratory experiments. The zones were originally thought to be a consequence of plastic deformation in a rigid crust. C. Known fracture zones in 1965. The long lines of great fractures in the eastern pacific were discovered in 1950-1952. D. Data that were interpreted in terms of the sequential development of mid-ocean ridges. The crust was thinned around ridge crests and the flanks moved outward in both directions from the crest, but no new crust was created. E. Profile and plan of differentially offset magnetic anomalies and a fracture zone in the sequential development hypothesis.

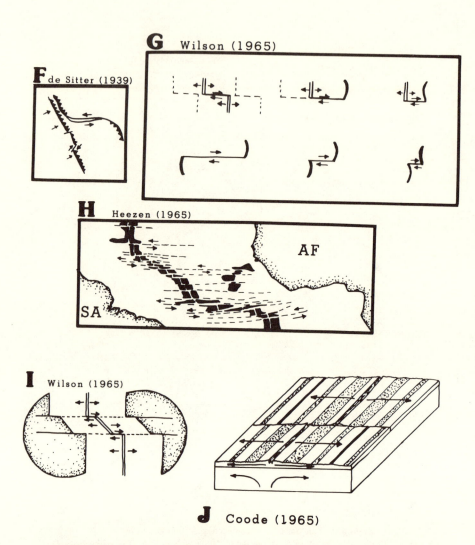

F de Sitter (1939)

G Wilson (1965)

H Heezen (1965)

AF

SA

I Wilson (1965)

J Coode (1965)

FIG. 9 Evolution of ideas on transform faults. F. Tear faults (lines flanked by opposed arrows) in the Jura Mountains associated with thrust faults (toothed pattern, bites down dip). Such tear faults were familiar features in structural geology. G. Wilson's (1965a) classification of transform faults. Double lines are ridges (R), heavy curved lines are trenches (T), concave toward the dipping Benioff zone, and connecting lines flanked by opposed arrows are transform faults. Upper line: left, an R-R transform; middle and right, R-T transforms. Lower line: T-T transforms; the left T-T transform is the exact analogue of a tear fault connecting thrust faults in the Jura Mountains. H. Heezen's interpretation of relative motion on the fracture zones of the equatorial Atlantic. The motion is left lateral on each fracture zone despite the fact that Heezen believed that the Atlantic was expanding away from the median rift. I. Wilson (1965a) realized that the motion in the equatorial Atlantic was right lateral if the sea floor is spreading. J. Coodes's remarkably advanced, three-dimensional illustration of a ridge-ridge transform fault and offset, symmetrical, magnetic anomalies.

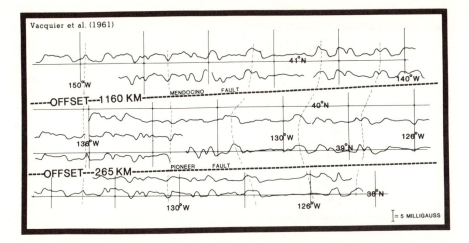

FIG. 10 Magnetic profiles in the northeastern Pacific showing their distinctive patterns and the ease of identification of offsets on fracture zones. The size of the offsets was quite unprecedented. The lack of distortion of the anomalies between fracture zones indicated a rigid sea floor.

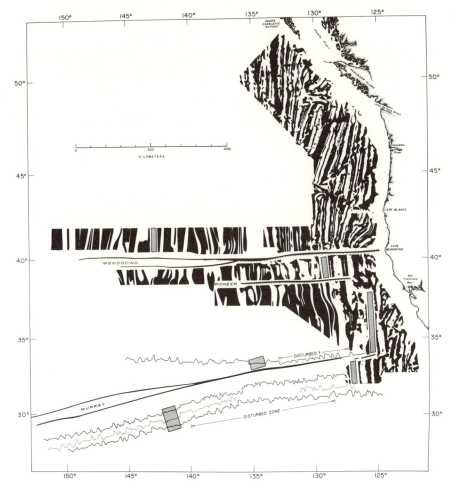

FIG. 11 Magnetic stripes discovered in the 1950s by Mason, Raff, and Vacquier, showing offsets on fracture zones. The survey was from south to north over a period of years.

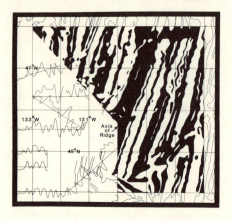

FIG. 12 Detail of the magnetic pattern, discovered by Mason and Raff, in the area of the crest of the Juan de Fuca Ridge. In this pattern, and without knowing the topography, Wilson and Vine perceived that the magnetic anomalies are symmetrical. The northeasterly and northwesterly offsets in the anomalies are a consequence of propagating rifts—a phenomenon that would not be identified for more than a decade.

A   Runcorn (1956) ideas as of June 1955

FIG. 13 Magnetic evidence for polar wandering and continental drift. A. In the spring of 1955 Runcorn had enough data to show that the magnetic pole had wandered (arrow path) relative to Europe (dots) and North America (triangles) during the last billion years (1956b). By summer he had four more points in North America (framed triangles) and could show that the apparent paths of polar wandering were different for the two continents (1956a). Thus the continents had drifted apart.

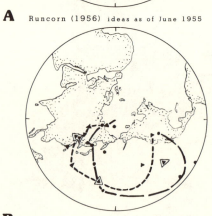

B   Runcorn (1956) ideas as of summer 1955

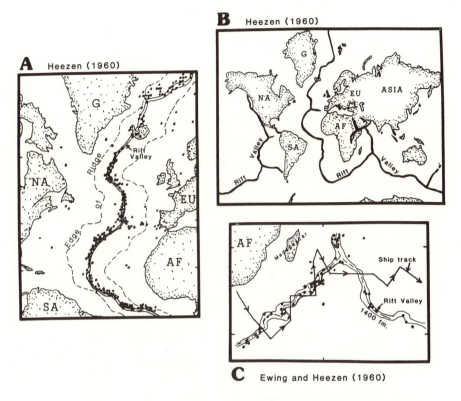

FIG. 14 Discovery of the mid-ocean rift. In the early 1950s, Marie Tharp and then Heezen realized that the earthquakes (dots) characteristic of the Mid-Atlantic Ridge might all be in the median rift as shown in A. In Fact, many earthquakes are on transform faults that offset the ridge. B. The global rift valley, whose locus was predicted in part from bathymetry but largely from earthquake locations. Note that the rift valley almost circles Africa and that it includes the San Andreas fault where the rift extends through California. C. The immediate and spectacular confirmation of the existence of the central rift valley in the southwestern Indian Ocean. A few zigzags discovered deeps (dotted line) on the seismically active (dots) ridge crest (parallel solid lines). Later exploration has revealed that the ridge is not sinuous but offset in a rectilinear pattern by transform faults and that many of the deeps are along the transforms.

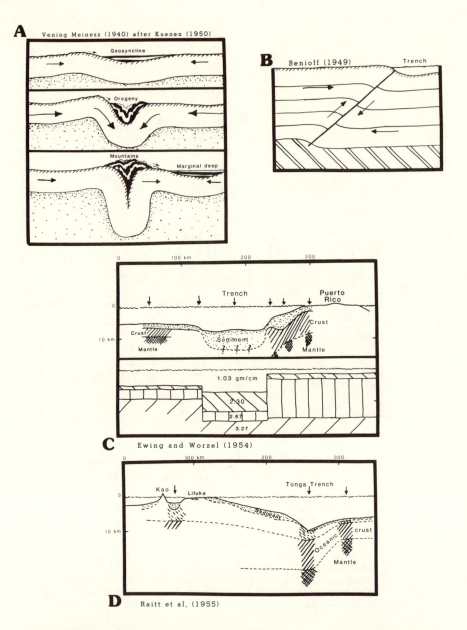

FIG. 15 Evolution of data and ideas on trenches and subduction zones. A. The tectogene as conceived by Vening Meinesz. Horizontal compression first produced a trench, then mountains with a low-density root as indicated by gravity anomalies. B. The Benioff zone as drawn by Benioff. An enormous dipping thrust fault has very little displacement. C. The Puerto Rico Trench as an example of all trenches according to Ewing and Worzel. Symbols are the same as in Figure 16. Arrows are seismic stations. Lamont had not detected oceanic crust under trenches but thought it was thin. D. Meanwhile, Raitt had established that the crust is, if anything, thick under the Tonga Trench.

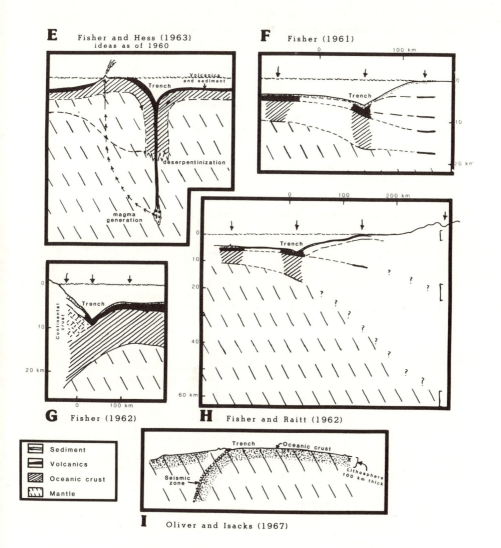

FIG. 16 More evolution of data and ideas on trenches and subduction zones. E. The tectogene as modified by Fisher and Hess to be a sink to dispose of all the crust created by sea-floor spreading. They ignored the Benioff zone. F. Fisher found that the oceanic crust under the Middle America Trench is slightly thicker than normal and appears to continue under the continental shelf, although the velocity structure was not exactly the same. G. Fisher interprets his Middle America data as showing an oceanic crust dipping under the continental margin. H. Fisher and Raitt make the same interpretation regarding the structure of the Peru-Chile Trench. I. Oliver and Isacks five years later identify the lithosphere, capped by the thin oceanic crust, dipping into a subduction zone.

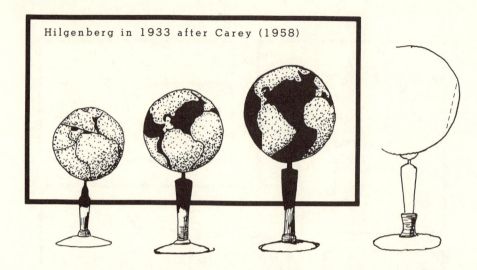

FIG. 17 As was shown repeatedly and independently, if the earth was once completely covered by a layer of continents, its diameter was only half of what it is now. The former existence of such a continuous layer was an attractive idea for many geologists.

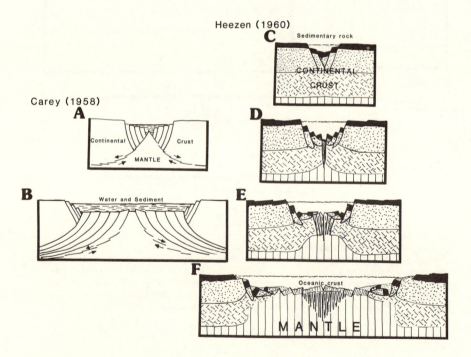

FIG. 18 Ideas on creating an ocean basin by expanding a rift. A. Initial collapse of continental margin according to Carey. B. Addition of new crust in the center of an expanding ocean basin. C and D. The much more detailed but similar concept of initial collapse according to Heezen. Note that the black layer is sedimentary rock on a continent before rifting. E. The collapsing blocks at the continental margin continue to rotate as the ocean expands, and dike injection begins in the mid-ocean rift. The lower continental crust thins. F. The center of the expanding ocean has only a thin oceanic crust over the mantle.

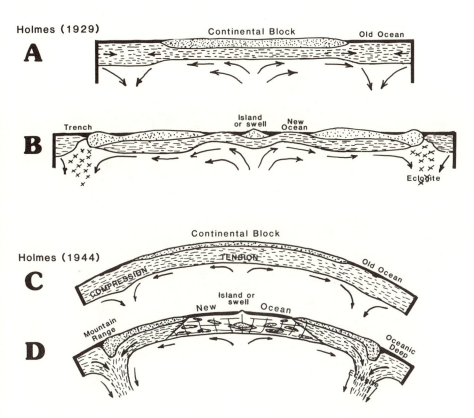

Fig. 19 Evolution of ideas on sea-floor spreading. A and B. The early ideas of Holmes. A continent was rifted by spreading convection currents, and dense eclogite formed by compression and sank in subduction zones. The "island or swell" is the Mid-Atlantic Ridge or Iceland, which rises from it. C. Holmes's later ideas. Mantle convection rises under a continent and sinks in compressed subduction zones. D. New oceanic crust forms, but mysteriously it contains fragments of continental crust, which presumably is foundering. The continental crust and underlying basaltic intermediate layer move as on a conveyor belt to subduction zones where the basalt is changed to eclogite and disappears into the mantle. The nonsubducting continental crust forms a low-density root under developing mountains.

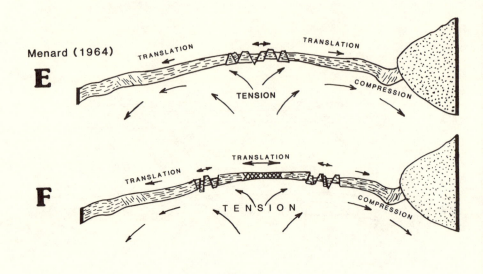

Menard (1964)

E

TRANSLATION

TRANSLATION

COMPRESSION

TENSION

F

TRANSLATION

TRANSLATION

COMPRESSION

T E N S I O N

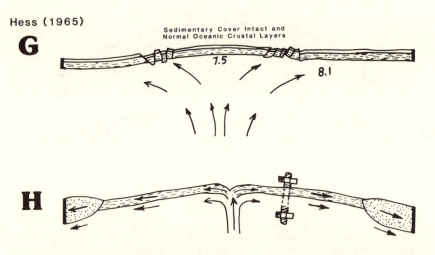

Hess (1965)

G

Sedimentary Cover Intact and
Normal Oceanic Crustal Layers

7.5

8.1

H

FIG. 20 More evolution of ideas on sea-floor spreading. E. Menard's concept of forces acting on the oceanic crust of the East Pacific Rise, which is unwittingly almost exactly the same as Holmes's idea in Figure 19 C. F. Menard's identification of rifting on the flanks instead of the crest of the East Pacific Rise (EPR). The physiographic evidence was correct but reflected ancient geological history. G. Hess's interpretation of F. There should have been earthquakes on the rifted flanks, but there were not. Note that the crest of the EPR is old crust covered with sediment. Thus a mid-ocean ridge could develop as a swell before spreading was initiated. H. The first drawing by either Hess or Dietz of sea-floor spreading, but even it did not include subduction.

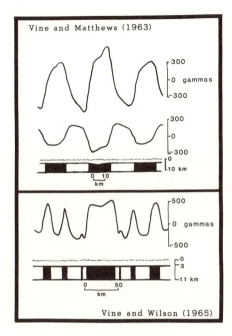

Vine and Matthews (1963)

Vine and Wilson (1965)

FIG. 21 (*Left*) Evolution of modeling magnetic anomalies. In the first published examples of simulated magnetic anomalies (1963), the model of normal and reversed polarities (black and white bars, respectively) was bilaterally symmetrical, but the resulting profiles were not because of the geograpical location. Two yeas later, in the northeastern Pacific, the symmetry in the profile was obvious.

FIG. 22 (*Below*) Eighty million years of magnetic anomalies. E1-19S is the western flank of the East Pacific Rise as shown by the famous Eltanin-19 profile. By varying the horizontal scale, it becomes readily apparent that the reversal profiles are similar in the North and South Pacific and the South Atlantic and therefore in the whole world ocean. The rate, however, could not be constant in all three regions. Heirtzler and colleagues at Lamont (correctly) selected the South Atlantic as a standard with constant spreading.

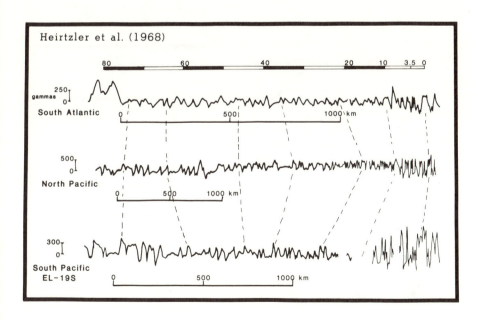

Heirtzler et al. (1968)

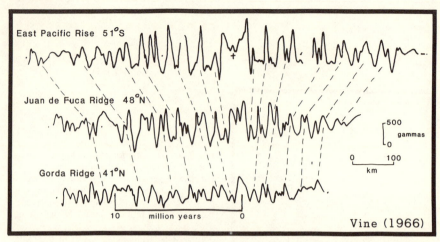

FIG. 23 Ten million years of magnetic anomalies. Although reversals were dated for less than 4 Ma on land, Vine could show that the rates of spreading were consistent in each of several regions for more than 10 Ma. He assumed constant spreading and had a time scale that could be modeled.

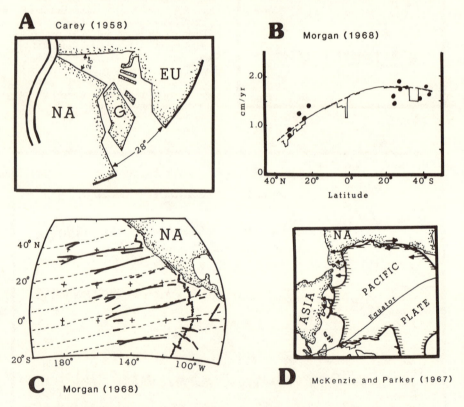

FIG. 24 Evolution of ideas on rigid plate rotations. A. Carey's demonstration that the North Atlantic had opened by a constant angle. B. Morgan's evidence that the spreading rate in the Atlantic varies with the Euler latitude. The latitude in the figure is geographical, not Euler. The dashed line shows the predicted spreading rate parallel to spreading. The solid line shows the predicted rate perpendicular to the crest of the ridge—the trend of which was not well known. C. Morgan's evidence that the old fracture zones that I had plotted (heavy lines) followed small circles (dashed lines) around a single Euler pole. D. Slip vectors of earthquakes (arrows) around the North Pacific presented on a computer-generated transverse. Mercator plot, so all the arrows can be seen to have the same trend. Evidently the whole North Pacific crust is rigid.

Chapter 10

OCEANIC TRENCHES

Each generation of scientists believes that its observations will be able to solve the great problems that eluded its predecessors. The prognosis is particularly optimistic if new kinds of instruments have been invented to shed light on the old problems. If the old observations could not really solve the problem, then the conflicting conclusions drawn from them may be neglected—even if by luck one of them will turn out to have been correct. So reasons the new generation. So Philip Kuenen, the new man in the early 1930s, wrote regarding the trenches and island arcs of Indonesia:

> The earlier conceptions were naturally based on scant data concerning the structure of the islands and a very poor knowledge of the morphology of the sea bottom. It is unnecessary to name all the authors separately, for the reason that their opinions are only of historical interest.[1]

The existence of deep-sea trenches was unknown in the middle of the nineteenth century, but almost all had been discovered by the beginning of the twentieth. They are difficult to miss, because they are very long and lie along the edges of the broad ocean basins and must be crossed to reach port. Not only were they discovered, they were explored and their continuity established because a network of deep-sea telegraph cables was being emplaced around the world. It was not desirable to put the cables in extreme depths or on faulted topography, so proposed lines for the cables were surveyed. If a deep was found, the surveying ship extended the soundings to each side seeking a suitable route, and the cable was redirected to stay in normal water depths. In modern terms, many cable routes cross the structurally continuous, subducting trenches on obducting groups of guyots.

The discovery of long, relatively narrow deeps at continental margins created great interest among geologists. James Hall and James Dwight Dana among others had established the existence of long, relatively narrow, thick bands of sedimentary rock, which were called "geosynclines." Most of the rocks were shallow-water facies, but some could be deep water, and the volume of the rock required deep erosion of a large nearby continent. The trenches appeared to be the only possible modern equivalents of geosynclines, so investigations of their dynamic geology and geophysics could provide essential information about the past. Likewise, if one was the fossil of the other, the history recorded in the geosyncline could help explain the origin of the trenches. The geological history was relatively accessible because geosynclines, after long subsidence and sedimentary filling, were uplifted to form normal continental mountain ranges.

The surficial sediment in trenches was sampled by the cable-survey ships and by HMS *Challenger* and found to be red clay and radiolarian oozes. This seemed promising because radiolarian-rich chert beds were a distinctive feature of some geosynclines. Geographical and geological exploration also showed that most of the active and inactive volcanoes of the world lie in arcs or lines near to and roughly parallel to trenches. This, too, was promising because rocks with volcanic sources are commonplace in geosynclines and less so elsewhere. The first maps of the distribution of earthquakes showed another interesting feature. The quakes were concentrated in a few long, narrow bands, and the oceanic edges of many of the bands corresponded to the line of oceanic trenches. The fundamental theory of geology is that the present is key to the past. Deformation of geosynclines into folded and faulted mountains had occurred frequently in the past. Similar deformation should be widespread now, and the only promising sites were the seismically and volcanically active lines of trenches and island arcs.

Confirmation of the extraordinarily dynamic nature of trenches was provided by gravity measurements at sea by O. Hecker.[2] From the boiling temperature of water the air pressure (P) can be calculated independent of anything else. The height of the mercury in a barometer is balanced against air pressure and is equal to the density of mercury (p) times height (h) times gravity (g). Thus $g = P/ph$. Hecker devised a method of measuring the barometer height and the boiling temperature with great accuracy and correcting for the motion of the ship. The error of the determination of gravity was about 30 milligals. Hecker made enough measurements to show that the Atlantic, Pacific, and Indian ocean basins are as close to being isostatically compensated as the continents. There could hardly be a more fundamental discovery, and it was famous in its time. For example, it was a foundation for Wegener's profound belief in isostasy and therefore for his understanding of the difference between continental and oceanic crust. Hecker made one other highly significant discovery in a crossing of the Tonga Trench region. Over the trench the free air gravity anomalies were negative and as great as -287 milligals. The deep-ocean basin immediately east of the trench had positive anomalies up to $+80$ milligals. The shallow Tonga region to the west had even larger positive anomalies. Trenches, or at least one trench, apparently were being actively stressed so intensely as to override isostasy.

We next come to the "peerless researches"[3] of Felix Andies Vening Meinesz (1887-1966; Fig. 33) who was born in the Hague.[4] He received an engineering degree at Delft in 1910 and a doctorate from the same university in 1915. He joined the Netherlands Geodetic Commission in 1911 and remained with it until 1927 when he became Professor of Geodesy and Geophysics at Utrecht, which is now the site of the Vening Meinesz Laboratory of Geophysics. Soon after starting work for the Geodetic Commission, he attempted to establish a gravity network in the Netherlands. Conventional pendulum gravimeters could not be used because of the unstable ground, so he invented one that could endure some motion. If some, why not more? In 1923 he found himself in a Dutch submarine ready to begin a brilliant career of measuring gravity at sea with a triple pendulum of his own devising. He must have had some misgivings. Harry Hess, who sailed with him on the U.S. submarine *S-48*, wrote in a volume honoring Vening Meinesz,

> No Navy would have accepted a man of his size for the submarine service. There was no place on the S-48 where he could stand upright. He slept in a bunk a foot shorter than he was, with only 18 inches to the bunk above. To turn over he had to climb out and get back in again. . . . Dives and gravity observations came every four or five hours around the clock. There was seldom opportunity for more than a couple of hours' sleep. . . . The real hazards of the early work in submarines of World War I vintage are not mentioned in the literature. For example the S-48's air vents . . . jammed open. . . . The ship dove 150 feet below her designed crushing strength. Only quick and drastic action. . . .[5]

Under these conditions Vening Meinesz went on numerous cruises from 1923 to 1938, from his 36th to 51st years. He sailed 125,000 miles and made 844 observations of gravity while submerged below the motions of surface waves.

> For so complete a conquest over the difficulty of designing adequate equipment, for his prolonged, self-sacrificing work among the dangers to life on an ocean-going submarine, and for his unquenchable ambition to allow for every conceivable error in measurement and in reductions of his observations, geophysicists and geologists cannot express too much admiration.[6]

So wrote Reginald Daly about 1939 as World War II began. In that year Vening Meinesz was elected a Foreign Member of the U.S. National Academy of Sciences. He was already a Foreign Member of the Royal Society of London and of the Académie des Sciences of Paris. Immediately after the war, late in 1945, the Geological Society of America gave him the Penrose Medal for "one of the outstanding contributions to geology in this century." Vening Meinesz acknowledged his debt to those American entrepreneurs of oceanography, Bowie and Field. They had helped him obtain time on U.S. Navy submarines, which were mercifully bigger than the Dutch ones.

Vening Meinesz had greatly strengthened the evidence for isostatic compensation of ocean basins.[7] He had proved that the Hawaiian Islands are a load that is supported in part by the rigidity of the crust. His greatest contributions, however, were associated with trenches, particularly those in the East and West Indies. He first discovered that Hecker's gravity profile across the Tonga Trench was typical of all trenches. Then he made an extensive survey of the trenches of the East Indies in 1929-1930. He next formulated a hypothesis of crustal deformation that sought to explain the gravity, vulcanism, topography, and geology. The narrow strip of intensely negative anomalies he attributed to a downward buckle of the crust (a "tectogene"; Fig. 15 A) where two convection currents converged. The buckling produced the trench. The active volcanoes all occurred on the concave side of a trench, and this was because the crust on that side was stretched laterally and faulted as it plunged into the tectogene. The tectogene was naturally buoyant, so the negative isostatic anomaly indicated that tectonic forces were active. Nonetheless, the tectogene would rise from time to time, and so would the mantle under it. Thus a downward flow would occur in the nearby mantle, and it would cause the observed regions of positive gravity anomalies. The steep gradients in the gravity field indicated that the tectogene was no more than 40 km deep. No large movement of crustal blocks was required or, perhaps, permitted. If the trench filled with sediment, the silts and shales would be squeezed and deformed into Alpine mountain ranges with overthrust faulting on two sides.

The tectogene hypothesis was speedily tested in the laboratory and the field. Philip Kuenen made a series of scale-model experiments on the deformation of a floating plate. He showed that under compression a plate first bends in broad waves and then one wave collapses into a tectogene. P. P. Bijlaard, Professor of Applied Mechanics at Delft, calculated that the lithosphere would actually deform plastically rather than buckling downward. The deformation, however, would occur as a thickening along lines with certain orientations. Between the lines of plastic thickening, the lithosphere would be relatively undeformed.[8] Geological mapping of Indonesia seemed to confirm this style of deformation. (So, later, did the bathymetric mapping of the fracture zones in the northeastern Pacific.) The few islands within the belt of negative anomalies were intensely deformed, whereas most other islands were not. Moreover, Timor, in the negative belt, had some extraordinary sedimentary rocks including red shales with manganese nodules. It appeared that Timor had been depressed into a deep trench and then elevated into an island.

Vening Meinesz's hypothesis of the tectogene was an immediate success. It was widely accepted that (1) Indonesia was a modern example of the formation of some kinds of geosynclines, (2) the negative anomaly belt was a consequence of a low-density root, and (3) the root was caused by and held down by converging convection cells or one cell pushing against an immobile continent. The hypothesis was attractive to many geologists because unlike most grand geophysical generalizations, it actually explained many details of geological history as mapped in the field. A measure of its acceptance is indicated in Harry Hess's first postwar paper on the trenches of the western Pacific:

> The trenches are the topographic expression of the dominant structural feature of the region, the downbuckling of the crust or tectogene. This structure . . . is the core and essence of mountain building, and all other major structures as well as the volcanic activity and the seismic activity of the region are subordinate to and related to it.[9]

Many other competing hypotheses, however, were offered, and no real consensus could be reached about the tectonics of trenches and island arcs. It was noted, for example, that many trenches are remarkably arcuate. In 1931, Arthur Lake attributed this arcuate shape to the intersection of a planar thrust-fault with the surface of the spherical earth. The tectogene hypothesis had no explanation for the regularity of the arcuate shapes. Several leading American geologists, including A. C. Lawson, W. H. Hobbs, and Walter Bucher, published their own hypotheses about the origin of island arcs and presumably impressed them on their students. Kuenen reviewed the hypotheses of Lawson and Hobbs and found them inadequate to explain the data available even in 1935, so they will not be discussed. The tectonic ideas of Bucher were another matter. He had assembled them in 1933 in a famous book, *The Deformation of the Earth's Crust*, as a logical code of scientific laws and opinions. The scholarship and rigor of the presentation made the book highly influential. Bucher proposed that the earth alternately expands and contracts, that trenches are cracks generated during expansion, and that mountains are folded upward during contraction. Thus trenches should have a thin crust in the early stages of their evolution. Bucher had a national and international influence upon geological ideas, but his impact must have been greatest at Columbia University. To that university came Maurice Ewing in 1946. Bucher was then 58, a member of the National Academy, and on his way to being president of the Geological Society and Penrose Medalist. He must have had a strong influence on Ewing's thinking on tectonics. Unfortunately, the perfect opportunity for Ewing himself to clarify this point in print was lost. He was awarded the Bucher Medal posthumously.

Immediately after the discovery of plate tectonics, Ewing came under criticism for his "opposition" to continental drift, and I shall consider this curious concept of the workings of science in due course. This "opposition" was yet to be perceived, as the Lamont geologists began their exploration at sea. It seems desirable at this point, however, to present Bucher's views on continental drift. They were the views that would at least be familiar to every geology student, and young geophysicist, at Columbia.

Bucher observed that the hypothesis of continental drift

gave promise of offering a solution for three difficult problems: (1) the excessive shortening that seemed indicated in the structure of Alpine mountain chains; (2) the correspondences in the configuration of opposite shores of continents and islands; and (3) the extraordinary distribution of ice during the Permian glaciation.[10]

As to the third point he showed that Pleistocene glaciation was so complex and ill-understood that it seemed extravagant to appeal to continental drift and polar wandering to explain Permian complications. Bucher, however, claimed no special competence in glaciology or meteorology. (Wegener was an expert.) Regarding the second point, Bucher considered the hypothesis sufficient to explain the correspondences, but it was hardly necessary. He showed that if the eastern United States were submerged 1,000 feet, the new shore lines would be correspondent like those of the northwestern Atlantic, so correspondence did not require continental drift. He did not realize that there would be no new continental slopes created by his exercise in submergence. Thus he could not show that when he submerged the eastern United States, the continental slopes would be correspondent as they were across the Atlantic. This was a long step backward from Wegener's perception of the differences between continents and ocean basins. But was Wegener correct? It was not certain. That was why Field and Bowie wanted Ewing to put to sea.

On the first point Bucher was an expert and he stated,

Opinion 23. The actual amount of crustal shortening involved in orogenic folding is much smaller than the resulting structures lead one to expect.[11]

In sum, Bucher did not believe that anything as dramatic as continental drift was necessary to explain the phenomena it was supposed to explain. Moreover, he showed that drift had little predictive power in explaining tectonic phenomena that had not been discussed by Wegener:

it is clear that the fundamental concept of continental drift, with a parent continent tearing along lines of weakness and buckling under the influence of friction is not sufficient to account for the existence of the large intracontinental mobile belts. A special auxiliary cause for the changes of relative speed of drifting must be introduced to account for the folding of the geosynclinal belts. To the writer's mind this is not the sort of clearing perspective that accompanies growing understanding.[12]

The last sentence would be echoed by Munk and MacDonald in 1960 with regard to the paleomagnetic evidence for continental drift. Bucher's statements illustrate the importance of basic assumptions in the thinking of a scientist. Some intracontinental mobile belts are still puzzling, but that fact does not now limit acceptance of plate tectonics. Bucher's statements also show that Ewing and the Lamont group were exploring a very different conceptual world in the 1950s from the one we all perceive now.

One of the new discoveries about trenches that was only beginning to be appreciated in the 1950s was the distribution of deep earthquakes. H. H. Turner reported in 1922 that some earthquakes were within the mantle, but the evidence was viewed as equivocal.[13] By 1928, K. Wadati demonstrated that earthquakes extended to depths of several hundred kilometers. Moreover, they occurred in a dipping zone that outcropped at the Japan Trench and deepened toward Asia. Similar dipping planes of earthquakes were found in association with other island arcs and trenches. Accordingly, Vening Meinesz had to elaborate his tectogene hypothesis. In order to explain the gravity anomalies he had proposed both a low-density root and mantle convection. The root extended down only 20 km or so, but the convection was much deeper. He proposed that the deep-focus earthquakes occurred on a zone of shearing caused by the different directions of motion in a convection cell.

In 1949, Hugo Benioff (1899-1968) developed a new way of analyzing earthquakes. He compared the creep-recovery of rocks in a laboratory with the strain rebound during a sequence of earthquake aftershocks and found them in close agreement. This enabled him to

> determine whether or not a given sequence of earthquakes is derived from movements of a single fault structure and so, in effect, to establish the existence of a fault which may have otherwise escaped detection.[14]

He, naturally, found some of the largest undetected faults on earth, according to his definition. The great fault plane associated with the Peru-Chile Trench (Fig. 15 B) was 4,500 km long, about 900 km wide, and dipped east from the trench to a depth of 650 km under South America. The fault related to the Tonga-Kermadec trenches was similar, although not so long. It dipped west, and subsequent investigations elsewhere showed that almost all these faults dipped away from the Pacific or another ocean basin.[15]

These great faults now appear to be tabular zones of earthquakes called "Benioff zones" by later investigators, although not of course by Benioff himself.[16] Benioff proposed that these zones develop because the boundary between a high-standing continent and an ocean basin is unstable. It should adjust by surface flow toward the basin and deep counterflow toward the continent. Consequently, the great boundary fault of the Benioff zone is tilted as observed. Despite the enormous area of the fault plane, Benioff did not visualize a large displacement. The lithosphere in the vicinity of the zone of deformation could support stresses for decades even at great depths. It was by no means plastic on such a time scale. Perhaps the bottom of the lithosphere was displaced no more than the top—a matter of 4-5 km as shown by the depth of oceanic trenches. The Benioff zones, as known in the mid-1950s, bore little resemblance to subduction zones as they are now visualized. They marked a boundary rather than a plunging cold slab. Marine geophysics, however, was even then beginning to reveal details of crustal structure that would outmode Benioff's tectonic models. Unfortunately, the scientists at Lamont and Scripps were to disagree increasingly about trench structure as they collected more data. For several years confusion would reign.

Robert Lloyd Fisher (1925–; Fig. 34) received a B.S. in geology from Caltech in 1949 after two years in the wartime Navy. He went on to Northwestern for a year and arrived at Scripps just too late to participate in the Midpac expedition. He missed few thereafter, and by Capricorn he was ready to begin the exploration that made him the foremost American investigator of deep-sea trenches.

Oceanic trenches are 6,000-10,900 m deep, and the oceanographic equipment of the 1950s was not designed or suitable for such depths. Bob's first task in investigating each trench was to make a detailed bathymetric chart. To do so it was necessary to reconcile the survey so that the depth was the same each time the ship crossed over its own track. The wide soundcone of the sounder gave misleading echoes from the steep slopes in the vicinity, so the depth recorded for a point was not necessarily under the ship. As a further complication, we operated the sounders in a gated mode for which they were not designed. We thereby obtained much more information, but it was difficult to identify the gate; errors of 400, 800, or 1,200 fathoms might be undetected for hours. Moreover, a ship rarely knew its position more accurately than 2-3 miles, so it was not certain where the intersection of two tracks occurred. From this kind of of material Bob patiently wove a consistent network of sounding lines by testing a range of assumptions about echo correlations and about current and wind drift.

Preparation of the final chart took months of work after an expedition, but a rough chart had to be drawn on shipboard in real time in order to sample a trench, scarp, or seamount, or prepare for a seismic station. Bob and I developed an elaborate system for eliminating errors while recording and plotting soundings under such circumstances. The worst trials, however, were in his trench work where everything was pushed to the limit.

We did not know these things when the Capricorn expedition began to survey the Tonga Trench in the Christmas season in 1952. We soon realized what was involved in taking a piston core at the end of 10,000 m of wire from a ship whose position was uncertain relative to the bottom;

> a core barrel with a heavy lead weight, which because of difficulties with the winch was dragged along the sea floor for several hours before it could be raised, came up badly battered by the rocks.[17]

Nonetheless, we did core and measure heat flow; Bob surveyed; Russ Raitt shot his seismic stations; and Ron Mason made some magnetic profiles before we headed toward Samoa. The first published results of this work were concerned with setting geographical records. The oceans were so little known that it was an inviting challenge to discover the deepest point.[18] It was not until 1955, three years after Capricorn, that the geology of the Tonga Trench began to appear.[19]

Raitt shot seismic profiles along the axis of the trench, parallel to it in the Pacific basin to the east, and over the Tofua Trough, which lies within the Tonga Islands and is also parallel to the trench (Fig. 15 D). He found that the basin structure was normal for the Pacific both in thickness and in velocity of the layers. The trench section was also relatively normal, although somewhat thick. Normal crust was merely depressed under the trench. The structure under the Tofua Trough, however, was different. The Moho was not detected, and it appeared that the crust was very thick.

Despite the great effort to make a bathymetric chart, the topography of the trench did not prove very informative regarding tectonics. It was notable that the V-shaped trench seemed empty of sediment, whereas the Puerto Rico trench had a flat bottom and thick sediment. The amount of sediment was interpreted as a possible consequence of a youthful age for the Tonga Trench because the many active volcanoes would supply abundant sediment to fill it.

Fisher and Revelle synthesized the tectonic implications of the Capricorn survey of the Tonga area in terms of mantle convection. The high heat flow under the ocean basins suggested rising convection. Somewhere there must be a return flow ''where cold rock moves downward.'' In fact, the heat flow in the Tonga Trench was lower than normal, and one interpretation of the magnetic profiles was that the upper mantle west of the trench was cool:

So it may be that relatively cool rocks are slowly moving downward under the trench

and

crustal material, including sediments, may continue to be dragged downward into the earth. This is suggested by the fact that the deepest trenches contain virtually no sediments, although they are natural sediment traps.[20]

This seems startlingly close to modern ideas of subduction, but it is easy, after the fact, to be misled. The illustrations in these papers show no lithospheric slab plunging deep into the mantle. The base of the oceanic layer is merely pulled down a few kilometers under the trench.

Meanwhile, back at Columbia where Bucher believed that trenches were produced by tension, Ewing, Worzel, Heezen, and colleagues were reaching the conclusion that the tectogene hypothesis was ill-founded. There was no low-density root under trenches. Brackett Hersey shot a seismic line near the Puerto Rico Trench in 1949, and six more were shot by Lamont in 1951. Like Russ Raitt in the Pacific, Ewing and colleagues apparently were too busy collecting more observations to work up those on hand very quickly. The seismic data finally were analyzed in relation to the gravity observations of the West Indies. In the *Bulletin of the Geological Society of America* for 1954, Vening Meinesz brought his tectogene hypothesis up to date in a lucid paper on pages 143-164. Beginning on page 165, Ewing and Worzel began to present the data that led them to conclude that

> The large negative gravity anomaly is attributed to a great thickness of sediments in the trench[21]

rather than to a sialic-root or tectogene "as formerly thought." They had not actually penetrated the sediment, but this in itself was deemed extraordinary and suggested at least 12 km of sediment to them (Fig. 15 C).

Worzel and Ewing published separately on oceanic trenches in the same volume in the following year and elaborated their analysis. There still was no seismic station to penetrate the sediment in the Puerto Rico Trench, but Worzel estimated the thickness was great and "perhaps a small thickness of oceanic crustal rocks" was below the sediment. He and Lynn Shurbet went on to analyze the structure of other deep trenches for which there were topographic and gravity profiles—but no seismic sections. They assumed, correctly, that the V-shaped Mindanao and Marianas trenches had no filling of sediment. They then made the bizarre assumption that the mantle had a uniform density below the Moho. If the sea floor was deep, the oceanic layer was thin. If shallow, the layer was thick. Consequently,

> On these assumptions it must be concluded that the crust is very much thinner under the great deeps than elsewhere, and so one must conclude that these great deeps must be the result of tension in the crust. . . .[22]

Three "musts" in one sentence.

Ewing and Heezen noted that the Cayman Trough was generally regarded as a tensional feature and that their data indicated it was an extension of the Puerto Rico Trench. Perhaps the latter was also tensional. In fact, they believed that the work of Worzel, Ewing, and Shurbet

> has demonstrated that a downbuckle ("tectogene") is not required to explain the gravity anomalies observed over the deep oceanic trenches . . . tension rather than compression may be the dominant force involved in the formation of the trenches.[23]

A mere renunciation of a scientist's crowning synthesis does not imply a diminished respect for a lifetime of doing what counts. Ewing, Worzel, and Shurbet were shortly acknowledging it a privilege to contribute to a tribute to Vening Meinesz.[24] Harry Hess also paid his tribute and took the occasion to pay his respects to the Lamont group while denouncing *their* science. If the trench was in tension, what could prevent it from rising until it was isostatically compensated and the negative gravity anomaly disappeared?

> It perhaps would not be worth discussing the opinions on this point published by the Lamont group, were it not for the high respect of the geologic profession accord[ed] this group for their past distinguished achievements. Many geologists with little facility in geophysics accept their statements at face value. Some brief counterstatement therefore is in order to indicate that the problem has not been satisfactorily solved by the Lamont group's postulates.[25]

Hess "felt certain that more seismic work would necessitate a change of mind," and as usual he was right.

The seismic workers were already at sea when the Lamont "tension" papers appeared and were ready to publish when Hess made his brief counterstatement. Very likely he knew that more seismic work had already been done. Charles B. (Chuck) Officer and associates shot 47 seismic lines in the Caribbean in January-March 1955, of which five were in the Puerto Rico Trench. "The crustal structure of the trench profiles is similar to that of the Atlantic." The oceanic layer was a few kilometers thick and sediment was up to two kilometers thick, but the former was not unusual and the latter was to be expected from the flat bottom. In short, Officer and others had virtually duplicated Raitt's results for the Tonga Trench. They had found nothing like the section predicted by the senior men at Lamont. To explain their data they noted that there were two hypotheses regarding trenches and island arcs. One was the tectogene hypothesis of Vening Meinesz. The other was a tectogene modified by a profound reverse fault as proposed by Ross Gunn. Both hypotheses assumed a primary compression. On the basis of the new seismic data,

> The writers conclude that Gunn's hypothesis is the primary explanation of gravity anomalies over the trenches and island arcs. Before these seismic investigations Ewing and Worzel (1954) had come to much the same conclusion. . . . They demonstrated that the gravity anomalies could as well be explained by a smaller thickness of lighter sedimentary material near the ocean bottom as by a large downbuckle of crustal material at greater depth.[26]

Walter Bucher might have been wrong in 1933 in proposing alternating periods of global tension and compression, but at least he could tell one from another. Apparently not everyone could. In a "refinement of previous work" in 1959, Manik Talwani and others continued to emphasize that the gravity anomaly could be caused by thick sediment but skirted the issue of tension or compression as the proximate cause.[27] The issue was also ignored in 1961 in a paper on the structure of the Tonga Trench by Talwani, Worzel, and Ewing.[28]

The final Lamont paper of interest at this point was submitted by Worzel and

Harrison in 1960 but was not published until 1963. It referred to Vening Mein-esz's "detailed and complicated theory" to account for the gravity anomalies. The available seismic observations showed no sign of a tectogene or thickening of the crust as defined by the Moho. If in any way abnormal, the crust tended to be thin. There was no mention of a proximate cause or an origin of trenches. At Lamont the tectogene hypothesis seemed to be dead, and a successor had yet to emerge. In any event, there was no certain evidence of compression in trenches. If Bruce Heezen wanted to create vast areas of new crust at ridge crests, he could not automatically dispose of it in tectogenes at trenches. He was in an environment where a hypothesis of global expansion might be the most natural one to evolve.

Scientists at Lamont had lots of gravity but hardly any seismic data. Those at Scripps had the opposite data sets, and they were growing along with the number of seismologists. George Gershon Shor, Jr. (1923–) came to Caltech in 1940 and moved into Dabney House just when I was in charge of its mild hazing of freshmen. George was a Naval Reserve officer from 1943 to 1946 and returned to Caltech for an M.S. in 1948. He spent 1948-1951 in seismic exploration for petroleum and in 1950 married Elizabeth Noble, daughter of James A. Noble, Professor of Geology at Caltech. Betty's work as a historian of science is cited elsewhere in this volume. George came to Scripps in 1953 and received his doctorate from Caltech in 1954. He lives on a short street at the opposite end from Russ Raitt. Fisher, Raitt, Shor, and I, all from Caltech, cooperatively directed a seemingly endless series of deep-sea expeditions.

Fisher, Raitt, and Shor produced a series of papers in 1961-1962, based on fieldwork from 1954 to 1958 (Fig. 16 F, G, and H).[29] They were concerned with the topography and structure of the Middle America and the Peru-Chile trenches. A total of seven seismic lines reaching the mantle were shot along the axes of the trenches, and most were parts of profiles transverse to the trenches. The oceanic layer under the trench axes had normal velocities, and the thickness ranged from normal to about 50% greater. The average thickness of the whole crust was 10 km compared with 6 km for the deep-ocean basins. "The feature in common" in five trenches with seismic crustal observations was a thickening of the crust, but not as much as "that implied in the tectogene hypothesis." Nor did the observations agree with the gravity interpretations by Worzel and others. In fact, they seemed to show that despite the volcanoes, earthquakes, and gravity anomalies, nothing very dramatic happened to the oceanic crust.

Thus at Scripps we were in an environment where the seismic refraction studies showed a crust that was present everywhere in the ocean basins. It thinned at ridge crests and thickened at trenches, but the changes were not enormous. Perhaps it was to be expected that I would think in terms of a solid lithosphere stretched and compressed by mantle convection. Hess and Dietz would have more of a problem. Without global expansion they would need to destroy as much crust as they created. The logical site for disposal was in Vening Meinesz's downbuckled crust, but just at the time when it was needed most, the hypothesis of the tectogene was dead. It appeared in a strange new form in 1963 when Fisher and Hess pictured a conveyor-belt type of serpentinized tectogene plunging straight down until it once again became deserpentinized mantle (Fig. 16 E). The tectogene had begun its apotheosis into the subduction zone.

Chapter 11

THE SEQUENTIAL DEVELOPMENT OF RIDGES

Before World War II the grand geological syntheses such as continental drift could not be tested. In the 1950s new tools and new men began to accumulate enormous quantities of entirely new kinds of observations. Paleomagnetics, explosion seismology, heat flow, marine magnetics, and echo sounding—all were yielding startling results. For the first few years, confusion grew as a few geologists attempted to absorb and evaluate the unfamiliar observations. Most geologists, of course, continued, unconcerned, with their research in petrography, stratigraphy, or other specialties. I do not mean to imply that the new men were dynamic and farsighted, as opposed to stodgy classicists. The new types of research were generally no broader or more difficult than the old. The new men, as this history shows, were rarely visionaries with a clear perception of the great promise of their work. It just happened that the new kinds of data would be suitable for a new understanding of the earth.

After a decade, by 1960, the ad hoc explanations for individual groups of observations were no longer very satisfying. It was time to try to integrate all the new data with the old and to generate testable syntheses. These syntheses took many complex forms depending on the type of data that were emphasized. For simplicity in the remainder of this book, I shall group them in three classes: (1) "sequential" hypotheses in which ridges are of different ages and convection acts at different times; (2) "expansion" hypotheses in which crust is created but not destroyed by the expansion of the earth; and (3) "sea-floor spreading" hypotheses in which crust is created and destroyed by mantle convection. All the phenomena can act at once. The earth can expand while convection subducts some crust into the mantle and ridges are created and destroyed. Most of the controversy in the 1960s, however, was in the form of criticism or support for one of these classes of hypotheses, so it seems reasonable to treat them separately.

Only one of the hypotheses, sea-floor spreading, originated in 1960-1961; but by a curious coincidence, or by the nature of the times, Bruce Heezen synthesized the new data in relation to the expansion hypothesis in 1960, and I did the same for the sequential hypothesis. Our attraction to the various hypotheses was an almost inevitable consequence of the results of our individual research. The research, in fact, was narrow, mostly marine geomorphology, but the areas were hemispheric and the conclusions correspondingly grand. Heezen worked in the Atlantic, an ocean created by continental drift according to Wegener and followers, but Heezen traced the median rift around Africa into the Indian Ocean. There was no intervening trench to absorb the crust being created at the rift. In the context of current thought, it was reasonable for Heezen to conclude that the earth was expanding and that the ridge was ancient and all of an age. I had accepted the conclusion that continental drift had opened the Atlantic. However, my own observations in the Pacific suggested that mid-ocean ridges were ephemeral, so I was attracted to a sequential hypothesis. Hess had not been to sea for about 15 years but had explored both the Atlantic and the Pacific long before. In any event, he was really trying to explain mantle phenomena and trenches at the same time, and it was almost inevitable that if he produced an integrating hypothesis it would be sea-floor spreading. Dietz was interested in marine geomorphology and rejected expansion, so he came to the same hypothesis. There really were not very many hypotheses that could even appear to explain the wealth of new data. In retrospect, it may seem surprising that any hypothesis but sea-floor spreading could be entertained. If that had been true at the time, however, only one hypothesis would have been entertained. As proposed, all the hypotheses had virtues and flaws. It would take different scientists five, or ten, or more years to become convinced that one or another was right. Some apparently are still doubtful.

This chapter is largely concerned with the sequential hypothesis and its status in 1960 as the tectonic revolution intensified. In the 1930s and 1940s even convection in the mantle was a radical hypothesis. Except for the few converts to continental drift, most geologists, especially in America, thought of localized crustal shortening in terms of tens of kilometers. Crustal stretching was on an even smaller scale. There was little reason to call for convection in the almost wholly unknown mantle except that Vening Meinesz and Hess had found gravity anomalies over oceanic trenches. Then Griggs and others gave some theoretical underpinnings to mantle convection and Holmes invoked convection as the cause of continental drift. In America that probably did not add to the credibility of convection. With this background it is surprising that mantle convection seemed to be such an attractive explanation for the phenomena discovered by marine geologists. Convection was offered as the cause of high heat flow, fracture zones, the elevation and subsidence of mid-ocean rises, and the formation of the Hawaiian Islands as well as oceanic trenches. In the late 1950s, however, no general synthesis existed to explain many of these phenomena at once. One was to arise from the Downwind expedition, which although largely confined to the southeastern Pacific would also lead to an explanation for the long mysterious slope between California and Hawaii. This expedition included the two principal Scripps ships, *Horizon* and *Baird*, which were aging but still serviceable, and lasted four and a half months.[1] Naturally it was during the southern hemisphere summer—from 21 October 1957 to 28 February 1958. It was the first of three Scripps expeditions that were part of the integrated International Geophysical Year under the auspices of the International Union of Geodesy and Geophysics, which was headed by Tuzo Wilson. Downwind began rather abruptly for me with a phone call to my home from Director Roger Revelle.[2] He had money and ship time and asked if I would organize an expedition. I did so by assembling the usual willing scientists to pursue their usual studies. We split the scientific party into two groups, one to work down to Valparaiso and the other back, but the work of the groups was to be similar. In retrospect, the expedition was a reprise of Capricorn with slight changes in cast, scenery, and plot: in order of appearance for each pair, Menard and Fisher, leaders plus dredging and bathymetry; Shor and Raitt, explosion seismology; Goldberg and Rakestraw, water chemistry; Riedel, sedimentary cores; von Herzen, heat flow; Shipek, bottom photography; supporting players, students, and technicians.

The basic objective of the expedition was to map the physical characteristics of the East Pacific Rise along two enormous transects with some work along the crest. The southeastern Pacific is the ocean most remote from land. From 40°S to 60°S latitude, one gale succeeds another, and enormous swells sweep across the empty sea. No trade routes cross the southeastern Pacific and World War II avoided it, so the only information available came from deliberate exploration by oceanographic ships. HMS *Challenger* discovered a broad swell on its doglegged track at about 45°S between Tahiti and Chile. It was named the "Albatross Plateau," in the absence of any land, because of the enormous wandering albatrosses that float like white ghosts above the curling swells. Half a century later, Alexander Agassiz, son of the great Louis, explored the region in the U.S. Fish and Wildlife Ship *Albatross*—which struck him as a pleasing coincidence. Agassiz established, however, that the broad swell extended north from *Challenger*'s track to Easter Island and on up to western Mexico.

In 1929 the ill-fated *Carnegie* sounded and cored the eastern Pacific and thereby collected the samples for Roger Revelle's doctoral thesis. Some cartographers renamed the swell the "Easter Island Plateau," but no one did much more fieldwork until after World War II. Then the U.S. Navy undertook a program of sounding the region, "Project Highjump," as a part of the annual shipments to support research in the Antarctic. Bob Dietz went on one of the first of these resupply expeditions, which were remarkably well conceived and executed. Even so, when we began Downwind there were no echograms of the deep sea floor in the southeastern Pacific or any other modern oceanographic observations except those taken on Capricorn.

We headed into this void with almost the same tools as on Capricorn, although all had been improved through evolution. We had no magnetometer.

One echo sounder developed a transducer leak when *Horizon* ran aground on the reef at Rapa. It was almost useless on the western flank of the southern transect. Much of the false decking on the fantail of *Baird* was smashed by heavy waves when I turned the ship to try to find the median rift predicted by Ewing and Heezen. We lost a boat and large stocks of core liner to the sea. The chief problem that arose was in the operation of the heavy winch on *Baird* that was needed for piston coring, dredging, and heat flow. The cold was unprecedented for our work, and the protective grease fell off in strips. The winch paid out wire without difficulty, but whenever it was stopped to begin hauling in, it threw a kink in the half-inch wire. The kink could not be removed without splicing, which was impossible at sea. The kink could be reeled in but could not pass through the sheaves into the water. After I understood the problem, I had to restrict usage of the winch to heat flow and make each measurement shallower than those before. In short, it was a normal expedition. I took time out to discover and survey a seamount on 19 December as a birthday present for my wife.

The preliminary results of Downwind were presented by Roger Revelle at the 1958 annual meeting of the American Geophysical Union a few months after we returned. Thus they were widely known to geologists during the next few years when the tectonic syntheses were being formulated. The first publication of the Downwind data, however, did not come until December 1960.[3] Our combined efforts had been remarkably successful in the sense that they were geographically consistent relative to the rise. For once in that decade of exploration, we had not generated more problems than we had solved. For once, everything had been as predicted or seemed consistent with predictions. If ever, it was a time for a broad synthesis.

No interpretation at that time, however, could be formulated in isolation. In order to explain what had been done we must turn back to 1959, to a paper describing recent marine exploration in the Atlantic.[4] Maurice Ewing had begun to publish about the Lamont expeditions with his younger brother John. The acknowledgments listed a familiar cast of participants: two Ewings, Bruce Heezen, Chuck Drake, Joe Worzel, two Katzes, and Jim Dorman. Not everything was the same; Frank Press had crossed the bar for the last time. The continuing seismic refraction studies had established that the deep Atlantic probably has the same structure as the deep Pacific except that the sediment is thicker, especially under abyssal plains. The second or volcanic layer, so evident in Raitt's data, was obscured by the thick sediment but appeared to be present.

The interpretation of the seismic stations on the Mid-Atlantic Ridge was complicated by the mountainous, rifted, fractured, volcanic topography. It appeared, however, that a layer of rock 3-5 km thick, with an average velocity of 5.15 km/sec, was characteristic of the ridge. It was taken to be volcanic. Below it was a layer with an average velocity of 7.21 km/sec, which was "tentatively identified as a mixture of mantle and oceanic" rocks. Normal mantle material with velocity 8 km/sec or more was not detected below the low-velocity mixture. The thickness was estimated by isostasy as 30 km because no gravity anomaly existed, and

> The formation of the ridge requires the addition of a great quantity of basalt magma and raises the question of its source. This is attributed to a rising convection current, as described by Hess (1954, p. 345). We suggest that a convection current system has contributed the basalt magma and has applied the extensional forces to the crust to produce the axial rift.

> The mid-ocean ridge system is worldwide in extent . . . and generally follows a course midway between continental masses, suggesting that it is as old as the continents themselves. . . . The thermal flux through the ridges is several times greater than the average for ocean basins (Revelle, 1958; M. N. Hill, personal communication). On the East Pacific Rise, which Revelle reports as a region of definitely high heat flow, no seismic velocities greater than 7.5 km/sec were recorded, in good agreement with the Mid-Atlantic Ridge seismic results.[5]

The Ewings were misinformed about the seismic results, which is not too surprising considering that they were not published in any detail for 12 years. Even so, the *Preliminary Report on the Downwind Expedition* in 1958 noted that the mantle was reached in some places on the crest of the rise. "Furthermore, it cannot be concluded that failure to observe the mantle wave is proof of its great depth. It may merely be a result of poor propagation."[6] At three of the six seismic stations on the crest of the East Pacific Rise, mantle velocities of 8.0-8.5 km/sec were reached at relatively normal depths. There was no sign of a thick lens of basalt.

When I tried to analyze the results of Downwind, I began by preparing a new bathymetric map and relating the physiography to it. The rise was like nothing else known on earth, a vast low-bulge 2-3 km high, 2,000-4,000 km wide, and 13,000 km long from Mexico to New Zealand. It was part of the continuous oceanic ridge system surmised from seismicity by Ewing and Heezen, so the length was not important. The height was also normal for the system, but the width and the gentle slopes of the flanks were quite unlike the Atlantic or Indian ridges. Moreover, the crest and the upper flanks of the rise were remarkably smooth. Bruce Heezen might believe there was a central rift, but no one else to whom I showed the records could see one. Thus both the bathymetry and the physiography were unlike the Mid-Atlantic Ridge.

I used the new bathymetric map as a base to plot all other kinds of data. We had known while on board ship that heat flow was from two to four times normal on the ridge crest.[7] What was unexpected was that heat flow was low on the western flank and probably low on the eastern flank as well. Everything seemed to be related to the rise.

Interpretation of the crustal sections required a redefinition of "crust" and "mantle." The definition in common use in seismology was that material below the Moho was mantle and material above was crust. The Moho was identified as the upper boundary of material with a velocity of 8 km/sec. It did not seem that these were very useful definitions for interpreting the seismic crustal sections of ocean basins. Everything of interest would be crust-mantle mix, and the top of the mantle would be unreachable. Considering the uniformity of the thickness and velocity of the oceanic layer, it seemed more reasonable to define it as the crust. That is, the bottom of the seismic crust is the bottom of the 6.8 km/sec material, and the material below it is the mantle, whatever its velocity. Either the oceanic Moho is not tectonically meaningful or it is the bottom of the crust. Often considerable advances are made in science just by changing a frame of reference or by redefining a variable as a constant or a constant as a variable.[8] The redefinition of the seismic refraction results made it possible to map variations in thickness of the crust and in velocity of the upper mantle. The crust is thinner under the ridge crest than under the ridge flanks or normal deep-ocean basins. Likewise, the mantle velocity is lower than normal under the ridge crest. This would be expected from the high heat flow without any mixing of mantle and crust. Ewing and Ewing could not reach such an obvious conclusion just because of the way they defined mantle.

Ewing and Heezen had concluded from the distribution of earthquakes that the mid-ocean ridge and its marvelous rift came ashore from the Gulf of Aden and the Gulf of California. From the latter gulf, the line of earthquakes extended along the San Andreas fault and went out to sea again off northern California. What did our new data show? After Downwind we had a series of short expeditions near Scripps including one in the Gulf of California and one off central and northern California. Von Herzen, Raitt, Shor, Fisher, and I collected more data like those from Downwind, and I had permission to synthesize the unpublished results. The belt of high heat flow was traced into the Gulf of California and at sea again off northern California. Likewise, the crust was thin and the mantle velocity low off northern California.

In a letter of 14 September 1960 I wrote in some excitement to Bruce Heezen:

> You may be interested to know that I am putting together a paper demonstrating that the East Pacific Rise sneaks under the western United States and then emerges off Cape Mendocino. I regret to say that I still cannot see your median rift as a prominent and reasonably continuous feature. However, I have become convinced that our Ridge and Trough province is the same gen-

eral horst and graben topography which you have described for the crestal belt of the Mid-Atlantic Ridge, and for the kernel of this idea I am indebted to your track of the rift.

He responded a few months later (in a letter dated 8 December) just before my paper was published:

I'm glad to hear that you have been converted to our view that the East Pacific Ridge connects through the western part of the United States via the Gulf of California and Cape Mendocino. I'm quite interested in knowing what additional information caused you to make this shift in your position . . . it certainly does appear that the Ridge that runs north-northwest from Cape Mendocino is similar to the Mid-Atlantic Ridge, although, of course, it is minute in scale in comparison.

Following the track of the earthquakes I had traced the crest of the rise by bathymetry, heat flow, and crustal characteristics both north and south of western North America. The high plateaus, faulting, and rifting in East Africa and western North America suggested that mid-ocean ridges extend under continents. The bathymetry west of North America slopes toward the west—just like the west flank of the East Pacific Rise. I realized (at last) that most of the northeastern Pacific *is* the west flank of the rise, even though part of the crest is missing.

The most conspicuous features of the northeastern Pacific are the great fracture zones. Are they related to the rise? The world-girdling seismic belt and rift as conceived by Ewing and Heezen had no offsets like those on the fracture zones. I had long before noted, however, that the eastern ends of the four great fracture zones are seismically active. How could they and the rift be active when they intersected at right angles? For the moment I was content to state that "it now appears that [seismicity of fracture zones] may be an incidental effect of the intersection of the East Pacific Rise seismic belt with transverse fault structures."[9] Not much later I would become more contentious in this matter.

Once I understood that the sea floor off California was the west flank of the East Pacific Rise I looked at the offsets on the fracture zones in a new light. The flanks in the South Pacific were remarkably smooth. The horizontal offsets of such a smooth plane would be indicated by simple offsets of the bathymetric contours. And so they were in the North Pacific. The 4,500 m contour was offset by almost the same amount as the mysterious magnetic anomalies on the Mendocino, Pioneer, and Murray fracture zones. It was "difficult to escape the conclusion that the sea floor was sloping before the offsets occurred. . . ." So much for the western flank, but what of the eastern flank?

The enormous displacements of marine crustal blocks are not matched on the adjacent continent. The Mendocino fracture zone offsets the continental slope 100 kilometers in northern California, but the displacement is right-lateral—that is, in the opposite direction from the displacement of the deep-sea floor farther to the west. . . . The Channel Islands and the transverse ranges of California appear to mark an extension of the [Murray] fracture zone on land. Wrench faults with the trend of the Murray fracture zone are common in these islands and ranges. Displacements are as much as 15 kilometers, but they are generally left-lateral—once again in the opposite direction from the displacement of the sea floor. It appears that some zone of discontinuity is parallel to, and near, the continental slope off California; that crustal blocks on opposite sides of the zone move in different directions; and that movement on the continental side is much less than on the oceanic side.[10]

That was the very region that Tuzo Wilson would be studying when the idea for transform faults flashed into his mind. Crustal blocks moving away from a central zone offset by active fracture zones is what transform faults are all about. Had I casually invented transform faults? There is one thing to be said for personal involvement in a history. One certainly avoids the error of adumbration. I didn't conceive of transform faults. Tuzo did.

What I did do was turn from the path that would lead to the ocean of truth to pursue the dream of all scientists, the dream that led Wegener astray, the dream of a quantitative confirmation of a prediction. I did not publish the prediction as such, but I was aware of it and it was implied by the mechanism I invoked to produce the East Pacific Rise, namely, mantle convection.

A rising hot material, marked by high heat flow, produces an upward bulge of the mantle because of thermal expansion or physical-chemical changes. The mantle bulge arches the overlying crust and forms a system of tension cracks parallel to the rise. Arching stretches and thins the crust, but the observed thinning is so great that translation of the crust toward the flanks of the rise is also required. Accordingly, the horizontal limb of the convection cell moves the crust outward and thins it at the crest of the rise by normal faulting along the tension cracks.[11]

I could calculate the translation from the thinning. The thickness of the crust was 4.9 km in the deep basins and flanks, but only 3.8 km in a band 2,800 km wide along the crest of the rise (Fig. 8 E). If the crestal crust was initially of normal thickness, its initial width was 2,200 km and the flanks had moved apart by 600 km. The translation, in short, was unexpectedly but predictably about equal to the observed offsets on the fracture zones. This struck me as a strong confirmation of my hypothesis—although, in fact, it was quite wrong. As to the cause of the offsets,

Blocks are displaced different distances by wrench faulting on fracture zones because of variations in intensity of convection along the rise. Displacements on fracture zones have opposite directions on the two sides of the crest because convecting material moves in all directions from a rising hot center. Farther out on the flanks of the rise the convection current moves the crust between fracture zones as a unit (the displaced magnetic anomalies are not distorted as they are near the crest). On the outermost flanks the sinking convection current, marked by low heat flow, defines the outer limit of wrench faulting and, presumably, the crust is thickened by thrust faulting.[12]

The conclusion of the paper addressed the questions of the evolution of rises and continental drift, which I took to be related. In the late 1950s, as ever, I thought about the subsidence of the guyots in the western Pacific. So did Harry Hess. Indeed, as the following correspondence shows, we had difficulty keeping our ideas straight. The letters were written in 1966. First to Hess, dated August 22:

> I am writing to point out an error in scholarship in your paper in the Colston papers. It doubtless arises from confusion in my previous papers, but in any event we might as well come out with the same version of the idea that mid-ocean ridges are ephemeral features. You state on page 327 of the Colston papers that "it was proposed by me in 1959 that mid-oceanic ridges were ephemeral features and Menard 1964 arranged them in an evolutionary sequence" and so on. This statement really is not correct. I had already proposed that mid-ocean ridges were ephemeral features in my publication in 1958, but *you* had proposed that they were ephemeral features in your publication in 1955. Thus there is not any question that you have priority on this idea but your references are wrong.

Hess responded promptly on September 6:

> I appreciate your corrections to my Colston paper. This was written during the first two weeks I was in England to meet a March 1 deadline and virtually without access to a usable library.

A typical battle over priority.

My publication of 1958, alluded to above, compared the position of the global "rift" of Ewing and Heezen with the median line of the ocean basins. The median line of the mid-ocean ridge corresponded very closely to the median line of the ocean basins. It really was a *mid-ocean* ridge. The correspondence continued rather dramatically in previously unsuspected places such as around the southern end of Africa and between Australia and Antarctica. The only median line of an ocean basin lacking a mid-ocean ridge was in the western Pacific. This suggested that the guyots in the region were submerged because they had been on a mid-ocean ridge that had subsided, which in turn appeared to confirm Hess's idea that ridges are ephemeral. I conceived of an age sequence. The East Pacific Rise is seismically active, exceptionally broad, un-rifted, and not centered in an ocean basin. Perhaps it was youthful; the Mid-Atlantic Ridge was mature; and the western Pacific was a dead ridge. The age sequence implied that ridges can be created anywhere, but in time they become centered in ocean basins between continents. How? Did the continents stay fixed and the ridge move, or vice versa? In 1958 I thought that the marine evi-dence was not pertinent to the hypothesis of continental drift because the ridge circled southern Africa and ridges intersected continents in East Africa and western North America. Everyone, or at least Ewing, Heezen, Hess, and I, thought the ridge was fixed. If so, Africa could not drift away from the ridge, and yet the continents should have been torn asunder at the intersections with the ridge.

By 1960, I had thought of a way to have both continental drift and fixed mid-ocean ridges. If different sections of the ridge were of different ages, perhaps they moved continents at different times:

Continental drift, as suggested by the parallelism of the Atlantic coastlines and the crest of the Mid-Atlantic Ridge, has been a very attractive concept for continental geologists, particularly since it was revitalized by the paleo-magnetic evidence for polar shifts and possible drift. Marine geologists, on the other hand, have been reluctant to accept the concept of continental drift because they find no evidence for it in the geology of the sea floor. Indeed, the existence of rises centered in the Indian and Pacific oceans seemed to eliminate the possibility that Africa and South America had moved away from the Mid-Atlantic Ridge. However, if a random distribution of rela-tively short-lived "oceanic" rises is accepted, the picture is entirely differ-ent. If all rises were in the center of ocean basins it would not be clear whether the convection current, or another agent, which produced the rise centered itself relative to the margins of the basin or created the basin. With rises bordering the Pacific and penetrating Africa, it appears more probable that most rises are centered because the margins of the basin have been ad-justed by convection currents moving out from the center. If so the African and East Pacific rises may mark relatively young or rejuvenated currents which have not yet had time to produce much continental displacement. Even so, East Africa is being torn by deep rifts and Baja California has al-most separated from North America along the crest of the East Pacific Rise.[13]

Chapter 12

THE EXPANDING EARTH

Geologists commonly report their results for some fieldwork, relate them to some phenomenon such as compression, and then refer to an "underlying cause" for the compression. The underlying causes have little to do with the results, which could be produced by a variety of causes, but such a reference distinguishes the broad research scientist from a mere field technician. The underlying causes are phenomena that affect the whole earth such as interaction with the moon or other celestial bodies, heating or cooling, or changes in volume, rate of rotation, or orientation of the axis of rotation. Little risk is involved in an appeal to these phenomena because little is known about them. Thus perhaps the most useful effect of the underlying-cause syndrome is to help the historian identify the basic assumptions underlying the authors' thoughts.

The fact that the geological history of global phenomena is inherently intractable makes it inevitable that all possibilities become functional hypotheses. So it has been with the volume of the earth. The common theory in the nineteenth century and the first half of the twentieth was that the earth is contracting.[1] In this theory the paramount fact to be explained was the crustal shortening in folded and overthrust mountain ranges, and all else followed. Wegener could explain regions not only of compression but also of expansion by the drift of continents. The geologists who later proposed large-scale mantle convection could also explain both phenomena. Consequently, "drifters" and "convectors" had little reason to consider the volume of the earth and generally assumed it was constant through geological time.

The possibility of an expanding earth was advanced at least as early as 1922; it still persists and was in vigorous competition with the constant-volume hypothesis just when the plate tectonics revolution occurred. Clearly, it was (is?) possible to accept the hypothesis of an expanding earth. The adherents, being converted, occupied themselves in solving problems related to and building on the hypothesis. The principal problems are the following: (1) elucidation of evidence supporting expansion in order to convert others, and determining, for the converts, the facts concerning expansion, that is, (2) the amount, (3) the rate and variations in rate, (4) whether it is uniformly radial or varies from place to place, (5) its cause, and (6) its effects on surface geology.

S. Warren Carey wrote a review essay on the history of ideas about the expansion of the earth that provides a classical example of multiple discoveries.[2] Carey began by observing that epeirogenesis and tension were always major elements in European tectonics, and therefore it was natural "that the idea of earth expansion was conceived and developed primarily in the German literature." In 1927, B. Lindemann wrote a book in which he attributed continental drift to expansion of the earth.[3] M. Bogolepow in a German paper in 1930[4] referred to three papers he wrote in Russian in 1922, 1925, and 1928[5] that proposed earth expansion. Neither Lindemann nor Bogolepow knew of the other's work. The next pertinent event was the discovery of the cosmological red shift by Edwin Hubble in 1929.[6] I have not found direct mention of it in geological literature, but from then on any scientist presumably knew that the universe is expanding.[7] Why not the earth?

O. C. Hilgenberg wrote a short book in 1933, dedicated to Wegener.[8] He constructed a small globe, fitted the continents from a large globe on it, and succeeded in assembling the jigsaw puzzle so as to cover the whole small globe with a layer of sial (Fig. 17). He required a 50% expansion of a globe from an initial radius of about 4,000 km to its present value of 6,380 km and proposed that the additional mass was acquired by the conversion of energy residing in the aether. Hilgenberg was not aware of the discoveries of Lindemann or Bogolepow. J. K. Halm, in 1935,[9] did not know of Hilgenberg or his predecessors, but he certainly knew about Hubble because he was an astronomer. His presidential address to the Astronomical Society of South Africa deduced an expanding earth from an "analysis of the evolution of celestial bodies," tied to the "variation of the effective size of atoms." The radius increased by 1,000 km, and here we see, for the first time, an astronomer advocating expansion but only on a smaller scale than that proposed by most geologists. Still, Halm believed that the expansion was adequate to explain Wegener's evidence for continental drift.

In 1937, P.A.M. Dirac wrote a letter to *Nature* that was to affect the expansion hypothesis of geologists 20 years later.[10] He was interested in the relation between the Eddington numbers that define certain properties of the universe in dimensionless form. The first number, the ratio of the electrical to the gravitational force between an electron and a proton, is 2×10^{39}. The second, the age of the universe expressed as a ratio of atomic constants, was also about 10^{39} (the present value is about 4×10^{40}). Finally "the ratio of the mass of the universe to the mass of the proton" is about 10^{78}. The square root of this is also roughly 10^{39}. Dirac has remarked, however, that the fundamental properties of the universe do not involve equations with fractional exponents. Even though dissimilar, all the Eddington numbers are enormously greater than normal physical constants, and Dirac reasoned that they constitute a special class of numbers with similar properties. Since one number varies with time (t) he further reasoned that all must do so according to some simple power such as t or t². This led him to a few "elementary consequences." The mass of the universe is increasing proportionally to t². The gravitational constant (G) decreases inversely with time. The paper established Dirac's priority, should the matter prove important, and nothing was lost thereby. Because he was a Nobel Prize winner, his letter was naturally printed first in that issue and therefore directly under the line:

> The editor does not hold himself responsible for opinions expressed by correspondents.

The letter was very brief. Dirac did not even write that it had not escaped his notice that the earth was expanding. Still, sweeping vistas were opened for other theoretical physicists. Another physicist thought that it would be truly remarkable if geology and astronomy could actually help solve a question of fundamental physics.[11]

In 1940, Keindl wrote yet another book that explained continental drift by expansion of the earth. Everyone had heard of Wegener, but Keindl did not know of any prior ideas on expansion and made yet another independent discovery. Like Hilgenberg, he started with an earth covered with a single sialic continent. Like Halm, he expanded the earth along with the rest of the universe. He added, however, a "super-dense metastable core in the earth" that gave additional expansion when a phase change was triggered.

The sequence of independent discoveries continued with Shneiderov in 1943 and 1944. He developed an independent formula of gravitation, and it led him to an alternation between large expansion and smaller contraction such that continents drifted during expansion and mountains were built during contraction.

The discussion among physicists continued with Edward Teller in 1948. The long persistence of life on earth provided a test for the changes in the Eddington numbers proposed by Dirac. Taking the age of the universe as 2-3 aeons, he calculated the effect of a decrease in G for the last 200-300 million years. Regardless of assumptions affecting solar radiation, life on earth would be destroyed.

Returning to geologists, P. Jordan in 1952 wrote yet another book about an expanding earth.[12] He used Dirac's decrease in G to explain the expansion of an earth originally covered with sialic continents. Walker and Walker were independent economic geologists who deduced the expansion of the earth from their own field observations and published one more book in 1954. H. C. Joksch generally followed Jordan's analysis but published in a normal scientific journal.[13]

L. Egyed probably was the penultimate independent discoverer. In 1956 he wrote,

Essentially there are two opinions concerning the volume of the Earth: (1) its volume is decreasing, that is, the Earth is shrinking; (2) its volume has remained practically constant during geological times.[14]

In 1620, Francis Bacon—in an attempt to explain the distribution of continents—advanced the possibility that the earth expanded more in the southern hemisphere than in the northern one.[15] Thereafter, there were at least eight independent discoveries of the hypothesis that the earth is expanding. Most of them were published in books and many were privately printed rather than distributed in the mainstream of scientific literature. Nonetheless, the discoverers generally became aware of the existence of prior literature a few years after they themselves published. Presumably, readers informed them. Thus it appears likely that, each time, the hypothesis seemed so radical that the possibility that it was already published was not considered. Perhaps, also, most of the discoverers published their seemingly radical ideas where they did because they doubted that scientific journals would accept them.

With Egyed the expansion hypothesis entered the normal literature of science. In his second paper, he cited many previous publications; Carey cited him shortly thereafter; and then the literature on expansion began an exponential growth. Egyed's first short note boldly stated that the volume of the oceans had grown with time, but by no more than 4% in the last 500 million years. Like most grand generalizations of planetary history, that was not actually known and probably is unknowable. Egyed assumed that the shape of the ocean basins had not changed during the same period, which is knowable and is wrong. He then drew on paleogeographic data from standard sources and determined that the area of continents covered by the sea decreased with time. Ignoring isostasy, he concluded that sea level had declined because the area of the oceans was expanding. The rate was 0.5 mm/yr, the amount 250 km in 500 million years, and the expansion a mere 4%. This quicksand of misinformation and assumption was to be the foundation for many a scientific hypothesis of noble (if not Nobel) proportions.

Egyed's 1956 note in *Nature* was just a ranging shot to establish priority. Only a year later he was ready to fire the full broadside of an integrated hypothesis. The form and style of the paper have an uncanny resemblance to Wegener's *Origin of Continents and Oceans* and Hess's geopoetry. Perhaps the way to write a revolutionary hypothesis is one of the things subject to multiple independent discoveries by scientists.

Egyed began by dismissing "the former Earth models" and quickly disposed of iron in the core and the role of heat in tectonic phenomena. The "new Earth model" had a homogeneous interior, so the boundaries of the core and inner core were phase changes and thus reversible. The core and inner core formed under much higher pressures than now exist. They were in an unstable "ultraphase" whose existence was related to changes in atomic nuclei. The former connection is implied by the evidence that the expansion was as old as the earth. The latter connection was required because the rate of expansion had been constant for four aeons. We may contrast this argument with the appealing simplicity of Carey's statement that geology proves expansion has occurred and it is up to the physicists to modify physics accordingly.

Having stated his model, Egyed proceeded to a discussion of the evidence that supported it and its logical implications. First, however, it was necessary to avoid a trip to the library.

A series of hypotheses were established as to the origin of the Earth's crust. Many of them are arbitrary and involve such striking mistakes that it is not necessary to discuss them.[16]

Like Wegener and Hess, he did discuss the structure and composition of the crust and the apparently bimodal distribution of levels of the earth's surface. Like Hess, he cited a name or two but no references to the literature. The continents all have the same composition and uniformly thick layers and were originally all of a piece. In contrast,

the coincidence of the percentuate depth-distribution curve of the Atlantic, Pacific, and Indian oceans, respectively—indicates that the formation of the ocean basins was brought about by the same mechanism.[17]

This mechanism was earth expansion with the creation of widening ocean basins. It is possible for an adumbrationist to read this latter sentence as an early statement that the sea floor subsides as it ages.[18]

The principle of conservation of angular momentum requires that if the earth expands without change in mass, the rate of rotation must decrease. This principle gave Egyed a quantitative, rather spectacular, and certainly highly reassuring confirmation of his hypothesis. He must have been as elated as Wegener when Koch confirmed the drift of longitude in Greenland, or as I would be by the correlation between fracture zone offsets and thinning of the crust of ridges. Egyed had developed five lines of geological evidence that, within the framework of his hypothesis, give the rate of expansion of the earth. The time intervals ranged from 5×10^7 to 4.4×10^9 years, but the average rates were within the narrow range of 0.4 ± 0.005 and 0.6 ± 0.005 mm/yr. With regard to the rotation of the earth, it was known from observations of eclipses that the angular velocity is decreasing. Egyed stated,

Formerly, the decrease has been attributed to the effect of tidal phenomena, but the latest investigations have shown that their effect is probably entirely compensated by the effect of the atmosphere.[19]

Thus all the decrease could be caused by expansion. Egyed calculated that an expansion of 0.5 mm/yr would slow rotation by 1.34×10^{-5} sec/yr. For the observed slowing he quoted no less a source than Sir Harold Spencer Jones, recently retired after 22 years as Astronomer Royal. Over an interval of a thousand years, rotation had slowed by about 1.0×10^{-5} sec/yr.[20] Egyed thus had six methods, five geological and one astronomical, that gave about the same average decrease in angular velocity for time intervals spanning six orders of magnitude.

The expansion hypothesis in simple form is incapable of producing compressional features such as folds and thrust faults. Egyed therefore had to introduce a complication. The strength of the crust was such that strain accumulated for periods of 50 million years or so before failure occurred and strain was released. Epeirogeny accompanied the accumulation of "tectonic energies"; and orogeny, ocean-basin formation, and continental drift occurred during the release of such energies. It is a corollary of this mechanism for producing orogeny that it occurs at uniform intervals. The standard geological thesis at that time was that orogeny is episodic and does occur at regular intervals. In fact, in the absence of datable fossils an orogeny located in one place was often assumed to have the same age as a dated orogeny elsewhere. It was also a corollary that if the continents fitted together on a small globe, they cannot fit together exactly on a larger globe because of the difference in curvature. Egyed utilized the surface compression that accompanies a decrease in bending of pieces of the crust to explain tectonic phenomena.

In sum, Egyed had a hypothesis for the origin of a wide range of geologic and tectonic phenomena and one that was quantitatively confirmed by the observed decrease in the angular velocity of the earth. It was to serve as a working hypothesis for a large group of influential geologists, but sometimes only after a radical metamorphosis.

The next major contribution to the expansion hypothesis was by S. W. Carey, and it is not easy to determine when he did it. Most of his ideas presumably were presented at his symposium in Tasmania in 1956. The printed version two years later, however, contained references to papers published in 1957 and 1958 including Egyed's in 1957.[21] Perhaps it will suffice to consider Carey and Egyed to be working simultaneously but independently. In any event, Carey's efforts were both more detailed and grander than Egyed's and earlier into print than Heezen's. We have already discussed many of his tectonic discoveries in Chapter 7. Here we are concerned only with their origin by expansion of the earth.

Carey began by discussing the evolution of a continental rift into a broad Atlantic-type ocean basin with a median rift in a mid-ocean ridge. His Figure 3 was a cartoon of the stages of evolution, which although purely conceptual, could have served as a model for Heezen's topographic profiles a few years later (Fig. 18). The mid-ocean ridges, however, did not merely indicate expanding rifts of uncertain origin. They almost mirrored the continental margin and almost surrounded southern Africa but were longer than what they mirrored. This was telling evidence for a lengthening ridge crest, which he attributed to earth expansion. It was not this line of evidence, however, that convinced Carey of expansion. "My conclusion that the earth expands is original, and follows directly from the orocline concept which I have developed."[22] His apparently was not the last of this series of multiple discoveries. Heezen was to follow.

Carey noted another virtue of the expansion hypothesis compared with continental drift. The latter required a place to drift to, namely, the Pacific basin, which therefore was ancient. Carey showed that the East Pacific Rise was as active as, and otherwise similar to, the Mid-Atlantic Ridge. Thus the Pacific was expanding, too. Moreover, the continents bordering the Pacific have all drifted apart; therefore the perimeter has increased and so has the area. There being no place to drift to, the earth was expanding.

Having established his hypothesis, Carey proceeded in normal fashion to develop it—more than warrants analysis here. He rejected Egyed's tectonics because the 50-million-year cycle neglected relaxation of strain by creep and flow. He proposed that geosynclines develop by sedimentation in what would now be called "failed rifts" or "aulocogens." This was certainly correct for one type of geosyncline, but he proposed that Alpine-type geosynclines are so produced and that the sedimentary rock was then mobilized by heating. Next, in the "regurgitation phase," the rock was squeezed upward by the expanding mantle and spread outward as allochthous mountains without any involvement of horizontal compression. This application of the hypothesis has not found much acceptance.

In his conclusion, Carey made a helpful statement of the physical implications of the hypothesis. The earth's radius had expanded twofold; the original density was 44 gm/cc; surface gravity was four times the present value. Expansion has accelerated and amounts to a radial increase of about 1,600 km in the last 200 million years. Thus from the evidence of drifting continents and abyssal morphology, the rate of expansion for the period has been 8 mm/yr or almost 20 times the rates calculated in six ways by Egyed.

Carey spent little space seeking a physical explanation; that was the problem of the physicists. By a miracle, of serendipity at least, the physicists were ready for the challenge. In 1957, R. H. Dicke, of Princeton University, with whom Jason Morgan would later work, first addressed Teller's arguments against Dirac's cosmology.[23] For a time in the early 1950s, the geologists' age of the earth was greater than the physicists' age of the universe, but in a decade the physicists tripled their age of the universe. Thus Dicke could calculate that the decrease in G would have had much less effect on solar radiation than estimated by Teller. Moreover, the temperature of the earth is strongly affected by the water vapor and carbon dioxide content of the atmosphere. Hence it was by no means evident that Dirac's cosmology would destroy life. Dicke went on to discuss the effects of a decrease in G upon the earth. The age he used was 3.25

aeons.[24] The radial expansion was about 320 km, and the average rate 0.1 mm/
yr. This affected the rotation and heat flow in ways that were not incompatible
with observations. Expansion would produce only small cracks in the moon,
but the compressibility of the interior of the earth would produce such effects
as widening cracks in the sea floor. Even if Egyed's phase changes could cause
the accelerating expansion required by Carey, however, the physicists had not
lent much support to the geologists' hypotheses.

The next scientist to publish about an expanding earth was Bruce Heezen,
and he was highly deliberate in his acceptance of the idea. I heard him give his
first published talk on the subject at Nice in 1958, and the impression I received
of his commitment is fairly summarized in his abstract:

> The major question involved in continental displacement remains the forces
> of displacement. If the original hypothesis of Taylor that all the continental
> blocks are moving toward the North Pacific is true, then the extensional fea-
> tures of the ocean basins' floors could be well explained, unless, of course,
> the North Pacific turned out also to be under extension. In this case we would
> be forced to assume an extension of the ocean floors in general. Since the
> North Pacific seems to differ in no fundamental way from other ocean floors
> it seems necessary to reject the Taylor thesis. Our other alternatives are to
> assume all the continents are being compressed to compensate for oceanic
> extension or to appeal to an expansion of the mantle perhaps by a phase
> change at the core mantle boundary. This latter hypothesis has been little
> considered by geologists and in view of the arguments presented in this pa-
> per seems worthy of more serious investigation.[25]

Heezen cited no references except Ewing or Heezen. Since everything quoted
above had already been published in very similar form by Carey, we can only
assume that Heezen was in the process of becoming yet another independent
discoverer of the expansion hypothesis. Even his main figure resembled Carey's cartoon of ridge development. Marie Tharp does not indicate that he had
reached any positive conclusions about expansion before the Nice meeting.[26]

My own memories are that Bruce was talkative about the possibility of an
expanding earth but that he lacked any strong convictions in the period from
1958 to 1960. I kept waiting for him to publish a proper scientific paper expos-
ing his hypothesis to critical review. Meanwhile, it was hard to take him seriously.

Heezen's second publication on expansion was an abstract for the International Oceanographic Congress in September 1959. His talk was bolder in advocating expansion, but the printed word said no more than the following:

> The location of the Mid-Oceanic Ridge, oft cited as a remnant of the original
> continental rift, opposes continental drift since it seems to require that the
> continents drift in several directions at the same time. A possible way out of

this dilemma is to postulate an expanding earth; but, in view of the meager evidence now available, this may seem too drastic and may itself have other more serious objections of astronomical nature.[27]

Thus, in New York, he was still tentative, had not yet written a word subject to prepublication critical review, and had yet to write anything not already published by Carey.

Tuzo Wilson soon published in *Nature* a proper scientific paper on expansion.[28] He cited papers by Dirac, Dicke, and Teller as well as by Egyed and other geologists. Unlike Carey, he accepted the limits on expansion set by the physicists. With typical Wilsonian ingenuity he proposed expansion of the earth along mid-ocean ridges, but on an acceptable scale. Perhaps the expanding ridges are not as wide as the basins:

> If, on the other hand, one postulates that only the mid-ocean ridges and not the whole ocean basins have grown, the expansion would have been much less. Their width varies and is not accurately known, but Heezen's estimates seem to range from 400 to 800 miles.[29]

The increase in area gave a radial expansion of 6%, and since "the formation of the ridges has probably required all geologic time," the rate of expansion was close to Dicke's. Wilson went on to develop some geological consequences of expansion. We now know that the whole sea floor is generated by spreading. Even at the time, Wilson's hypothesis seemed to abandon the former connection between continents, which was the primary evidence for continental drift. The paper, however, addressed problems and contributions in the literature and was itself in a form that could be evaluated. Thus it was science even if it was wrong.

Bruce Heezen's next publication that mentioned an expanding earth was in the *Scientific American*. He described the sinuous, world-girdling rift and paleomagnetic evidence of large-scale continental drift. He then made an even-handed analysis of the compatibility of the data to various hypotheses of global tectonics: classical continental drift, mantle convection, and earth expansion. Regarding the last,

> In an attempt to overcome this dilemma I have recently suggested that the earth is neither shrinking nor remaining the same size; rather, it is expanding. If the earth were expanding and the continents remained the same size, additional crust would have to be formed in the oceans. This is apparently just what is happening in the mid-ocean rift valleys.[30]

As to the amount of expansion, Heezen referred to Dicke and to Wilson's recent paper in *Nature*. In the text he discussed Wilson's quantitative analysis stating that the mid-ocean ridges fill only part of the ocean basins and he does not reject it on the basis of either abyssal morphology or my data on the width of the East Pacific Rise (note the continental crustal fragments well into the Atlantic in Fig. 18 F). Heezen gave equal credence to the possibility that phase changes could cause a large expansion and that continents "once covered the entire surface of the smaller earth with an unbroken shell of granitic material."

The *Scientific American* has an extensive system of internal review but accepts scientists' manuscripts without external peer review. Heezen still had not published a proper scientific paper, crediting sources and written in a form for scientific criticism. I was still waiting for him to be serious. Nonetheless, the *Scientific American* paper would provide the published basis for Heezen's position in the tectonic debate during the next five years. Among converts to the hypothesis of an expanding earth the principal questions for discussion were those raised by Egyed and Carey. In the debate, however, Heezen's prominence gave him a particularly important role. When he merely spoke about the existence of the world-girdling rift at Princeton in 1957, Harry Hess said, in essence, "You have shaken the foundations of geology."[31]

Chapter 13

SEA-FLOOR SPREADING

In 1959, Harry Hess published his last ideas on tectonics before embracing sea-floor spreading. For publication, he was reduced to the final resort of a busy man of affairs: an abstract for a meeting and a bit of science spliced into a progress report of a committee.[1] Fortunately, this style of publication is printed very rapidly and gives an accurate impression of his thinking at the time. The tectonic ideas in his abstract[2] differed little from those in his papers of 1954-1955. He merely added Heezen's rift to the list of important discoveries and made a few calculations to the effect that the excess heat flow of ridges can be caused only by upward convection in the mantle. In conclusion, "subcrustal drag of the horizontal flow produces extension and the graben on the crest." He still adhered to standard views in which mantle convection merely stretched or compressed a relatively unbroken crust. In the committee report he went a step farther because he became more concerned with the implications of the very thin sediment in ocean basins. Rejecting the possibility of ancient, unmeasured sedimentation rates that were very different from the recent ones subject to measurements, he identified alternative explanations for thin sediment:

1. The oceans are relatively young.

2. The pre-Cretaceous sediments have in some manner been removed, for example, by incorporation into the continents by continental drift.

3. Nondeposition of any sediment over much of the sea floor was a common attribute of the past.[3]

The point he was discussing was that, if drilled, the Mohole could not possibly produce a continuous section of sedimentary rocks dating back to the creation of the earth. The point that Hess himself apparently absorbed was that the implications of the first two alternatives looked exciting. Perhaps his thinking about epeirogeny for the past five years was wrong. Perhaps the concepts of vertical movement triggered by his discovery of guyots 15 years before had been misleading. Perhaps the sea floor drifted sideways. Thus was the good committeeman rewarded. As Roger Revelle remarked in other circumstances, "You just can't keep some people from doing research."

The Mohole progress report was published in December 1959. Just a year later Hess began to circulate a manuscript titled "The Evolution of Ocean Basins'' prepared for *The Sea, Ideas and Observations*, a planned new series of volumes edited by Maurice Hill and others. It was an elaborately presented manuscript, with a grey cover and a spiral binder, because it had the double purpose of being a preprint and a progress report for an Office of Naval Research contract. It received an exceptionally wide official distribution in addition to the normal circulation of a few copies to colleagues with similar interests. Thus began the unusual prepublication history of Hess's paper. The new series on *The Sea* was delayed, as new series are, and volume three was not submitted for printing until after May 1962, or printed until 1963. Meanwhile, *Petrologic Studies: A Volume to Honor A. F. Buddington* was in preparation, and Harry certainly wanted to honor his teacher, colleague, and friend.[4] Moreover, Bob Dietz had published on sea-floor spreading in *Nature* in June 1961, and created a priority muddle.[5] Hess shifted his manuscript to the Buddington volume, which was moving along faster than *The Sea*. It was printed in November 1962, almost two years after it was written. It may seem surprising that Hess himself did not rush to publication in *Nature* or *Science* with his revolutionary ideas. The explanation may be complex, but the ideas were just ideas— of which he had many. At the time he may not have thought of them as anything other than the "geopoetry" he called them. The more revolutionary they were, the more he might feel the desirability for review by selected colleagues.

Hess's paper hardly changed at all during the review process.[6] He originally acknowledged his debt to Carl Bowin for a critical evaluation of his ideas. The printed version also acknowledged comments from Rubey, Bass, Helsley, Engel, Burk, and myself. The lengthy letters we exchanged (25 May 1961 and 12 June 1961) may well have been typical of the extent of the comments and his response. A line by line comparison of original manuscript and ultimate publication, however, shows little more than a change in the title, a change from third to (mostly) first person, and a revision of the discussion of sedimentary thickness that did not change the meaning. Nothing is more repugnant to a writer than the thought of altering a single painfully created sentence.

Before considering the content of the new ideas, let us turn to Bob Dietz and his work while the sea-floor spreading concept emerged. Dietz returned only briefly to the Navy Electronics Laboratory from his Fulbright fellowship in Japan (Figs. 36 and 37). Then he departed for London where from 1954 to 1958 he was occupied in scientific liaison for the Office of Naval Research. During this period he issued a large number of mimeographed reports titled "Oceanography on the Iberian Peninsula," "Geological Oceanography in France," and so on. He set new standards for this genre, but he was not doing what counts. Nor was he doing so writing articles for the *New Scientist* and the *London Times*. He collaborated with Jacques Piccard, who was developing the bathyscaphe invented by his father, August. That work led to a book, *Seven Miles Down*, that was translated into French, German, and Japanese, but it was largely a tale of invention and adventure; the science with research submersibles did not come until much later.[7]

From age 40 to 44 Dietz was essentially detached from creative scientific research. He did not suffer for it; he lacked original data, but apparently his mind was in ferment and the absence of new details fostered a broad view. In any event, not long after he returned to the Navy Electronics Laboratory in 1958 he began to pour out papers. Between June 1960 and December 1961 he was sole or first author of eight scientific papers, seven abstracts, a reply to a discussion, and an article in the *Scientific American*, as well as the book with Piccard. Half the publications were divided among marine geology and the concept of sea-floor spreading, but the other half were about his major research topic, meteorite impacts with the earth.

Dietz's interest in what he later called "astroblemes" began when he was a graduate student at the University of Illinois. He offered to submit a doctoral thesis on earth and lunar impacts but was rejected. Only by default did he begin work in marine geology. He was an Army Air Force pilot in World War II after receiving his doctorate in 1941. Lacking original data but not the urge to do research, he took to studying the recently published large-scale photographs of the moon. By 1946 he published on "The Meteoritic Impact Origin of the Moon's Surface Features."[8] In 1947 he focused on the implications of peculiar structures called "shatter cones," which resemble the fractures produced by shooting a small bullet at thick glass. He plunged soon after into a lengthy attempt to establish the abundance and importance of astroblemes. These efforts were much more prolonged, controversial, and hotly contested than his battle for acceptance of sea-floor spreading. Despite any impression that might be given by this book, not many people were concerned with or opposed to sea-floor spreading—simply because the origin of marine topography did not touch their research directly. Dietz's ideas on astroblemes proposed radical new explanations for well-known geological features—many with major mineral deposits. Whole generations of geologists had worked and published on those structures, and many individuals rose in outrage to defend conventional views.

Dietz's outpouring of papers in 1961 was not too unusual for scientists. Edward Drinker Cope seems to hold the record in geology.[9] Between 1885 and 1918 he published 1,395 papers on vertebrate paleontology. That averages one paper every 8.6 days for 33 years. I don't know about Cope's colleagues, but the marine geologists around San Diego in early 1961 began to weary of editing and reviewing Dietz's manuscripts.[10] Dietz, however, never wrote a dull sentence in his life, and perhaps with resignation, I began to read yet another, mercifully short manuscript sometime in the spring. I was dumbfounded. Hess had sent me a copy of "The Evolution of Ocean Basins" soon after he bound the manuscript. Dietz's manuscript, "Continent and Ocean Basin Evolution by Spreading of the Sea Floor," was amazingly similar in more than the name.[11]

I remember what then happened 23 years ago with perfect clarity. (Dietz remembers it exactly the same way.)[12] I phoned him with the news. He expressed surprise because he had not received or seen Hess's manuscript. He came over to my office to read the manuscript for the first time.

The number of identical points in the two manuscripts was very large. At the time this seemed astonishing. I don't believe any of us was aware of the frequency of simultaneous discoveries in science. In the comparison that follows I shall attempt to assess whether the individual ideas were original or were generally being discussed informally at the time, and whether they were in press or even published. The assessment will then provide the basis for evaluating priority of ideas and the historical significance of these revolutionary syntheses.

For convenience in discussing the papers it is useful to recapitulate pertinent information that was generally known in 1959 when the ideas in the papers were gestating. In legal terms these would be the facts stipulated by both sides in a controversy. Of course many other facts were known that seemed equally important in 1959, but given a single insight, would one not automatically select these stipulations?

Data and ideas related to ocean basins and in common circulation in 1959 were as follows:

1. Convection currents probably exist in the mantle and may cause continental drift, mid-ocean ridges, island arcs and trenches, high oceanic heat flow, and fracture zones.

2. My ideas on convection currents, presented in Chapter 11, had been circulated in preprint.

3. Heezen's ideas on earth expansion, presented in Chapter 12, were widely known from his talk at the International Oceanographic Congress meeting. Carey's papers were little known.

4. The crests of mid-ocean ridges are seismically active and most are in mid-ocean.

5. The structure of continental and oceanic crust was established.

6. The Moho is a physical rather than a chemical boundary. Not everyone accepted this as a fact, but everyone did accept it as a possibility.

7. Isostasy prevails in any large area.

8. There are alternative explanations on every point, but several lines of evidence suggest that the sea floor may be young. I did not believe that it was, but nonetheless I made the first summary of the evidence in 1959 as follows:

(a) sediment is thin (Hess proposed it was swept away by continental drift in 1959).

(b) all dredged fossils are relatively young. "If present sampling is considered to be adequate, the Pacific Basin is no older than Cretaceous."

(c) seamounts are relatively few; "Within the accuracy of the estimates, it appears all the large volcanoes of the Pacific Basin could have been produced since Cretaceous time at the present rate of vulcanism."[13]

The two papers began in characteristic style. Hess wrote,

The birth of the oceans is a matter of conjecture, the subsequent history is obscure, and the present structure is just beginning to be understood. Fascinating speculation on these subjects has been plentiful, but not much of it predating the last decade holds water.[14]

One is spared many trips to the library, but in this paper at least, Hess actually made formal citations to printed references. He cited a "brilliant summary" by J.H.F. Umbgrove in 1947 as an example of what was no longer pertinent and "like Umbgrove, I shall consider this paper an essay in geopoetry."[15] Thus the true heir of Wegener revealed himself.

Dietz had in mind a short paper in *Nature* and plunged right in:

Any concept of crustal evolution must be based on an Earth model involving assumptions. . . . The concept proposed here, which can be termed the "spreading sea-floor theory," is largely intuitive, having been derived through an attempt to interpret sea-floor bathymetry. . . . Since the model follows from the concept, no attempt is made to defend it.[16]

At one place or another each writer advocated or assumed the stipulations on convection currents, ridge crest location and seismicity, structure of the oceanic and continental crust, isostasy, the reversible nature of the Moho, and the three lines of evidence for the youth of the ocean basins. Viewed in the abstract, the long list of stipulations inevitably makes the papers seem remarkably similar, but it required neither original data nor wit to make the list at the time. There were some differences of opinion. Both writers mentioned that the Moho was either a serpentine-peridotite boundary or a basalt-eclogite phase change and cited the same references. They differed in that Hess, who was the expert, incorrectly chose the former, and Dietz chose the latter; but as Dietz wrote, "its exact nature is not vital to our concept."[17] It was enough that the crust and mantle be interchangeable, so the sea floor was effectively the mantle and could be created in one place and destroyed in another.

Given the stipulated facts, all that was required was the single step of combining Heezen's arguments that the sea floor splits apart at ridge crests with my arguments for convection under the crust. How Hess made this step can be extracted from the caption for Figure 10 in his paper (Fig. 20 G). In defiance of accepted editorial standards, especially for publications of the Geological Society of America, there is no mention of this figure in the text, so it stands alone. It is quite possible that as his ideas evolved, Hess half forgot the implications of the figure.

Figure 10. Approximate outline of the East Pacific Rise, which possibly represents an oceanic ridge so young that it has not yet developed a median rift zone, and pre-Rise sediments still cap most of its crest.[18]

In short, Hess accepted both my ideas and Heezen's ideas on the origin of sea-floor topography on the grounds that they merely described different stages in a sequence of development. Then both he and Dietz rejected Heezen's mechanism of an expanding earth by contending that it was improbable at best. That left them with the sea-floor spreading hypothesis. According to Hess,

> The mid-ocean ridges could represent the traces of the rising limbs of convection cells while the circum-Pacific belt of deformation and volcanism represents descending limbs. The Mid-Atlantic Ridge is median because the continental areas on each side of it have moved away from it at the same rate—1 cm/year. This is not the same as continental drift. The continents do not plow through oceanic crust impelled by unknown forces; rather they ride passively on mantle material as it comes to the surface of the crest of the ridge and then moves laterally away from it.[19]

According to Dietz,

> The gross structures of the sea floor are direct expressions of this convection. The median rises mark the up-welling sites or divergences; the trenches are associated with the convergence or down-welling sites; and the fracture zones mark shears between regions of slow and fast creep.[20]

The similarity seems more than coincidental, but compare Dietz with Arthur Holmes in 1944:

> Most of the basaltic magma, however, would naturally rise with the ascending currents of the main convectional systems. . . . Thus, in a general way, it is possible to understand how the gaps rent in the crust come to be healed again; and healed, moreover, with exactly the right sort of material to restore the basaltic layer. To sum up: during large-scale convective circulation the basaltic layer becomes a kind of endless travelling belt on the top of which a continent can be carried along until it comes to rest (relative to the belt) when its advancing front reaches the place where the belt turns downwards and disappears into the earth.[21]

In turn, compare Holmes with Dietz, in 1968, who did not know about the Holmes paper in 1944 and was referring to one in 1931:

> In any event, Holmes' concept . . . is not really sea-floor spreading. He resorts to thinning the crust rather than creating new oceanic rind at the mid-ocean swell and destroying it in trenches in the conveyor-belt fashion of sea-floor spreading.[22]

It appears that when two or three people present a concept, the language may be very similar simply because the concept lends itself to such language. The same may be said of illustrations of a concept; see, for example, Figures 9, 18, 19, and 20.

I have attempted to show that the remarkable similarities between the papers by Dietz and Hess were almost inevitable if they made only one connection between two documented concepts that were just then being widely discussed. Thus it would have been likely that two or more people conceived of sea-floor spreading independently. But did they? I certainly cannot be sure, and I am the only person thanked by both writers for critical discussions of the manuscripts before publication.

The published record appears to imply that the work was not independent. In 1968 Hess wrote, rather casually it appeared, in a one-page response to a criticism by Meyerhoff,

> The cogent term "sea-floor spreading," which so nicely summed up my concept, was coined by *Dietz* (1961) after he and I had discussed the proposition at length in 1960.[23]

Dietz in 1962 wrote, also rather casually it seemed, the following:

> *Note added in proof*—The writer's attention has been drawn to a preprint by H. Hess also suggesting a highly mobile sea floor. Full credit of priority is to be accorded to him for any merit which this suggestion has.[24]

Dietz wrote this remarkable note on the strong urging of myself, Ed Hamilton, and, I believe, others. I remember that, at the time, we thought only one person could have priority and that, by circulating his manuscript, Hess had it. We also thought the similarities in the papers would prove embarrassing unless priority was openly ceded. These urgings were despite the fact that I was certain, and said so, that Dietz had written his paper before he saw Hess's manuscript. I assumed that somehow Dietz had heard of Hess's ideas at some meeting or cocktail party and then forgotten about it. This is a common occurrence. I once thought of a new explanation for the origin of the continental slope and wrote it up in a few frantic days only to recall, in time, that I had heard it all from Dietz. Hess's statement, much later, about a lengthy discussion in 1960 seemed to confirm this assumption, but I recently discussed these events with Dietz. He was unaware that Hess had published such a statement and could recall no such meeting. He had not seen Hess for some time before writing the paper. I can only speculate that relatively early in 1960 Dietz and Hess talked, as we all did, about the stipulated data and current ideas about their meaning. If Dietz had thought that Hess was proposing a revolutionary hypothesis, he surely would have remembered it when I told him his manuscript was so similar to Hess's. On the other hand, when Hess learned of Dietz's manuscript, he might well have attached more significance to their conversation than was warranted.

Dietz never meant to imply that he did not think of the concept of sea-floor spreading independently. He just meant that Hess thought of it first. Priority, however, is normally based on publication in a scientific journal, and there Dietz was first. Taken all in all it appears to me, for the first time in almost two decades, that there were two independent discoveries of this idea that Hess considered to be geopoetry and not worth rapid publication and that Dietz wrote up as a potboiler.[25]

I cannot say that Hess had a responsibility to publish an abstract (or deposit his ideas in a sealed box with the Royal Society) if he wanted to establish priority. A quick abstract, however, was conventional and would have saved some bother. In my rather lengthy comments to Hess by letter on 25 May 1961, I made no mention of Dietz's paper due to be published eight days later. I can only assume that I knew that Hess was well aware of the pending publication. If so, he had some opportunity for a preemptive note to *Science*.

If there was any priority, it may have belonged to Arthur Holmes for the ideas in his textbook of 1944 (Fig. 19 C and D). Hess until his death and Dietz as late as 1984[26] were familiar only with Holmes's paper on convection and drift in 1931. In the first paper Holmes merely stretched the crust, a view that was unwittingly duplicated by me in 1960 with regard to the East Pacific Rise. In the 1944 book, Holmes proposed that stretching was merely an initial stage, a view that was unwittingly duplicated by Hess in 1962. Regarding the next stage, Holmes's views are confusing, but he writes of the newly created basaltic crust moving like a traveling belt away from the torn and outstretched crust of the disruptive basins left behind. He injects intrusions and extrusive lava flows to replace the disruptions. All that is different between Holmes's hypothesis and sea-floor spreading is the median rift.

Although they both introduced the concept of sea-floor spreading, Dietz and Hess each discussed other ideas that were to influence the further development of tectonics. To begin with Hess, a major change had occurred in his interpretation of the origin of the broad relief of mid-ocean ridges—a subject curiously ignored by Heezen and vaguely attributed to rising convection by Dietz and to thermal expansion or phase changes by me. Hess had previously proposed an oceanic crust of basalt over a peridotite mantle that expanded by serpentinization to produce the broad bulge of a ridge. A new appreciation of the data of oceanic heat flow had caused him to shift the 500°C isotherm up to the Moho. That put the serpentine in the crust and made crust and mantle interchangeable, which is just what he needed for sea-floor spreading to work. By the same token, however, he no longer could produce the bulge by serpentinization below the Moho. Thus,

The topographic rise of the ridge must be attributed to the fact that a rising column of a mantle convection cell is warmed and thus less dense than normal or descending columns.[27]

With this statement Hess had lost his mechanism for causing the ridge to subside by mantle deserpentinization. The only serpentine was in the oceanic crust, and it was preserved as the crust drifted. Hess did not discuss the point, but the only remaining way a ridge could subside was by cooling. In this fashion, by default or agreement, a consensus was reached in the literature that ridges are basically thermal phenomena. Hess turned to the implications of drifting upon the subsidence of guyots. Active volcanoes occur near the crest of mid-ocean ridges. They have a relief of about 3-3.5 km when truncated by waves. Then sea-floor spreading would take them across the sloping flanks of a ridge and thereby submerge them, and the guyots on the distant flanks would be older than those nearer the ridge crest. This was an idea that intrigued Hess and Tuzo Wilson for some time and one that I believed I could disprove. By now we know that guyots are submerged both on ridge flanks and on the slopes of mid-plate swells.

Dietz stated that his ideas on sea-floor spreading had evolved in an attempt to interpret bathymetry, and appropriately, he devoted much of his paper to that attempt. Unlike Hess or Heezen, he followed me in associating fracture zones and magnetic anomalies with the evolution of mid-ocean ridges. At that time, Hess viewed the relation of the great fracture zones to the ridges as a "bone of contention,"[28] although much later he would come to agree with me.[29] I had said that fracture zones were the boundaries between regions that had moved different distances from a ridge crest (Fig. 8 E). Dietz rephrased this in terms of sea-floor spreading with fracture zones marking "shears between regions of slow and fast creep."

Dietz also went far beyond anyone else in observing that

Much of the minor sea-floor topography may be even directly ascribable to spreading of the sea floor. . . . Can it not be that these expanses of abyssal hills are a "chaos topography" developed as strips of juvenile sea-floor (by a process which can be visualized only as mixed intrusion and extrusion) and then placed under rupturing stresses as the sea floor moves outward?[30]

This is a far more detailed description of spreading center processes than any offered by Hess or Heezen. It was also the first, and correct, proposal that the endless fields of abyssal hills are created at ridge crests (Fig. 7).

To complete his creative burst, Dietz turned to spreading and magnetic anomalies. "Mason, R. G., and Raff, A. D., in press" had found a north-south lineation that seemed to "reveal a stress pattern," and

Such interpretations would fit into the spreading concept with the lineations being developed normal to the direction of convection creep. The lineation is interrupted by Menard's three fracture zones, and anomalies indicate shearing offsets of as much as 640 nautical miles in the case of the remarkable Mendocino escarpment. Great mobility of the sea floor is thus suggested. The offsets have no significant expression after they strike the continental block; so apparently they may slip under the continent without any strong coupling. Another aspect is that the anomalies smooth out and virtually disappear under the continental shelf; so the sea floor may dive under the sial and lose magnetism by being heated above the Curie point.[31]

That was a paragraph that launched a thousand studies: subducted fracture zones, magnetic evidence of subduction, the generation of abyssal hills, and, most fertile, an association of the destruction of crust with the destruction of magnetic anomalies. Given sea-floor spreading, it would be a very small step to the creation of magnetic anomaly stripes along with the "strips of juvenile sea floor."

Chapter 14

1960

In tectonics the decade of the 1960s was characterized by the chaos that precedes general acceptance of a theory. Some geologists believed in continental drift and sea-floor spreading. Others believed in an expanding or a contracting earth, and so on. Those who took a stand on global tectonics as suggested by one type of information hardly communicated with other geologists with other kinds of information. The groups of specialists tended to talk only to each other with little effort to convert the ignorant, and the vast majority of geologists were not involved with global tectonics and remained quite neutral—at least in their publications. This chapter and the succeeding ones follow the developments of the decade more or less chronologically rather than by topic as heretofore. The objective is to try to capture contemporary ideas in context and to identify when and why concepts were accepted or rejected by different groups of specialists.

Before beginning the chronology it may be useful to compare this decade with those before and after, and to identify the high points within the decade. The 1950s were characterized by the collection of massive quantities of new kinds of geological information on a global scale. The discovery of fracture zones, magnetic stripes, and the median rift made it increasingly evident that a new theory was needed. Several were offered.

In the decade of the 1960s the acquisition of data accelerated but fewer new kinds of phenomena were discovered, so concepts were tested more by the generation of hypotheses than by the collection of data. The hypotheses were proposed in three steps. First came sea-floor spreading, the Vine-Matthews hypothesis, and a few corollaries in 1961-1963. Then came transform faults, magnetic symmetry, and their global confirmations in 1965-1967. Plate tectonics *sensu stricto* followed in 1967-1968, as did the involvement of the uninvolved, at a time when the theory of the new global tectonics was promulgated (it is hard not to say ''merchandized'') in a series of symposia and road shows, as the participants called them. This was the decade when the scientific revolution was most intense.

The decade of the 1970s was much more polarized than that of the 1960s. Indeed, it was more like the 1920s in the intensity of debate. A small group of skeptics rallied to question the evidence of sea-floor spreading and continental drift. A debate with some of the intensity of the nineteenth-century debates on evolution ensued; it largely involved the uncommitted whose observations were not closely linked to global tectonics. Meanwhile, the marine geologists, the paleomagicians, and those confronted with previously inexplicable observations began to confirm and develop the new theory.

A simple way to put global tectonics in context during the decade of the sixties is to examine the indexed references as a function of time. In the *Bibliography and Index of North American Geology* the pertinent papers were indexed under "Continental Drift," which included "mantle convection" and "global expansion" among others. The table shows that the number per year increased five- to sixfold during the decade. Perhaps symbolically in relation to the present study, the ancient *Bibliography and Index of North American Geology* was combined with the newer *Bibliography and Index of Geology Exclusive of North America* in 1969, so the number of references increased by a factor of 1.7. Correcting for this factor the references more than doubled in 1970, and by 1972 they were about 16 times as many as in 1960. This seems to indicate an enormous expansion in geologists' interest in continental drift. To test this speculation I have also tabulated the number of papers on conodonts that are indexed. Conodonts are rather mysterious fossils that are unknown parts of extinct animals of some sort. They can be utilized as stratigraphic markers, and their color is an index of the thermal history of the surrounding sediments. Walter Youngquist, with whom Bruce Heezen wrote his first paper, and Anita Harris, protagonist of a series of articles titled "In Suspect Terraine" in the *New Yorker* magazine, both work with conodonts. These fossils have little, if anything, to do with continental drift, sea-floor spreading, or plate tectonics. The table shows that there were more indexed papers about conodonts than about continental drift during the sixties, and the number per year increased dramatically. Apparently the number of specialists actively engaged in research on conodonts was about the same as the number doing research on continental drift. In fact, at least part of the apparent increase in papers per year may merely reflect more intense indexing. It was not until the early seventies that more people became involved with continents than with conodonts, and by that time the revolution was over. Perhaps, in the sixties, the vast majority of geologists were hardly concerned or even aware of the publications and events that loom so large in this history.

Another way to put global tectonics in context is to examine the papers in some of the principal geological journals. For this purpose I shall analyze six journals that publish papers on all aspects of geology as well as on a range of special interests.[1] In addition, I shall analyze papers in other journals and books that are related to global tectonics in order to follow the particular topic of interest here. As to the interest of geologists in global tectonics, the six journals published more than 200 articles in 1960. Only 12 contained a significant component of global tectonics, and even fewer were actually about the subject.

One of the recurring themes is the paleomagnetic evidence for continental

TABLE 2. Indexed References, by Year, to Global Tectonics

Year	Continental Drift (CD)	Conodonts (C)	CD/C
1960	10	11	0.9
1961	8	15	0.5
1962	23	36	0.6
1963	30	23	1.3
1964	19	24	0.8
1965	23	43	0.5
1966	34	35	1.0
1967	35	37	1.0
1968	54	48	1.1
1969	56	91	0.6
1969	95 (56)	162 (91)	0.6
1970	216 (127)	149 (83)	1.5
1971	256 (151)	149 (83)	1.7
1972	284 (167)	162 (91)	1.8

Note. Figures above the dividing rule taken from the *Bibliography and Index of North American Geology*. Figures below the rule are combined references with those from the newer *Bibliography and Index of Geology Exclusive of North America*, the portion from the older *Bibliography and Index* appearing in parentheses.

drift and polar wandering. In the six journals, in 1960, are three papers that are based on the concept of drift. A.E.M. Nairn noted that "the separation of North America and Europe would appear to date from Permian time." David Collison and Keith Runcorn utilized new observations to confirm Runcorn's earlier demonstration of polar wandering and continental drift by means of paleomagnetism. Opdyke and Runcorn show that paleowind patterns support the paleomagnetic data. These papers had gone through the standard review process and appeared in irreproachable journals—or at least journals not reproachable on grounds of frivolity or sensationalism. Apparently continental drift was an acceptable concept for discussion. It was not, however, universally accepted even by specialists in paleomagnetism.

Allan Cox and Richard Doell, who later would contribute so much to establishing a magnetic reversal time scale, published a "Review of Paleomagnetism" that included a table with all known paleomagnetic measurements. They summarized as follows:

> The post-Eocene results are impressive and offer very strong evidence for the dynamo theory and against substantial Tertiary polar wandering or continental drift.[2]

This statement would have been more in line with the limitations of the data if it had been noted that longitudinal drift was in no way inhibited by paleomagnetic observations, so the opening of the Atlantic was not opposed by the observations. This point, however, was not noted, and the apparent absence of late Tertiary drift and polar wandering encouraged a search for alternative explanations for the observations on pre-Tertiary rocks reported by Runcorn and colleagues. After a thorough analysis of all the data, they concluded that

> Paleomagnetic results for the Mesozoic and early Tertiary might be explained more plausibly by a relatively rapidly changing magnetic field, with or without wandering of the rotational pole, than by large-scale continental drift.[3]

RATES OF SEDIMENTATION

One of the influential studies in marine geology published in 1960 was by Ed Hamilton, who had turned from guyots to the physical properties of deep-sea sediments. While I was still at the Navy Electronics Laboratory we had begun a collaborative effort of applying the techniques of soil-mechanics to field and laboratory investigations of sediments. I had learned about soil mechanics in my first professional job, which was selecting sites for earth-fill dams in the high Sierras in the summer of 1946. Ed had a bachelor's degree in civil engineering and felt right at home. We pioneered in using scuba gear to measure physical properties *in situ*. The Navy loved this work, and Ed developed and enlarged it until he became one of the world's leading experts on the subject. Consequently, his views received widespread attention among geologists when he related soil mechanics to the problem of the age of ocean basins.[4]

Hamilton began with basics. It was widely assumed that the seismic second layer was volcanic because the velocity was appropriate and the ocean basins were full of volcanic seamounts. As Russ Raitt had stated, however, all that was known was that the layer was material with a certain velocity and variable thickness. Hamilton simultaneously addressed two problems: the nature of the second layer and the paucity of sediment in ocean basins. He proposed a bold solution. If the second layer was largely consolidated sediment, there was no paucity of sediment.

The two principal types of abyssal sediment are red clay and carbonate ooze. As successive layers accumulate above them, they consolidate, and with time they alter to the rocks shale and limestone, respectively. The velocities of seismic waves in these sedimentary rocks are in the observed range in the second layer. The amount of red clay necessary to produce a given thickness of consolidated clay or shale increases with depth and can be surprisingly large. It takes 2,230 m of unconsolidated red clay to produce a layer 1,000 m thick after consolidation. Likewise 8,360 m of clay consolidates to 3,000 m.

If the second layer is largely sedimentary rock, only rates of sedimentation are needed to calculate the age of the ocean basins. Hamilton made corrections for the physical effects of coring and relatively rapid sedimentation in Pleistocene time. The indicated rates for unconsolidated sediment in pre-Pleistocene time ranged from 0.5-13 m per million years. When consolidation was taken into account, the ocean basins appeared to be hundreds of millions to billions of years old. Hess, Dietz, and Heezen would attribute the paucity of sediment to the youth of the ocean basins. Thus they could retain a volcanic origin for the second layer. For those, such as Ewing and myself, however, who were not inclined toward sea-floor spreading, Ed Hamilton's analyses eliminated the absolute necessity for young ocean basins.

TECTONIC HYPOTHESES

This was the year (1960) in which Munk and MacDonald expressed doubts about the validity of the increasingly complicated interpretations of paleomagnetic observations. It was also the year when Blackett and others published a collection of new observations and presented them ''in a way that clearly indicated the reality of continental drift.'' Marland Billings in his Presidential Address to the Geological Society of America on ''Diastrophism and Mountain Building'' expressed the general reaction at the time:

> Geologists cannot help viewing with extreme skepticism most of the conclusions reached so far from these paleomagnetic studies. For one thing, the various workers within the field do not agree on their interpretations. Moreover, if slipping of the entire crust is not sufficient to explain their observations, some investigators also invoke drifting and rotating continents. . . . This, of course, is an unbeatable combination; anything can be explained.[5]

This address was in the classical style of the decade of the fifties in that it reviewed all proposed causes of diastrophism, could eliminate none, and appeared to lead nowhere. Perhaps it was frustration with this fair, reasonable, and scholarly style that led to the more polarized papers of the sixties.

Billings concluded that

> A successful theory of diastrophism and mountain building must explain among other things (1) the horizontal compression that is essential to form belts of folded and thrust-faulted strata; (2) extensive vertical movements, with or without high-angle faulting, and unrelated to folding; and (3) extensive strike-slip faults.[6]

Dietz would publish such a theory just a year later, and within a decade most of the phenomena would be quantified. Billings' emphasis on strike-slip faults shows the remarkable change that had occurred as a consequence of global mapping in the previous decade. He noted that "it is now the fad to assign all kinds of faults and even nonexistent faults to the strike-slip category," but large displacements had occurred on many faults. Even larger ones were in the process of being discovered by Vacquier and colleagues on the Pacific fracture zones. I found this emphasis by Billings to be particularly noteworthy because I had taken courses in tectonics from J. P. Buwalda in California and then from Billings in Massachusetts. Each professor had stressed the tectonic style characteristic of his region. Buwalda talked about strike-slip faulting, and Billings talked about thrust faults and plutonic intrusions. Now, in 1960, an increasingly global viewpoint required that all types of faults be considered together.

One of the hypotheses of diastrophism discussed by Billings involved the conversion of sial to mantle or "oceanization" as proposed by the leading Russian tectonicist, V. V. Beloussov. Beloussov published two papers on tectonics in 1960 in the American journals I have monitored. His book, *Basic Problems in Tectonics*, was published in English translation in 1962, so his views were familiar to concerned geologists. He became an outspoken critic of the concepts of sea-floor spreading and plate tectonics and helped to retard tectonics in Russia by a decade or more. Those developments, however, came later. In one of the papers in 1960 he outlined an interesting idea that explained many of the same observations as other new concepts of global tectonics. The formation of the oceanic crust began in Mesozoic time, and oceanization was progressively enlarging it. The ocean basins were in different stages of oceanization; the northern Atlantic was youngest, and the Pacific oldest. Oceanic trenches and related phenomena were a consequence of imbalances at the boundary of oceanization. The prime characteristic of Beloussov's concept was that almost all tectonic phenomena resulted from direct or differential vertical movement of crustal blocks. It was a concept not inappropriate to a geologist living in Moscow on the great stable platform of Eurasia.

As to other hypotheses of global tectonics, 1960 was the year that Bruce Heezen published his paper in *Scientific American* on the global rift and suggested that it was caused by global expansion. P. H. Reitan also investigated the possibility of expansion in relation to the thermal history of the earth. He calculated that the thermal expansion was negligible; the mere existence of the paper, however, demonstrated that expansion was a legitimate subject for research. The same could be said for mantle convection. I published in *Science* on convection under the East Pacific Rise, and A. L. Licht made calculations on the rate of convection. Likewise, there were covert acceptances or rejections of continental drift. Frank Dixey believed that the gap between Africa and Madagascar was the site of a great syncline rather than of crust created by drifting. In 1944, G. G. Simpson had explained the obvious similarities between the faunas on opposite sides of this gap by a random or sweepstakes dispersal. In 1960, A. H. Cheetham believed that faunal patterns around the North Atlantic "harmonized best" with the sweepstakes hypothesis instead of with continental drift. In contrast, D. H. Swartz and D. D. Arden, Jr., proposed that the Red Sea opened by blocks moving apart, perhaps in the same way that median rifts form on mid-ocean ridges.

In short, chaos prevailed in global tectonics and everyone who was interested had his say, but few people changed their minds as a consequence. Nonetheless, some of the most important events in this year were the conversions of two critical people to continental drift. One is already a familiar figure in this book, Teddy Bullard, who was convinced by Blackett's presentation of paleomagnetic data. The other was already a major international figure in geology by 1960, but one who was little related to marine geology and who only now enters this history. Henceforth, however, John Tuzo Wilson (1908; Fig. 38) will dominate the scene.

TUZO WILSON—EARLY YEARS

Wilson was born in Ottawa and educated there at Ashbury College School before entering the University of Toronto seeking first-class honors in physics and chemistry.[7] In high school and as a freshman, however, he had obtained summer jobs in forestry and prospecting camps and discovered that he liked life in the field. After his freshman year he applied for a change in his major but was informed that it was too late to take straight geology. He was directed to take a joint major in physics and geology. Apparently this was a serendipitous directive because 20 years later he would coauthor a book called *Physics and Geology*. Meanwhile, as an undergraduate he was able to perceive that

> physicists and astronomers . . . seemed to try to understand the basic principles first and later fill in the details, whereas geologists were so obsessed with details and had so little knowledge of the interior of the earth, that they seemed to have no chance of understanding the earth if they only worked upon that and the geology of the surface.[8]

Wilson graduated in 1930, receiving the Governor General's Medal and Massey Fellowship—the first of a long series of honors that still continues. He went to Cambridge where he fell under the influence of Harold Jeffreys, who looked upon continental drift as nonsense. In the small geological community he met Teddy Bullard's first wife Margaret and heard what he was doing, but Teddy himself was away. Tuzo received an M.A. in 1932.

From 1933 to 1936 Wilson was at Princeton where he studied with Harry Hess and met Maurice Ewing. Upon receiving his doctorate he became an Assistant Geologist with the Geological Survey of Canada and worked extensively in the extremely stable continental shield. At that time he found himself in a vexing position for a young scientist about to begin publishing. The name "J. T. Wilson" was already held by James Tinley Wilson, who worked for the Geological Survey of Canada and published on both the seismology of California and the geology of Canada. Two different J. T. Wilsons, interested in physics and geology, published for a while, but then "Jack" became "J. Tuzo" by which he has been recognized professionally ever since.

In 1938 Tuzo married Isabel Jean Dickson. A year later, at age 31, he went on active duty with the Royal Canadian Engineers that was to continue until 1946. Thus, like Wegener, Dietz, and Hess, he spent much of his thirties detached from science. Can it be that the great generalizers benefit from being forcefully prevented from early and complete commitment to a specialty? Tuzo spent 1939-1943 in regimental and staff appointments in Britain and Sicily, and from 1944 to 1946 he was Director of Operational Research at the National Defense Headquarters in Ottawa. He left the army as a colonel with an O.B.E.

In 1946, age 38, he became Professor of Geophysics in the Department of Physics at his old school, the University of Toronto. He would continue as a Professor until 1974, but he would be engaged as much in the activities of scientific organizations as in research and teaching. In this, as in the war, his career would parallel that of Harry Hess. Most research scientists would lack the stamina, let alone the desire, for visiting 200 universities in more than 100 countries, but Tuzo lists such visits as a "leisure activity" along with sailing his Hong Kong junk in the Great Lakes. Can it be that the speculations of great generalizers in geology benefit from interruptions in the collection of data because of frequent flights and committee meetings? "Apart from the fact I travelled a great deal and saw people in meetings, I really rather worked in isolation," he later wrote to me (a letter dated 1 February 1984). Committee meetings, or at least the accompanying cocktail parties, provide an opportunity for airing speculations, but writing them down, as Hess and Wilson could do, requires a rare discipline.

Tuzo began his dual career in 1946 by becoming chairman of the Canadian National Committee of the IUGG (International Union of Geodesy and Geophysics) and leading Operation Musk-Ox, a 3,400-mile scientific trip through the Arctic. A few years later he began to publish on global tectonics and major lineations in the Canadian shield. He was elected a Fellow of the Royal Society of Canada in 1949. In 1950 he was a Visiting Professor at the Australian National University, and he gave the R. M. Johnston Memorial Lecture "On the Growth of Continents" in Tasmania. His work, which will be discussed shortly, had made him famous among geologists, and we began to correspond in 1955 after I sent him some reprints on fracture zones and archipelagic aprons. He was due to lecture in Los Angeles in October and proposed that he visit me in San Diego on a Saturday to look at my unpublished charts. The Naval Electronics Laboratory, however, was not a real research facility, and it closed on the weekend. I took the charts up to Los Angeles.

In 1957 Tuzo was elected to a three-year term as President of the IUGG, which spanned the great growth in geology during the IGY (International Geophysical Year). He was Past President from 1960 to 1963 and thus was intensely involved in the ferment of international meetings during a critical stage of the evolution of sea-floor spreading. From 1957 to 1963 he was a member of the National Research Council of Canada, and from 1958 to 1964 was on the Defense Research Board. In 1958-1959, as the IGY came to a close, he received honorary doctorates from Cambridge and two other universities and another medal.

Tuzo's activities are reminiscent of those of Harry Hess and so are his publications, which include many abstracts and progress reports on Canadian geophysics. Also like Hess, he apparently was sometimes overcommitted to invitations to meetings, because he published papers in Tasmania and Canada with whole pages that are almost identical. Unlike Hess, however, he also published books, namely a coauthored textbook on geology and two popular books based on his experiences as President of the IUGG, *One Chinese Moon* in 1959 and *IGY, Year of the New Moons* in 1961.

Tuzo also published a remarkable series of prophetic statements about the status and future of geology, which made him an ideal spokesman during the IGY. Even before the war he perceived that a basic new understanding of the behavior of the earth was in the wind, but it was not until a decade later that his ideas on the subject began to appear in print. From a talk in 1950:

> The methods of field geology have been adequate in order to describe the earth's surface as it is and to tell a great deal about its history, but they have not disclosed the cause of the earth's internal movements and there in no indication that they are likely to do so. In order to discover the processes that have given the earth a history, a consideration of physical laws is necessary.[9]

It was quite a statement from a man who had spent 14 consecutive summers in the field before the war. I wonder if he would have made it if he had been allowed to pursue a normal major in geology at the end of his freshman year at Toronto. He was optimistic, however, about the prospects for revelations from new data:

> More recently . . . has come available . . . a body of geophysical observations which promise to bridge the gap and make possible the eventual transfer of geology into the group of precise sciences with a sound theoretical basis. These observations include measurements in different parts of the earth of gravity, radioactivity, heat flow, location of earthquakes, age of rocks, layering of the earth's interior.[10]

That is not what was being taught in the schools I attended a few years before, but just as he gave his talk a lucky few of us were immersed in such observations.

WILSON'S ECLECTIC TECTONICS

What of the personal research on global tectonics by this visionary geologist? We might expect that it would be on the main line leading through mantle convection and a phase-change Moho toward the hypothesis of sea-floor spreading that Wilson would accept as soon as Dietz published it. That expectation, however, would be disappointed. He kept generating exciting and widely read papers, but he started with a regrettable handicap:

> a firm grounding by both the geophysicist, Harold Jeffreys, and by the North American geologists at Toronto and Princeton, coupled with my own work on the extremely stable Canadian Shield, convinced me that Continental Drift was not the answer.[11]

On this slippery grounding Tuzo built a tectonic edifice that he elaborated for a decade.[12] He began with his own research contributions to the mapping and dating of the Precambrian rocks of the Canadian shield. The core was oldest, and at increasing distances from it the provinces were progressively younger. J. D. Dana in 1849 and A. C. Lawson in 1932, among others, had noted that the Cambrian and younger rocks of North America were oldest in the interior and progressively younger toward the shoreline. They proposed that the continent was growing by marginal accretion of sediment eroded from the interior. Tuzo concluded that the growth had begun at the dawn of geological time. The continents were, in part, metamorphosed accumulations of great blocks of andesite and granite created in island arcs and, in part, sedimentary rock eroded from the growing interior. He showed from published data that the present rates of andesitic volcanism and erosion were adequate to generate the volume of continental crust above the Moho. He thereby opened the study of such rates as a subject for further research, and I, for one, would pursue it.

A key element in Wilson's synthesis was the origin of island arcs, and he sought a physically satisfactory explanation by collaborating with the mathematician A. E. Scheidegger who was also at Toronto. They developed an origin that rested upon the contraction theory as developed by Jeffreys. If the earth is cooling at the surface by conduction and radiation, the interior "core" is presently unaffected and has a constant volume. Above it is a layer that is cooling, shrinking, and therefore under tension because it rests on the core of constant volume. At the surface is the cold, solid crust that is under compression because the cooling layer below is shrinking away from it. Thus at the boundary between the layer under compression and the one under tension was a layer of no strain. This "layer of no strain" also occurs when layered rocks are folded. The concept was of great interest in structural geology. Marland Billings asked me a question on the subject in the oral qualifying examination for my doctorate at Harvard in 1948.

Given the above state of stress in the earth, Wilson and Scheidegger noted that

> according to O. Mohr's theory, conical fracture can occur and that these cones should dip at about 20° to 30° if caused by compression and dip at about 60° to 70° if caused by tension.[13]

Jeffreys put the level of no strain at roughly 100 km and the limit of cooling at 700 km. Consequently, a change in dip of the perhaps conical fault planes under the arcuate trenches and island arcs should occur at 100 km—which is just where Benioff showed it to occur in some places. The analysis looked pleasingly scientific, even mathematical, and Wilson displayed it in exactly the same illustration from 1950 to 1959. Unfortunately for his progress in global tectonics, in relation to the earth it was much ado about nothing. The hypothesis of global contraction led him astray on another score. Calculations of the effect of cooling indicated only a 50 km decrease in circumference, which was too little to be satisfactory. Wilson, however, had shown that the total volume of the continents could have been generated by volcanism derived from the interior. The base of the continents, from which the volume was calculated, was the Moho, so the Moho "represents the original surface of the earth."[14] The displacement of rock to the surface must have caused a shrinking of the circumference of the Moho by 100 km. This gave a more reasonable total of 150 km of shrinkage. "This indeed is the closest estimate which we can make at the present time to the probable behavior of the earth." That was 1959.

As the new data came in from the oceanographic ships, Wilson began to consider fracture zones and mid-ocean ridges. He consistently rejected any possibility of mantle convection or continental drift to explain any of the newly discovered phenomena. As to the mid-ocean ridge,

> The lack of abandoned ridges and the slow rate of the ridge's volcanism suggest that it has been in its present position for a very long time, perhaps most of the Earth's history.[15]

With regard to his main theme, the growth of continents, J. Tuzo Wilson was about as far as one could be in 1959 from the sea-floor spreading hypothesis he espoused in 1961. He believed that the earth was contracting, that the mid-ocean ridges were primeval lenses of basalt and fixed in position, and that the Moho was as unlike a phase change as it could possibly be. In sharp contrast, as we shall now see, he was very close to the concept of transform faults that he would not propose until 1965. During the fifties Wilson had had glimmerings, almost revelations, about a secondary theme, the junctions of tectonic systems. In 1951 and repeatedly thereafter he published and discussed an illustration of "the shearing expected where two orogenic systems intersect at right angles."[16] It is a diagram of what we now call a "triple junction." Two plates are moving at right angles to each other toward a fixed plate. He explained that the boundary between the moving plates has to be a shear zone. In 1957 he considered the interaction of mid-ocean ridges and island arcs, which by then he accepted as the two primary orogenic systems of the earth. The ridges had only shallow earthquakes. What was most significant:

> Besides these two systems of fractures, one shallow and stationary and the other deeper and migratory one would expect ancillary fractures springing from them and connecting them. Such may include graben and fault scarps such as those recently described by Menard.[17]

I don't know whether it is possible to be guilty of adumbration when considering the work of a single scientist, but it appears to me that Tuzo conceived of transform faults in that passage in 1957. He connected spreading centers with subduction zones by ancillary fractures of which the Pacific fracture zones were examples. He didn't move anything very far, but in 1957 the enormous offsets on the fracture zones remained to be discovered.

What enabled Tuzo Wilson to shift from his long-time, active espousal of the contracting earth paradigm to the concept of sea-floor spreading so quickly? Perhaps it was because he went through a transitional stage. In 1957-1958 he gave some Sigma Xi lectures founded on a contracting earth. They were published as the first article in *American Scientist* in 1959. The second article was titled "Gravitation—an enigma"; in it R. H. Dicke proposed that the earth was expanding, as discussed in Chapter 12. A year later Tuzo was publishing on the geological consequences of such an expansion (as discussed in the same chapter). He found that he could explain global tectonics about as well with an expanding as with a contracting earth, and if both of them why not something else? He still was dubious about convection currents, although he noted that they could explain the high oceanic heat flow. Even continental drift on a small scale appeared likely. He also went another step toward the idea of transform faults.

> If [the East Pacific Rise] is widening . . . and if . . . there is an existing line of fracture due to buckling, then a northwest shearing motion of the Pacific Ocean relative to the Americas could take place. . . . Similar motions in the past could account for other shifts, such as those along the Great Glen fault or the Murray scarp.[18]

Tuzo was moving even faster than the publication date indicates. "About 1960 I became a convert to continental drift," he wrote in 1972.[19] In March 1963 he wrote to me,

A year or so ago I noticed that there is a striking resemblance between the Great Glen Fault in Scotland and a system of faults which I found I could trace in the literature from northern Newfoundland to Boston.

This aroused my interest in continental drift. . . .

Apparently that was the triggering event, but it may have occurred a bit earlier than he casually recalled. It must have been before sea-floor spreading or he would have said it was after. Whatever the exact date of his conversion, it was the most important event in global tectonics at the time.

Chapter 15

1961-1962 THE REVOLUTION BEGINS

Half a century after Wegener's opening salvo, the second phase of the revolution in the earth sciences began quietly in 1961. Bob Dietz published his sea-floor spreading paper in *Nature* in June, and there was some follow-up in that journal later in the year. J. D. Bernal commented on 14 October, Dietz responded.[1] Bernal rebutted, and Tuzo Wilson identified corollaries and expanded on Dietz's hypothesis. Those were the only publications regarding the new hypothesis that would ultimately carry everything before it. The existence of any follow-up publication at all, especially in such a journal and by such famous names as Bernal and Wilson, was enough to ensure that sea-floor spreading had an exceptional chance of attracting attention. The places where the hypothesis was of most immediate concern were the oceanographic laboratories that had the most data—Lamont and Scripps. Thus it is reasonable to begin to trace the development of the sea-floor spreading revolution in those laboratories.

TECTONICS AT SCRIPPS

To begin with Scripps, nine days before Dietz's paper was due to be published. I wrote (in a letter dated 25 May 1961) to Harry Hess about sea-floor spreading. The letter, however, was wholly about Hess's manuscript, which would be long in the press, and did not use the term "sea-floor spreading" or refer to Dietz's work. It must have been one of the first occasions in which Dietz's paper was effectively ignored by the geological establishment in which I include myself (although I certainly did not have such a perception at the time). Those who knew about Hess's manuscript before Dietz's paper appeared generally cited Hess rather than Dietz. Scientists who were less closely associated with marine geology or global tectonics tended to cite Dietz on the normal grounds of priority of publication. In any event, reading my own letter I find I ignored Dietz's manuscript.

I had much to say about Hess's ideas, particularly about observations that I thought were pertinent but that he had neglected to explain or had misinterpreted. Among them were the following:

1. His ideas about the occurrence of guyots in tropic waters would not explain why guyots and atolls were intermixed in three regions—of which only one was known to him. The intermixing, in fact, would not be explained by ridge-crest volcanism as he proposed. It requires, in modern terms, overplating by mid-plate volcanism.

2. Like Wilson, Hess had some misconception that mid-ocean ridges are of different widths that are unrelated to the total width of an ocean basin. He did not realize that contour maps can be very misleading in this regard.

3. He did not explain the observed variation in thickness of the oceanic lithosphere. I attached great importance to this point because I could explain the variations, but they turned out to have no significant bearing on the hypothesis.

4. He did not explain why the fracture zones had been active at the same time as the mid-ocean ridges, or at least the East Pacific Rise.

5. He did not explain the observation that the East Pacific Rise ended at about 45°S. I had mapped a very large offset of the ridge crest along what is now called the Eltanin fracture zone. I took it that the Pacific-Antarctic Ridge, at the other end of the fracture zone, was an old ridge whereas the East Pacific Rise was younger. Hess also believed that mid-ocean ridges had different ages, which was correct. In this case, however, the important point was that there was no continuous world-girdling rift as publicized by Ewing and Heezen. In the South Pacific there were two rifts connected by an enormous fracture zone. Thus had Hess wished to elaborate his hypothesis, he would have been confronted with the need for a ridge-ridge transform fault in 1961. In his response (a letter dated 12 June), however, he was "not quite prepared to accept your views but do not have a solution of my own" regarding the great fracture zones.

Harry did not mention Dietz's paper either, although *Nature* should just have reached Princeton before he wrote to me. Presumably, there was no immediate outcry in Guyot Hall. It is a curious fact that none of my objections to Hess's version of sea-floor spreading applied to Dietz's version. Bob included fracture zones, variations in crustal thickness, and everything else I had published in my 1960 paper on the East Pacific Rise. Moreover, the origin of guyots was not central to his hypothesis. Long before, he had guessed that they were Cretaceous islands, and he quit guessing about them while he was winning.

What of other responses to sea-floor spreading at Scripps in 1961? I cannot remember a thing. Sea-floor spreading was just one more hypothesis, like the expanding earth, or mantle convection.

TECTONICS AT LAMONT

Turning to Lamont, it appears that responses to sea-floor spreading varied widely. Bruce Heezen wrote to me on 2 February 1961 and did not mention it even though he would have had a copy of Hess's preprint. His letter reached me when I returned from the Monsoon expedition from New Zealand to Tahiti by way of the Antarctic ice. I responded on 4 April to describe our discoveries on the Pacific-Antarctic Ridge. I was more encouraging about the existence of a median trough than I had been when I wrote to him after Downwind. I wrote not a word about sea-floor spreading. He wrote in August, I in October, then he in December. We mentioned soundings, charts, Roger Revelle's departure for the Department of the Interior, our separate meetings with Gleb Udintsev, Heezen's discovery—at last—of fracture zones in the Atlantic, my article on the East Pacific Rise in *Scientific American*, and plans for future exploration. Not a word about sea-floor spreading.

Another indication of response to Dietz's paper is suggested by the memories of two newly arrived students in 1961.[2] Walter Pitman was a Lamont technician from December 1960 to September 1961 when he became a graduate student. Denny Hayes also registered at Lamont in 1961. Neither of them remembers any discussion, let alone excitement, about the paper. Both are reasonably certain that they were unaware of sea-floor spreading for some years, and Pitman did not actually read Dietz's paper until after he had discovered the symmetry of magnetic anomalies in the Eltanin 19 profile in December 1965. At Lamont, it appears that for most people sea-floor spreading initially was just another unproved and unprovable hypothesis.

The Director of Lamont came to be identified as a scientist whose mind was made up on the issue of continental drift.

"Ewing was an avid anti-drifter and the leader against mobility, both intellectually and emotionally. Jim Heirtzler and Xavier Le Pichon were similarly disposed," recalled Neil Opdyke.

Walter Pitman, too, remarked that "Ewing, Heirtzler, and Le Pichon did not believe in drift and sea-floor spreading."[3]

These were memories as of 1979. Opdyke came to Lamont in 1964; Pitman was unfamiliar with Dietz's paper until even later.

At least two scientists, who had then been at Lamont for several years, still recall Maurice Ewing's reaction soon after Dietz's paper appeared in 1961. It has been some time and on one point their memories differ about what occurred, but there is no question at all about Maurice Ewing's actions or reactions. Jim Heirtzler remembers that M. Ewing called him in to have a discussion with John Ewing as a third party. Manik Talwani remembers exactly the same thing, only about himself. John Ewing recalls no such meetings, although the ideas credited to his brother are those he does remember.[4] Can it be that the three people who met were Maurice Ewing, Heirtzler, and Talwani? At the meeting it was agreed that Dietz's paper was important and significant. Ewing stated that he thought sea-floor spreading was probably right and wished he had thought of it. Lamont, however, like Scripps, was organized to investigate problems rather than to take a stand based on available data. Ewing decided to direct a large portion of Lamont's efforts toward testing the hypothesis in the field. The ideal testing device for Lamont was the seismic reflection profiler, which had been brought to a new level of efficiency by John Ewing in 1960. The thickness of pelagic sediment should generally increase with the age of the underlying crust. If the sea floor was spreading, the reflection profiler would discover that sediment was thin or absent on the mid-ocean ridge crests and thickened toward the flanks. It would be jumping ahead of our chronology to describe at this point what happened in the next few years at Lamont as a result of Ewing's immediate decision to test Dietz's hypothesis. Suffice it to say that *Vema* and later *Conrad* began to course the world ocean like hounds on the scent.

This seems a sensible place to introduce one of the most important of the new men around Maurice Ewing at Lamont, his younger brother John. John I. Ewing (1924–) was born in Lockney, Texas, like Maurice. He worked at Woods Hole with Maurice in 1941 when he was only 17 and then served in the Army Air Force until 1945. I first met him when he was a sophomore at Harvard in 1948. I was a teaching fellow and had to approve his application for late admission to the introductory course in geology. I asked him if he was related to the Ewing whom I had seen on *Atlantis* at Woods Hole. "Brother," he said. I allowed as how we could squeeze him in. He graduated in physics in 1950, moved to Lamont, and never bothered to receive a doctorate. In this he was at an extreme, only approached by such student leaders of expeditions as Bruce Heezen and Bob Fisher. John spent about three months at sea per year for the next 20 years and became head of the marine seismology group in the mid-fifties. In that capacity he developed the seismic profiler system that was the marvel of the time.

REACTION TO SEA-FLOOR SPREADING

So the oceanographers continued their eternal exploration with the correct expectation that more data would be needed to understand how the world works. Meanwhile, what of the discussions ashore? What was the published reaction to sea-floor spreading? J. D. Bernal was one of the most famous scientists in England as well as an outspoken Marxist and scientific historian. Thus his reaction to sea-floor spreading was based on an exceptional breadth of knowledge. He characterized Dietz's article as

a marked contribution to what might be called the oceanographic revolution in geotectonics. One gets the impression that with it a stage has been reached like the last but one in fitting together a jigsaw puzzle, with the various parts of the puzzle having been laboriously assembled, piece by piece, at last seeming to look as if they are all part of one single picture.[5]

Dietz had effectively combined the recent discoveries of "mid-oceanic rift ridges and transcurrent faults" with older observations. Bernal, however, saw two major difficulties. First, deep-focus earthquakes implied rigidity to 800 km, so the upper mantle must drift with the lithosphere. This would require an embarrassingly large flow of material toward the spreading rift. Second, Bernal believed, along with Tuzo Wilson, that continents grow around their margins. This was not easily reconciled with Dietz's idea of underplating of continents along Benioff zones.

Dietz was

pleased to learn that [Prof. Bernal] finds the concept of ocean floor spreading, marking the top of mantle convection cells, at least generally palatable.[6]

He believed that Bernal's objections were surmountable. Deep-focus earthquakes were not proof of long-term strength, and Tuzo Wilson's annular growth of continents was more of a concept than a fact.

The next, lengthier, note was by Wilson himself. Like Bernal, he accepted the basic hypothesis of sea-floor spreading, but then he drew on his great knowledge of global geology to expand and develop Dietz's ideas. He began by citing Hess's manuscript as well as Dietz's published paper and said that they made "similar proposals" about the existence and consequences of mantle convection rising and spreading under mid-ocean ridges. Wilson then addressed the fact that the median rift nearly circles Antarctica and Africa. It was this fact that caused Carey and Heezen to call for an expanding earth. Not at all, wrote Wilson; concerning Antarctica:

To escape the dilemma one must suppose that, instead of a current flowing south from the ridge, the ridge itself is migrating northwards and that the ring which it forms around most of Antarctica is expanding in radius. If so, the current flowing northwards on the other side of the ridge is moving at twice the normal rate. . . .[7]

The migrating ridge would sweep the continents before it and thereby produce the northerly drift recorded by paleomagnetism. The ridge around Africa was also migrating, and in the Pacific two ridges were migrating without an intervening trench. This caused the very rapid movement on the San Andreas fault. Thus Tuzo first explained in terms of convection many of the phenomena that would enter into plate tectonics. He also concluded from the arguments that Hess and I had advanced that convection patterns could change.

Despite his considerable elaboration of the sea-floor spreading hypothesis,

Tuzo did not find it necessary to abandon every aspect of the global tectonics that he had previously espoused. The shapes of island arcs still needed explanation. "Thus some features of the contraction theory are preserved." Sea-floor spreading would sweep island arcs into continents, so continents would grow at their margins. Another fact that needed explaining was the persistent freeboard of the continents.

> Another possibility is that the earth is slowly expanding. A rate of a few mm/yr would suffice to maintain continents above sea level. This would not interfere with and indeed might assist the other orogenic processes described.[8]

Tuzo Wilson had made some brilliant elaborations of sea-floor spreading and, for the moment, managed to incorporate all fundamental hypotheses of global tectonics into a single synthesis. That was the last paper on sea-floor spreading for 1961. The year, however, did not lack for other, less eclectic, papers on global tectonics. Heezen and Ewing published about the mid-ocean ridge in the Arctic and came to their parting of the ways regarding the origin of the ridge. Heezen, of course, was in favor of an expanding earth, and

> both writers believe that the Arctic Basin between the Lomonosov Ridge and the Barents Shelf is currently growing wider and that a mid-oceanic ridge is growing along the line of parting marked by the epicenter belt.[9]

I published an updated version of the sequential-development hypothesis for the East Pacific Rise. Vacquier, Mason, and Raff finally published three notes on the magnetic anomalies of California and the offsets on the great fracture zones. Fisher and Raitt published on the structure of the Middle America Trench and eliminated the tectogene hypothesis. Talwani and others published a description of the structure of the Mid-Atlantic Ridge based largely on gravity. There were many papers on the crustal structure of continental regions. Beloussov published two more papers on basaltification and vertical tectonics in American journals, and Irving and others published several papers on paleomagnetism.

One paper in 1961 deserves mention on two grounds. It was the first of a notable sequence of papers on continental drift by Warren Hamilton, and it may have been the first paper "authorized by the Director, U.S. Geological Survey" that advocated drift.[10] Hamilton took the new geological data on the offset of the San Andreas fault in California and showed that the offset could have opened the Gulf of California by 300 miles. Wegener and Carey had already proposed such an origin for the gulf without knowing the offset on the San Andreas, and in retrospect, Hamilton believes that he should have given more credit to Carey.[11] Nonetheless, it was a radical paper for an American journal and is as good an indication as any that ideas were changing.

DIETZ'S ELABORATIONS

It is difficult to try to unscramble the development of the concept of sea-floor spreading between June 1961 and the end of 1962 because of overlapping meetings and publication delays. In any event, it is probably pointless because all the pertinent publications were by Dietz, except for the long-delayed paper by Hess. Other important developments occurred in global tectonics, but in 1962 the only elaboration of sea-floor spreading was in a burst of, largely, invited papers by Dietz. Criticism began to appear, but no one else joined the bandwagon started by Bernal and Wilson in 1961. In many ways, 1962 was a year of publicizing rather than developing sea-floor spreading. It was Harry Hess who would win the Penrose Medal and the Feltrinelli Prize four years later for his "History of Ocean Basins" published in 1962. It was Dietz, however, who not only conceived of a superior hypothesis of sea-floor spreading but also carried out the essential step of advertising it.

The papers in *Nature* 1961 generated enough interest to merit a popular account, and Dietz published one in the *Saturday Evening Post* for 21 October 1961. In 1962 he republished his basic concept in an abstract at the Pacific Science Congress in Hawaii, in a commemorative volume in Japan, and in a book titled *Continental Drift*, edited by Keith Runcorn. He also was invited to give his ideas wide circulation as a Distinguished Lecturer of the American Association of Petroleum Geologists. On and on he went for most of a decade, first a period of triumph and acclaim, then a spirited defense against the embattled skeptics. Bob expresses surprise that so many young geologists clearly recognize his name when he meets them. It would be a wonder if they did not.

In addition to the repetition typical of invited papers, Dietz's three papers in 1962 contain a few new points.[12] First, he added his note in proof giving priority for the idea of sea-floor spreading to Hess. Second, he made the mistake of accepting Hess's idea that the oceanic crust is partially serpentinized ultrabasic rock from the mantle. In 1961, Dietz had emphasized that sea-floor spreading would work regardless of the nature of the crust, provided the mantle and crust were interchangeable, but he had preferred a basalt-eclogite phase change at the Moho. In 1962 he changed his mind because he was beginning to integrate his research on continental margins with the concept of sea-floor spreading. He distinguished Atlantic or passive continental margins and Pacific or active margins. The distinction went back to Edward Suess, but Dietz recognized that a passive margin meant that a continent was moving with the sea floor, and an active margin meant that the oceanic crust was plunging under the continent.

Geologists had developed an elaborate classification of geosynclines depending on such characteristics as their shape and the occurrence of volcanic rocks. "Eugeosynclines" were long and narrow with abundant volcanics, and they commonly became deformed into mountain ranges. Eugeosynclines were often paired with parallel "miogeosynclines" that lacked volcanic rocks and were less deformed. Dietz accepted Tuzo Wilson's concept that continents grow at the margins, but this growth occurred

> by the plastering of eugeosynclinal prisms against the continental block. But unlike the usually held belief that eugeosynclines are developed on the continent, the writer supposes that they are formed off the continent and marginal to it at the base of the continental slope. The eugeosynclinal prisms are thus equatable actualistically with the continental rises now being laid down by turbidity currents which carry down material eroded from the continental block.[13]

The adjacent miogeosynclines were equivalent to the thick modern wedges of sedimentary rock under the eastern continental shelf of the United States.

A passive margin could, indeed probably would, develop a miogeosyncline and eugeosyncline, but it had to become active to transform them into mountains. At active margins Dietz visualized the formation of island arcs that might eventually be added to a continent by continental drift. Seamounts would also be swept into continents. So would pelagic sediment and the volcanic and oceanic layers of the oceanic crust. Such distinctive materials from the deep sea would be intermixed with the quite different continental rocks and the sediments derived from them. Alpine-type mountains built of geosynclinal rocks were already known to be characterized by three exotic materials: serpentines, partially metamorphosed basalts, and radiolarian cherts. They were "portions of the oceanic rind caught up in the orogenic folding." The radiolarian cherts and the basalts were expected components of the sedimentary and volcanic layers regardless of the nature of the oceanic crust. The extensive intrusions of serpentine, however, required a voluminous source. Therefore Dietz came to accept an oceanic crust composed of partially serpentinized peridotite. It was a pretty example of the continuity of a scientist's research. Hess had been thinking about serpentine since he was a graduate student, and Dietz had been concerned with the origin of the continental terrace for at least 15 years.

Dietz did at least one more thing that one can easily conceive of as significant to the future of the geological revolution. In *Continental Drift* he titled the last section of his paper "Spreading and Magnetic Anomalies." In the following year, Vine and Matthews would publish their inspired "Magnetic Anomalies over Oceanic Ridges," and Morley would try to publish his similar paper.

CRITICISM

Bruce Heezen's paper in *Continental Drift* contained the first published criticisms of the sea-floor spreading concept. The paper was a broad synthesis of discoveries in marine geology and geophysics that was reminiscent of the sea-floor spreading papers by Dietz and Hess and my papers on the East Pacific Rise.[14] Coming later, it naturally exposed weaknesses in previous syntheses because its author came to a different conclusion:

> It is just possible that the evidence for continental drift could instead be interpreted as evidence for continental displacement (without drift) due to internal expansion of the earth accompanied by the growth of simatic oceans through emplacement of mantle material in the floors of the mid-oceanic rifts.[15]

The basic difference between the expansion hypothesis and the others is that it eliminates the need for crustal plates to converge. For this reason Heezen focused on apparent aspects of convergence that he believed could not be explained if they were real.

He began with the pattern of the mid-ocean ridge, which

> when examined on a globe presents a formidable difficulty to conventional continental drift, for the Ridge tends to get in the way of the drifting continents.[16]

He considered the problems (apparently) posed by the fact that the ridge virtually encircles Africa and Antarctica and reduced the implications to the absurd:

> if one considers the drift of Antarctica relative to the Mid-Atlantic, Mid-Indian and Easter Island portions of the ridge, one must only conclude that Antarctica has shrunk, for the pattern of the Ridge would indicate that Antarctica must have drifted towards its geographical center.[17]

This striking argument does not easily lose its appeal. Consider, for example, S. Warren Carey in 1976:

> But what about Antarctica? Its encompassing rift surrounds it on all sides enclosing more than twice the area of Antarctica, again Quaternary at the rift, giving place in turn to the successive epochs of the Tertiary, with still a wide zone to have been formed during the Mesozoic. . . . Somewhere within Antarctica the theory demands a central sink, which has swallowed an area equal to the area of Antarctica. Where is it? It does not exist. Escape may have been possible for Africa alone, but not for Africa and Antarctica. The plate theory is false.[18]

A pity that Heezen and those who followed did not absorb Tuzo Wilson's paper on sea-floor spreading in *Nature* in 1961. Heezen read the adjacent paper by Bernal because he cites it. If he had read and remembered Wilson he would have realized that within the developing theory, no problem existed. The spreading rift merely drifted away from Antarctica. This may not have been easy to accept because, in the model of Dietz and Hess, the giant convection cells in the mantle would have to drift with the rift. I was among the majority who believed at the time that this was a major weakness in the convection-driven sea-floor spreading hypothesis. It was only after the hypothesis was generally accepted in 1966 that this apparent problem was solved. If the spreading was a fact, the spreading center *had* to drift, and the problem shifted to how the convection cells went with it or, a startling thought, whether the cells really existed. Meanwhile, there was Wilson's alternative explanation for the distribution of mid-ocean ridges that did not require an expanding earth. The most troublesome paradoxes of the old paradigm become the obvious corollaries of the new one. It is a pity also that Bruce did not have access to the seismic profiler data that were pouring into Lamont. As we shall see, they would have strongly reinforced the hypothesis of an expanding earth.

Heezen went on to discuss what he perceived as the principal weakness of the convection hypothesis for the origin of ridges, trenches, and fracture zones as proposed by Dietz, Hess, and myself. Once again the geometry of the convection cells seemed improbable. He also questioned the existence of the zone of compression that I had proposed would take up the stretching at the ridge crest. These were quite reasonable criticisms.

Few other people ventured to compare or evaluate the competing concepts in 1962. Geologists just went on developing the different hypotheses that each found attractive. In *Continental Drift* Runcorn and Opdyke further developed the intricacies of drift as determined by paleomagnetism and paleoclimatology. Vening Meinesz elaborated on new evidence of mantle convection derived from a spherical harmonic analysis of the topography of the earth. Chamalaun and Roberts discussed the mathematical stability problem of convection in spherical shells. Dietz spread, Heezen expanded, and confusion reigned.

BASALTIFICATION

Most distant from other ideas on global tectonics were those of Vladimir Vladimirovich Beloussov (Fig. 39), Corresponding Member of the Academy of Sciences of the USSR, and President of the International Union of Geodesy and Geophysics since 1960 when he succeeded Tuzo Wilson. We have noted his frequent publications in American journals in the early 1960s, and in 1962 the English translation of his great work, *Basic Problems in Geotectonics*, appeared. It was grand in scope, based on extensive reading and his personal mapping in the high mountains of central Asia, and espoused a fundamental belief that all of tectonics was a unified consequence of vertical motion. Even in its time it seemed an alien work with strange reasoning based on exotic examples and a philosophy far from the thinking of most of the people in this book. It was, however, logical, important, and influential, and much of what seemed strange about it derived from Beloussov's opposition to what Sam Carey called the "English-language obsession" with the role of compression in orogenesis:

> The Russians and many European geologists have long recognized that orogenesis is a diapiric, gravity-driven process, in a dilative environment, in which the upper part of the rising tumour spreads laterally under its weight. . . .[19]

This was Beloussov's thesis. Geologists marked the vertical oscillation of continents by the occurrence of rhythmic successions of deposition underwater and erosion above. From these motions and the associated orogenesis, both the history and the cause of geotectonics could be established.

Beloussov paid his respects to other hypotheses but found them wanting. The contraction theory did not explain the oscillatory movements and their association with orogeny, and it "hardly explains the familiar periodicity of the epochs of folding." It was rejected on these "purely geologic considerations." It was fully a dozen years earlier that James Gilluly had published his address as retiring president of the Geological Society of America. He said that he would have titled it "To question the periodicity of diastrophism" had F. P. Shepard not long since preempted that title. Periodicity was central to Beloussov's hypothesis, but it was discredited by others.

Beloussov turned to the "hypothesis of isostasy." Forty years after Wegener had recognized that isostasy provided the key to global tectonics, it was still merely a hypothesis to be rejected:

> In regard to folding, the isostasy hypothesis repeats and augments the mistakes of the contraction hypothesis: if the earth's crust had been so pliable as to warp easily under a load of sediments, it could never have transmitted tangential stresses over long distances. . . . Furthermore, this hypothesis separates folding and oscillatory movements, attributing them to different causes and thereby destroying the unity of the tectonic process, which, however, is evident from all observations.[20]

If not contraction or isostasy, what was left at the beginning of the twentieth century?

> The failure of these geotectonic hypotheses, especially the contraction theory, brought forth, chiefly in the period of 1910-1930, a flood of various hypotheses, the vast majority of which must be regarded as fantastic and having nothing to do with science.[21]

At the crest of the flood was the hypothesis of continental drift, which seems to draw such language from its critics. Beloussov was resident in Leningrad in 1941 when von Loeb's Army Group North punched up from Poland to the outskirts of the city and the horrors of the siege began. He developed his geotec- tonics in 1942-1943 in Moscow. It is a credit to the claim that science knows no nationalities that Beloussov, unlike Bailey Willis, demolishes continental drift on the grounds of evidence and logic rather than because it was a German fairy tale. After a scholarly analysis, however,

> one perceives even more clearly the total vacuousness and sterility of the hy- pothesis. Oscillatory movements are the fundamental type of tectonic move- ment. Where are the movements in Wegener's hypothesis?[22]

Beloussov considered several other hypotheses including mantle convec- tion, which he thought interesting and useful, but being ''based only on geo- physical considerations,'' it was far from reality. Thus he eliminated compet- ing hypotheses for various reasons but chiefly because they did not explain the unity of tectonics that he perceived. If there was no such unity, however, the whole structure of his criticism collapsed. He noted that

> M. A. Usov pointed out (1936-40) that the attempts at geotectonic general- izations outside the Soviet Union verged on metaphysics.[23]

Perhaps this was a nationalistic view, but I prefer to take it as an easy way to describe two quite different theories. In any event, it is clear that if it is applied to the two theories, Usov's statement works both ways.

Beloussov's hypothesis of the evolution of ocean basins derived from his basic concept of unity in geotectonics, but it could have stood on its own as a competitor with sea-floor spreading in 1962. He observed that there were no pelagic sediments, typical of the deep ocean, on the continents. This would shortly be subject to reinterpretation by Dietz, but it was the standard view at the time. Beloussov concluded that there had never been deep oceans ''on the present site of the continents.'' In contrast, geologists had found that many of the margins of continents include sedimentary rock that was obviously eroded and transported from the present sites of deep oceans. Likewise, Permian gla- ciers had flowed onto the southern continents from what are now deep-ocean basins. Wegener had interpreted these facts as evidence of continental drift, but he had not known of modern discoveries in marine geology.

Darwin's theory of the origin of atolls had been proved by drilling, the dis- covery of guyots by Hess, and the dredge hauls of the Midpac expedition. Clearly large areas of the sea floor were sinking. Darwin, who only deduced the existence of guyots, believed that some regions sank and some were ele- vated. By 1960, however, guyots were known to be very widely distributed, and Beloussov concluded quite correctly that the whole sea floor was sinking. The continents were permanent features and the deep-sea floor was sinking; what of the boundary between them? The most recent drilling and geophysical studies by Ewing and colleagues confirmed the geological observation that sedimentary formations thicken outward from land under some continental margins. The formations were deposited in shallow water, therefore the boundary between continents and ocean basins was sinking. Moreover, as the foremost authority, Shepard, had been saying for decades, the great submarine canyons of the continental slope had been cut by rivers and then submerged. Beloussov integrated all these observations related to ocean basins and concluded logically,

> There is reason to believe that the surface of the earth in earlier times contained no deep oceans and that their place was occupied by shallow marine basins smaller in area than the present oceans. With the passage of time the gradual subsidence of the earth's crust, occupying progressively greater areas and not being matched by deposition of sediments, led to the formation of the ocean basins.[24]

The geological evidence that Wegener interpreted in terms of continental drift showed that the subsidence had occurred chiefly in Mesozoic and Cenozoic time. Its cause was "basaltification" by massive intrusions and extrusions of dense basalt that pulled down the former continents. The former continents or identifiable fragments, however, still existed.

> The vestiges of both geosynclines and platforms may be found, submerged and basalticized, at the bottom of the oceans.[25]

Beloussov's hypothesis was quite unlike sea-floor spreading, but it explained many of the same facts and some others, such as submarine canyons, as well. Thus it was only reasonable that if, or when, a challenge to sea-floor spreading should arise it would come from Beloussov. The defender, however, would not be Dietz or Hess but J. Tuzo Wilson, and the debate would pair two men who were remarkably evenly matched, being presidents of the international organization that united geophysicists, equally famous, and even the same age—but the event would occur after the time limit of this book.

MORE IDEAS ON TECTONICS

In 1962, Tuzo Wilson published another note in *Nature* that was mainly concerned with the existence of a great wrench fault in eastern North America. He also expanded his concept of continental drift and sea-floor spreading. He noted that proponents of drift had weakened their arguments by adhering rigidly to particular paleogeographic reconstructions, whereas,

> If . . . they had assumed that continents and fragments of continents had always had a random distribution and motion, now joining together in one

place and now rifting and separating elsewhere, they could have explained the good fits of some coasts without the necessity of introducing hypothetical and unnecessary distortions.[26]

There was the kernel of the Wilson cycle and the accretion of small geological terraines. Wilson also considered the whole global distribution of ridges and continental mountains for the first time. He abandoned the mystical "T-shaped" distribution of modern mountain systems that had intrigued him for decades and made two hemispherical plots of "continental and mid-ocean mountain systems in early Tertiary time showing directions in which sub-crustal currents may have moved." The implications of this figure would concern him for some time. This figure may be taken as symbolic of a widespread search for patterns of mantle convection. With convection having been accepted to explain global tectonics, it was only reasonable to try to map it. Vening Meinesz, Runcorn, and others began to study the geology and geophysics of the whole globe in ways that would be increasingly fruitful as satellites started to collect global data.

The geological journals were exceptionally free of speculation or even pertinent data on tectonics in 1962. This might be attributed to doubt and indecision generated by the spectacular concept of sea-floor spreading the year before. Considering the normal delay in publication, it seems more likely that speculation based on some kinds of observations was beginning to appear futile. An article appeared on the paleogeographic development of South America that certainly justified a discussion of continental drift, but it contained not a word on the subject. In five journals for the year there was only one paper on the mechanism of continental drift and another on polar wandering.

Three papers in other journals attempted to extend fracture zones and median rifts into continents. Gilliland believed that he could trace the Mendocino fracture zone through or under the North American continent along the 40th parallel. I had proposed that the southern boundary of the Cascade volcanoes might mark such an extension. He found geological and geophysical anomalies, including the extraordinary east-west trend of the Uinta Mountains, all the way across the continent. If the extension was real, the fracture zone was Precambrian because that was the age of some of the anomalies.

Ronald Girdler analyzed the initial breakup of continents over newly created convective divergences. Keith Runcorn had indicated that if the mantle convected and the core grew with time, the pattern of convection would change. With time, the decreasing depth of the mantle would cause ever smaller convection cells to arise. Some divergence would thus be expected under continents from time to time. Girdler identified two: the Red Sea and the Gulf of California. He showed that, although narrow, they had the geophysical characteristics—oceanic crust, high heat flow, seismicity—of ridge crests. This analysis was a great step toward understanding the initial opening of older ocean basins.

The third paper, by K. L. Cook, was concerned with the significance of the low velocities in the upper mantle under the crests of mid-ocean ridges. Ewing had proposed that they were a consequence of a physical mixture of mantle and crustal rocks. I had conceived that they were a result of heating or alteration of normal mantle. Cook analyzed newly acquired data and opted for a crust-mantle mix. Ewing and Heezen had traced their rift into the western states by seismicity. I had proposed that the East Pacific Rise extends under western North America because the region is elevated and the rise extended into the Gulf of California and reappeared off Oregon and Washington. Cook and his associates at the University of Utah had been investigating the gravity and structure of the western states. He could show that the region had all the known characteristics of the East Pacific Rise with a granitic continental crust on top. It was uplifted, rifted, block faulted, seismically active, and had high heat flow, profuse volcanism, and abnormally low velocities in the upper mantle. I, at least, found this very persuasive evidence that the rise did extend under western North America, and thus the mantle convection that caused the rise did so well. Cook prepared an excellent illustration of the convection and its tectonic effects in the western states, and he presented his findings in several professional meetings. The idea that oceanic structures were influencing American continental geology was becoming widely disseminated.

Two highly significant papers appeared that were concerned with magnetic observations unrelated to continental drift. In the first of these, R. D. Adams and D. A. Christoffel published the results of their surveys between New Zealand and Antarctica. They recognized that they probably had found linear magnetic anomalies similar to those surveyed by Mason and Raff off California. They were, in fact, an older sequence of anomalies that were west of the *Pioneer* survey, but the widespread occurrence of such linear anomalies was established.

The second paper of note was by Cox and Doell who last appeared in this book two years before when they questioned the paleomagnetic evidence for continental drift. They published a description of the magnetic properties of basalt recovered from the Mohole test east of Guadalupe Island. The test hole was deliberately sited over a linear magnetic high, so the results will be reserved for the discussion of the Vine-Matthews hypothesis in the chapter after next.

Chapter 16

1963-1964
INCREASING TENSION

Certainly the most important geological hypothesis to emerge in 1963 was that magnetic reversals are recorded by the cooling lavas at a spreading center. That fact, however, was not appreciated at the time, and the idea languished—largely ridiculed or ignored except in one visionary paper by George Backus. On the other hand there was a continuing development of the broader hypotheses of global tectonics in 1963-1964, and they are the subject of this chapter.

The chaotic state of global tectonics prevailed through 1963-1964, becoming, if anything, even more polarized than before. Lacking critical new data, and lacking appreciation of a critical new idea, geologists continued to talk past each other. Very likely they would have done so for decades, but the end was in sight. During the next year (1965) Vine and Wilson would realize that old data were quite adequate to prove that the disdained new idea was correct. Meanwhile, geologists were busy developing ideas and implications related to sea-floor spreading, the sequential evolution of mid-ocean ridges, global expansion, and continental drift. Meanwhile, also, a powerful new geophysical argument appeared to eliminate the possibility of drift or spreading. It seemed as though the events of the 1920s were to be repeated. Just when the field evidence for global mobility seemed overwhelming, a brilliant young geophysicist emerged to say it could not be so. The latter-day Harold Jeffreys was Gordon J. F. MacDonald.

SEA-FLOOR SPREADING

Dietz, Hess, and Wilson were the only geologists converted to sea-floor spreading and publishing about it at the beginning of 1963. All published more on the subject in the next two years, but hardly any followers appeared as a consequence of their efforts. In the three years after it appeared, the concept of sea-floor spreading seemed to be just another wild idea. Nonetheless, it was not known to be wrong, and the converts felt justified in elaborating it. R. L. Fisher and Hess published a paper on trenches in *The Sea* Volume 3 (1963), that they had submitted in May 1961. Therefore it was written just when Hess conceived of sea-floor spreading and was a corollary of his original idea, quite unrelated to the later ideas of Dietz and Wilson. Perhaps Hess conceived of this trench paper and his "Evolution of Ocean Basins" as complementary. They had been intended to appear in *The Sea* together, and he gave few details about how he would dispose of the oceanic crust in the "evolution" paper. Fisher and Hess accepted a standard Hessian crust with sediment and volcanic debris over partially serpentinized mantle (crust) over peridotitic mantle. This assemblage plunged into the mantle where it was subjected to increasing temperature and pressure (Fig. 16 E). At the 500°C isotherm the crust was deserpentinized, and the hot water released there began to rise. The peridotitic mantle, sediment, and volcanics continued to plunge until the last two components were "partially fused" and melted. The melt rose and reacted with the water of deserpentinization to form the distinctive volcanic rocks of island arcs.

Volcanic arcs are always on the concave side of arcuate trenches, and this distribution causes no problems provided that the plunging plate follows the dipping Benioff zone. If so, the dewatering and melting are inevitably under the concave side of the arc. Regrettably, Fisher and Hess had the misfortune to write their paper at the only time for several decades when the significance of the Benioff zone was in question. J. H. Hodgson had vastly improved the consistency of analyses of the first motion of earthquakes.[1] According to his interpretation at the time, the great earthquakes of the Tonga and Peru-Chile regions were on vertical faults and the motion was strike-slip. The faults were either parallel to the trench or perpendicular. It appeared that there was no confirmation of dip-slip along a sloping plane as conceived by Benioff (Fig. 15 B). Fisher and Hess therefore proposed that mantle convection cells converged at a trench and the crust plunged vertically between them. It was almost exactly the same idea (and illustrative diagram) as Vening Meinesz's concept of the tectogene except that the crust vanished downward instead of merely bending. With vertical plunging, however, Fisher and Hess had to explain why the volcanic arc formed on only one instead of both sides of the trench. They appealed to the explanation offered by Vening Meinesz two decades before. The crust was compressed as it moved in to fit the convex side of the arc, but it stretched as it expanded to fit the concave side. The stretched side had tensional faults, and they provided easy access for magma rising to form volcanoes. Thus the sea-floor spreading hypothesis required only one important alteration of the tectogene hypothesis to dispose of old sea-floor. Fisher and Hess were able, somehow, to ignore or minimize the significance of the seismic studies in trenches that showed no extraordinary thickness of oceanic crust. Of course this attitude is quite normal; most geological hypotheses, even if correct in detail, have to accept the existence of some inconsistent data. It takes genius or luck to guess which to disregard. Fisher and Hess, this time, were unlucky in accepting Hodgson and disregarding Benioff and the seismic studies of Raitt, Shor, and Fisher himself.

Tuzo Wilson was off in a new direction. Many of us were collecting or had easy access to new observations of the age and nature of the oceanic crust. He had no ships, but it occurred to him that oceanographers were not taking advantage of the existence of islands that might be samples of the sea floor. Not that we did not sample islands; no geological oceanographer ever passes one by if he can help it. Moreover, nineteenth-century geologists had repeatedly visited almost every island in the deep-ocean basins. With the expansion of oceanography, however, the islands had been relatively neglected, and many had not been studied in the twentieth century. Wilson obtained a research contract from the U.S. Air Force and set out to compile geological information about the hundreds of oceanic islands. Volume 1, *The Atlantic and Indian Oceans*, was mimeographed and bound in January 1963, and the Pacific volume followed a month later. Tuzo kindly sent me copies, and it became almost impossible to talk to him without studying these books. Geologists are expected to know vast amounts of geography, but until Tuzo no one I knew had ever been so casual about references to Kerguelen or Tristan da Cunha or Fernando Po. Perhaps Daly was that way; he had mapped St. Helena and Ascension. Perhaps, come to think of it, I was that way, too, regarding the Pacific Islands.

Wilson focused his compilation on observations of potential interest to tectonics. He adopted Charles Darwin's definition of ocean basin islands, namely those rising from the deep ocean that are volcanoes or coral reefs resting on volcanoes, such as atolls. Other islands that contained granite or metamorphic rock generally did not qualify. They might, for example, be tiny, drifting fragments of continents. Such are the granitic Seychelles on the Indian Ocean, although this interpretation of their origin was not established when Wilson made his compilation. Few facts about volcanic islands and coral reefs are of much tectonic interest, so the compilation was on a manageable scale. Most important was the age because it was also the minimum age of the sea floor. Second was the nature of the volcanic rocks and any repetition or cycling of volcanism. Could a drifting volcano erupt in more than one cycle? Third was any evidence of uplifted marine erosional terraces or coral reefs that would demonstrate uplift of the islands and presumably of the underlying sea floor. The basic difficulty was that most of the observations were of dubious value. Potassium/argon dating was available in only a few places, and stratigraphic ages were based largely on casts of sparse fossils of little-known species. Likewise, the heights of uplifted reefs and terraces were usually uncertain at best, and the observations were often hurried or incidental. Nonetheless, no compilation of comparable utility had existed before, and Wilson wrote three papers to discuss different aspects of his discoveries. They were published almost simultaneously, so they will be considered in the order of their writing as deduced from the content: first, vertical motion; second, horizontal motion of ridge-crest volcanoes; third, horizontal motion of mid-plate volcanoes.[2]

In the first half of this century it was known that sea level fluctuated eustatically as a consequence of the waxing and waning of continental glaciers. Reginald Daly had established this fact and had proposed that a widespread insular terrace was a consequence of a former stand of the sea 3 m above the present level. The melting of Antarctic glaciers, however, would raise sea level by about 46 m, so it was reasonable to assume that insular terraces much higher than 3 m had the same cause. Harold Stearns, one of the pioneers of Pacific insular geology, discovered widespread evidence of elevated marine erosion and coral reefs.[3] These he attributed to eustatic changes even though some were hundreds of meters above sea level.

Tuzo Wilson's thinking was conditioned by his realization that the sea floor is tectonically active, splitting at the median rift, faulted by fracture zones, and drifting sideways. It was therefore likely that the islands moved up and down even when sea level was constant. He found that uplifted islands did not occur everywhere, as they should if caused by eustatic changes, but only in certain tectonic provinces. He divided these into five groups. One group was on the convex side of island arcs in the western Pacific and Indian oceans and at a distance of 300-700 km from the trenches. This was a remarkable generalization considering the facts then available. Echograms showed that a feature, now called an "outer arch," exists off many trenches, but this fact was scarcely known except to a few marine geologists. The arch is caused by flexing of the rigid lithosphere as it bends down into a trench. The amount of upward arching increases with age, and in the western Pacific and Indian oceans it produces the elevation of islands by 100-200 m as Wilson observed. Like many an innovator he went too far and included Ocean, Nauru, and Marcus islands in this group. The arch extends out only 200-300 km, and the islands are farther from the trenches.

Wilson's other principal generalization was that many uplifted islands lie along a great circle approximating a line of latitude across the South Pacific. He observed that along the eastern extension of this line Fisher and I had described many fault scarps and seamounts. He deduced that the islands were uplifted by faulting and that the great circle was like the ones I had described in the North Pacific and "perhaps forms one of the great fracture lines of the earth." The "line" was a coincidence. The uplift of islands in this region is also a consequence of plate rigidity but was not explained until 1978 when it was correlated with loading by young volcanoes.

Wilson's second paper in this series was concerned with insular evidence that the sea floor spreads.[4] This was an obvious subject for research, considering that the truncation and drift of islands down the flanks of ridges were discussed at length by Hess in his evolution of ocean basins. Wilson, however, made several marked advances in thinking about vulcanism at spreading centers. Unfortunately, he also made a number of refutable errors, and at the time I thought that he weakened the hypothesis that there was any spreading. At the least, I was later unprofitably distracted into publishing a refutation of the significance of his data.

The first conclusion Wilson drew was that almost all ocean basin islands, as he had defined them, are young.

> One immediate reaction to this apparent youth of the islands is to suggest that the ocean basins are nevertheless permanent and old, but that the study of islands is a poor method of sampling the floors. Another possibility [is that] the youth of the islands reflects the youth of the sea floors.[5]

He proposed to test these possibilities by assuming that islands are formed at spreading centers where volcanism was obviously occurring at the deep-sea floor. This assumption was manifestly incorrect. With the exception of Iceland there are hardly any islands at ridge crests, and active volcanoes show that they can grow anywhere in the ocean basins. Nonetheless, Wilson's data set, as plotted, gave him a correlation between island age and distance from the ridge crest; that certainly improved the correlation. We all have our moments of blindness, and those who see farther are probably more subject to them. Even so, his correlation rested on the existence of three very old islands very far from the ridge crest. What he could not know was that one of his ancient islands, the Bahamas, was not an ocean basin island by his definition; that another, Fernando Po, was in fact quite young; and that the third, the Cape Verde Islands, actually included fragments of uplifted ancient sea floor. The Cape volcanoes are all relatively young.

All in all, Tuzo Wilson would have done better, with regard to this line of evidence, to pay closer attention to Harry Hess. The reason volcanic islands are young is that old ones sink beneath the sea surface and become guyots or atolls. Their existence as volcanic islands depends on their size,[6] but hardly any last for more than 20 million years. In short, Wilson's conclusion regarding his evidence, "the farther an ocean island is from a mid-ocean ridge, the older it is likely to be," was both vexatious and wrong.

In contrast, Wilson's speculations about the consequences of (in modern terms) the intersection of a hot spot and a spreading center were gloriously correct. He was misled to the extent that he thought of Hessian convection but not so with regard to the topographic consequences. As a plate moved above a fixed hot spot, it would produce a line of volcanoes or possibly two symmetrical lines, one on each flank of a ridge. This would account for the [somewhat] symmetrical Walvis and Rio Grande ridges in the South Atlantic.

All such branches might be flow lines:

The well-known herringbone pattern of the [Atlantic] ridges could be an expression of the fact that Africa and North America did not move directly apart, but that each had a northward component of motion as well.[7]

Wilson already had the fixed hot spot concept well in hand for the Atlantic, and in the last of these three papers he developed the implications for the Pacific. The first scientist to visit several of the linear island groups that are characteristic of the Pacific was James Dwight Dana on the U.S. Exploring expedition in 1838-1842. Charles Darwin had crossed the Pacific a few years before but stopped only at the nonlinear Galapagos and at Tahiti. Dana observed that the Society, Samoan, and Hawaiian islands are all in lines trending northwest, and each shows a progression in age. He first explored the Society Islands and recognized that Mehetia on the southeast end of the group was a young volcano although not active. By the depth of erosion he judged that Tahiti and Moorea, the next pair in the chain, were older. Farther to the northwest were barrier reefs and beyond were atolls, which he took to be indications of even greater age. His excitement can be imagined when the expedition arrived in Australia and Dana learned from a newspaper that Darwin had announced his now-celebrated theory of atoll origin in London. Dana had found the whole Darwinian age sequence, in one linear archipelago. He observed an even more complete age sequence from active volcano to atoll, in the Samoan Islands and once more, and even more obviously, in the Hawaiian Islands. The first and last of these groups aged to the northwest, but the Samoan group aged to the southeast—or so it certainly appeared. Dana proposed that the linear groups were along faults, presumably produced by some widespread stress. It was not clear whether the islands were (1) successively active or (2) all active at once, and successively extinct.

Most of us who published on linear Pacific archipelagoes accepted Dana's underlying fault. Tuzo Wilson, however, took a new approach. He rejected the idea that the islands had all been active at the same time; that had been tenable only when none were dated, and I had dated one. I had conducted a one-day cruise off Honolulu in 1961 to show foreign oceanographers at the Pacific Science Congress how we used our ships. I had the luck to dredge fossil reef corals at the edge of the insular shelf at 500 m. My paleontological colleagues thought the fossils were probably of Late Miocene age. Thus, reasoned Wilson, the age progression was demonstrated. He further rejected the idea that the age progression was caused by volcanism at the tip of a propagating crack. The crack was not needed once he proposed the correct origin, namely, that the archipelagoes are produced by drifting the oceanic crust over a relatively fixed source of magma in the mantle. He began by explaining how a source could be fixed when Hess's model required a convecting mantle to cause the drift. The sources were within immobile cores above which moved relatively thin horizontal surface flow and below which moved a horizontal return flow.

Many scientists are content to propose a radical idea without elaborating it. Some scientists prefer not to develop detailed corollaries in the same paper as a radical hypothesis because it gives potential critics a larger target to attack. Tuzo Wilson characteristically plunged ahead. The Ewing and Heezen plot of the mid-ocean rift showed an inverted ''Y'' junction near Easter Island. Wilson recognized the need for complex mantle convection to produce this junction and, incidentally, identified what are now the Pacific, Antarctic, and Nazca plates. He concluded that the Nazca plate was drifting east because of a line of older islands and guyots extending east from Easter Island. The Pacific plate was drifting northwest as shown by the trends both of island groups and the chain of seamounts in the Gulf of Alaska. It is noteworthy how thoroughly Wilson had absorbed the results of the published oceanographic exploration of the past decade. In this sequence of papers he was using some of Dietz's research to elaborate Dietz's hypothesis. In these bold generalizations, however, he had left a few facts unexplained. What of Dana's southeasterly aging sequence in the Samoan Islands? How could that be produced by a plate drifting northwest? And what of the admixtures of guyots and high volcanic islands? Could they be produced without simultaneous volcanism along the linear archipelago?

Wilson published two summary papers in 1963 that incorporated all his new ideas from the three detailed papers into the basic hypothesis of sea-floor spreading. They included a few unpublished ideas as well. He observed that the paired aseismic ridges of the South Atlantic not only were generated by a single hot spot at the ridge crest, but their outer ends approached matching continental margins. Thus the aseismic ridges showed the path of drifting of Africa and South America (just as Carey had said). He also changed his mind on one point. He rejected the idea of Ewing and Heezen that a continuous mid-ocean rift existed. Instead he accepted my interpretation that ridges end and are connected by large transcurrent faults. Presumably, that took him one step closer to the concept of transform faults. Certainly it was a necessary step.

In contrast to Tuzo Wilson, Bob Dietz ignored the open ocean basins in this period and concentrated on astroblemes and marine geology. The marine geology, however, was partially linked to sea-floor spreading and is pertinent here. Dietz expanded his ideas on the origin of geosynclines in three papers that were not without publicizing repetition but showed an evolution and clarification of his thinking.[8] He was bold as usual; ''This essay is rife with speculation admittedly, and with little apology.'' If he had written all three at once he might have presented his analysis as follows. Marshall Kay had observed that eugeosynclines and miogeosynclines were common in the geological past, but it was questionable whether there were any modern analogs. In 1950, Philip

Kuenen had asked, in effect, "Where are the continental terrace wedges of pre-Mesozoic time?" Considering oceanographic discoveries in the 1950s, Dietz indicated that an even more important question was "Where are ancient continental rises?" He concluded that the solution to all these questions was that continental terraces and rises are the modern equivalents of miogeosynclines and eugeosynclines respectively.

Given this hypothesis, Dietz was in a favorable position to fire broadsides at several generally accepted geological ideas. Regarding the evidence for permanence of continents derived from the rarity of abyssal sediments on land,

> It is widely believed that deep-sea sediments are virtually or entirely absent from the continental blocks. . . . But if the eugeosynclinal deposits are former continental-rise prisms of turbidites, then the deep-sea deposits on land are of enormous extent and volume.[9]

As with Harry Hess and the occurrence of serpentinized oceanic crust on continents, it was all a matter of definition.

Dietz also had new insights on the accretion of continents and on the apparently fundamental difference between continental and oceanic crust:

> The basement beneath eugeosynclines is only rarely unequivocally exposed; for example, no sialic basement can be identified for the Franciscan and Knoxville formations of California and Oregon. This supports the view that these sediments were laid down as a continental rise [on oceanic crust].[10]

Thus he was diametrically opposed to Beloussov. Dietz produced continents from ocean basins; Beloussov did the reverse.

What of the widespread hypothesis that oceanic trenches are the modern analogs of empty geosynclinal troughs ready to be filled? This clearly was impossible if the sea floor spread and the earth had a constant radius. Any sediment that fell into an active trench would be obducted or subducted, and an inactive trench would rise to restore isostatic equilibrium. Consequently,

> Island arcs, borderlands, trenches, and tectogenes play no part in this actualistic version.[11]

Dietz also noted that he was able to give strong support to one side of an argument that he had not initiated:

> *Subsidence by compression or sedimentary loading?* The acrimonious debate between Hall and Dana is a classic of geology that persists today.[12]

One side said that horizontal compression was necessary to produce a deep trough to be filled. The other side said that all the sediment in miogeosynclines and eugeosynclines was deposited in shallow water, so there was no deep trough. The load of sediment apparently deepened a shallow trough so that it could hold a thick accumulation. In rebuttal, however, it was possible to show that the deepening by loading could hardly be more than two or three times the initial thickness. Dietz realized that his concept was capable of accommodating all the arguments of both sides during a century of discussion; ''subsidence was entirely a result of loading.'' The continental rise originated in water more than 5 km deep, so it could easily accumulate 10-15 km of sediment by loading. Moreover, the criteria for shallow water sedimentation in eugeosynclines were wrong. Turbidity currents were now depositing sediment with similar criteria on continental rises in deep water. In contrast, the criteria for shallow water in miogeosynclines were mainly correct, so no deep trough had existed under them. The critical fact was that the shallow continental terrace was connected to the deep continental rise by a rigid crust, and the subsidence by loading of the rise would pull down the terrace. Therefore the shallow miogeosyncline could accumulate thick sediment. The subsidence of the rise is now attributed more to cooling than to loading, but the essential role of flexural rigidity as conceived by Dietz is wholly supported.

MARINE GEOLOGY

Dietz, Fisher, Hess, and Wilson discussed marine geology in terms of sea-floor spreading, but most marine geologists did not. In part, this was because of delays in publication, particularly in *The Sea*, Volume 3, which finally came out in 1963. It is striking, however, that this thick volume, the first to summarize the geological oceanographic exploration of the fifties, is almost devoid of any mention of any hypothesis of global tectonics. Marine geologists and geophysicists were not so much talking past each other as not talking at all on tectonic speculation. There were 11 papers on geophysical exploration with not a word on tectonics. Nine papers were about topography and structure, and much was written about turbidity currents and abyssal plains, which had once been controversial in marine geology; but there was nothing much about tectonics. The paper on trenches by Fisher and Hess has already been discussed. It accepted sea-floor spreading. A paper on mid-ocean ridges by Heezen and Ewing briefly repeated their disparate views on the origin of the features. Heezen and I published an overview of abyssal geology without any discussion of tectonics.

Volume 3 of *The Sea* was largely a nonspeculative summary of techniques and observations that were startling in the fifties when discovered, but which were already absorbed into hypotheses of global tectonics. It also, however, contained observations that were still anomalous. Among them was the fact that

Available dates suggest a constant rate of subsidence of 2 cm per 1,000 years for the last 100 million years.[13]

The observation was based on the depth of datable horizons on atolls and guyots. Harry Ladd and I were unable to determine whether the sea floor had gone down or the sea level gone up as Roger Revelle had proposed. The fact that a rate could be established, however, boded well for future understanding of oceanic tectonics.

In contrast to their amorphous composite effort in *The Sea*, individual marine geologists continued to maintain strong and opposed views on global tectonics. At Lamont the controversy was causing increasing strain because of the personalities involved. I have noted that Bruce Heezen was a different person in the 1970s than he was in the 1950s. His problems, however, initially arose not so much because he changed but because he did not change while Lamont changed around him. He was temperamentally suited for the individualistic oceanography of the early fifties. The Chief Scientist was in charge at sea and had the power and responsibility to change schedules and programs at will. At Lamont he was among a few strong-willed pioneers who collaborated but who had more than enough work to keep all of them busy. Times changed; by the end of the decade Lamont had many more scientists, much more money, and an enlarged administrative structure. The new men began to expand their research, and the data that had once been used exclusively by Bruce were now divided among them. The greatest change, however, was that rules were established and seagoing programs passed out of the control of the Chief Scientist. Bruce could brook none of this restraint. He was a maverick who refused to renew his driver's license or his car registration. He would not acknowledge limits to his freedom to use any data at Lamont, and therein ultimately came his downfall.

The strain developed gradually. When Bruce insisted on doing what he wanted at sea, he was denied access to Lamont ships. He went to sea at least once a year from 1947 through 1958, but then there was a four-year gap to 1962. Bruce and I were traveling to Europe together almost every year during that period, and although he never said so, it was my impression that he had had a heart attack and was barred from sea for that reason. Marie Tharp confirms that he had a heart attack just a few weeks before the International Oceanographic Congress in 1959, but no indication of heart damage was found during his three weeks in the hospital. It was deduced that he might have had a psychosomatic attack because his father had died of a heart attack shortly after Bruce had visited him. Marie recalls that Ewing was very solicitous and telephoned the hospital every day.[14] Bruce, however, emerged from the hospital, gave 13 papers at the Congress, and changed a flat tire one night while driving home. So the heart attack did not keep him from the sea in 1960 and 1961. He was entrusted with *Vema* from New Zealand to Tahiti in July-October 1962, but once again he broke the rules. He disappeared for three weeks after leaving the ship in Tahiti—not the first sailor to do so. Ewing was furious,[15] and for whatever reason, Bruce never again had access to a Lamont ship while Ewing was Director.

Bruce's perception of what then happened was recorded about 1966;

At Lamont students have played a decreasing role in (marine science) as the operations developed into an efficient survey program manned by a corps of efficient but often unimaginative technicians who could be more easily controlled than the more gifted but often rebellious graduate students.

Such a course may provide an efficient data collection scheme for an aging scientist but. . . .[16]

Bruce made one last cruise soon after Talwani succeeded Ewing, but that was many years ahead.

Bruce had scientific as well as administrative setbacks in this period. One of his earlier triumphs was his demonstration, along with Ewing and Ericson, that the sequence of cable breaks off the Grand Banks in 1929 was caused by a turbidity current. Indeed, that was why Bruce was so jubilant while smashing core barrels, as Manik Talwani recounted. Bruce had proclaimed his success and declaimed upon the properties of this turbidity current for most of a decade. The most surprising pronouncement was that the current had moved extremely rapidly on the continental slope. This was greeted with skepticism, and Bruce was typically savage in response to his critics. That was the background when the continental slope in the region was surveyed with the seismic reflection system. Heezen joined Chuck Drake in presenting the results at the annual meeting of the Geological Society of America in 1962. The talk was titled "Grand Banks Slump." The whole slope had slumped at once, and the upper cable breaks were not caused by a superfast turbidity current. Presumably, the audience included some of those whose questions had received scathing responses. Certainly the audience laughed. This account of the slump that was not a turbidity current was published in 1964.

In some important ways, 1964 was a good year for Bruce. He was awarded the Henry Bryant Bigelow Medal by Woods Hole Oceanographic Institution for his work on the physiography of the sea floor. Considering his international fame and this award, his status as an untenured Assistant Professor seemed highly anomalous. I so informed Maurice Ewing in a bit of unsolicited advice as we were strolling the streets in Washington. Tuzo Wilson gave the same opinion when queried by Ewing. In any event Bruce was promoted to an Associate Professor with tenure in 1964. He came home and told Manik Talwani that he would never again allow Ewing to put his name on one of his (Bruce's) manuscripts. The alienation deepened, and although Bruce had tenure, that is not everything in the world of big research. Tenure guarantees employment but not research funds, supporting staff, laboratory space, or access to major facilities. Bruce was cut off from the Lamont ships, and worse was to follow.

We last noted the Lamont ships undertaking the test of sea-floor spreading as soon as Ewing read Dietz's original paper in 1961. By 1963, Ewing could report that 150,000 km of profiler records were in hand, and by 1964, 200,000 km. To a singular degree, Ewing and Lamont focused on a field test of the sea-floor spreading hypothesis. The ships ran a grid, and Ewing examined the data that were flown back from every port. This is not an unusual procedure; I fly data back myself on multiport expeditions, but the control exercised by Ewing from Lamont was exceptional.

One of the principal figures in this intense exploration was Manik Talwani, who received his doctorate at Columbia, through Lamont, in 1959 after graduating with honors from Delhi University in 1951. Much of his early work was on gravity anomalies, but it soon spanned the full range of marine geophysics and geology. He was Chief Scientist on *Vema* or *Conrad* on at least one cruise in every year from 1961 through 1977, and one year he sailed on *Eltanin* as well. He was to succeed Ewing as second Director of Lamont-Doherty Geological Observatory in 1972, without interrupting this long series of expeditions. Manik remembers Ewing as engrossed, even obsessed, by the profiler records in the early sixties and believes that he withdrew from active research in most other fields.

Manik shared a house with Bruce Heezen in the early sixties, but he collected data for and would publish with Ewing, so he saw both sides. He describes Bruce as ''frantic to see'' the records from which Ewing had excluded him.[17] I can imagine Bruce's frustration to be shut off officially from the data critical to his hypothesis of global expansion. He still brooked no restraints, however, and he had his ways. Marie Tharp smiled when she mentioned ''midnight requisitions'' during this period.[18] Still, Bruce could not publish the data.

Ewing was candid about his objectives:

> Some of us hope to find a record of much earlier times, believing that the rough surface of the solid basement rocks beneath the deep-sea sediments may be billions of years old and, in fact, may be the original surface of the planet.[19]

He acknowledged, however, other valid hypotheses:

> Other scientists believe that the slate was wiped clean, by one means or another, one or two hundred million years ago, and that no record of earlier times has been preserved on the sea floor.[20]

He went on to discuss, in a quotation from Hess, ''what could have 'wiped the slate clean. . . .' '' Either continental drift or sea-floor spreading by convection was a possibility, and surely they "could not have occurred without leaving clear evidence for the student of sediment distribution."

Maurice Ewing, recently described as a "fixist," wrote in 1963,

> There is strong evidence in support of each of these divergent viewpoints—continental drift with disposal of pre-Cretaceous sediments on the one hand and stability with a long-continued collection of sediments on the other. At the present time, new evidence is arriving at a great rate, and each bit of it must be weighed against two hypotheses.[21]

That sounds like a proper scientist and not any other kind of "ist."

What had been found on the first 150,000 km of seismic profiler track? First, the sediment was very thin except bordering some continents. This was a detailed confirmation of Raitt's discoveries in the Pacific a decade earlier. Ewing concluded the obvious: "Either the time of accumulation is far shorter than the age of the oceans, or the mean rate is far less than the present one." There were no obvious conclusions about the variations of sediment thickness with distance from the median rift. The central facts to emerge were that near the ridge crest the sediment was in flat-bottomed pockets between rocky outcrops of basement, and on the flanks there was a blanket of relatively uniform thickness draped over the basement. The existence of the pockets was known to Tolstoy and Ewing in 1951, and they had speculated among themselves about the origin of the flat-surfaced sediment. It was a time when abyssal plains were being discovered and explained by deposition from abyssal turbidity currents. In 1963, Ewing appealed to the same cause. Turbidity currents redistributed sediment at the ridge crest because slopes were steep. On the gentler flanks the currents lacked competence to erode and transport. Ewing was strongly convinced by this argument, perhaps because it supported an idea that had gestated a decade. Manik Talwani recalls a meeting he had with the two Ewings to examine and discuss the first profiler records across the Mid-Atlantic Ridge. They studied a map of sediment thickness. Sediment was absent or very thin near the ridge crest. M. Ewing said this was a consequence not just of local turbidity currents but of an extensive, integrated system of such currents. He said that he and Tolstoy had mapped such a drainage system. Then he insisted on finding their old maps, even though Talwani remembers the search as inordinately lengthy.[22] The chart was found. It did show such a drainage pattern—a decade before the sediment distribution had any possible bearing on the validity of the hypothesis of sea-floor spreading.

That hypothesis not only predicted thin sediment on ridge crests but also disturbed sediment in subduction zones. The far-reaching Lamont ships had already crossed the trenches of the eastern Pacific. Ewing found that the trenches were forming faster than sediment could fill them and that the sections contrasted in every way with those in the Atlantic. The number of trench crossings was small.

But on the assumption that these samples are typical, we may note an absence of indicators that the continent has been plowing through, or overriding, or is being undercut by the oceanic crust and any body of supposed sediments.[23]

There the matter stood in 1963. Ewing concluded by summarizing the four hypotheses that might possibly explain the observations. In addition to permanence, continental drift, and sea-floor spreading, the older sediments might be buried by extensive lava flows. There is a curious omission. Ewing had found thin sediment or bare rock at ridge crests, suggesting spreading of some sort. He had not found any disturbance of sediment in trenches, suggesting (incorrectly) that it was not being pushed into the continent. Bruce Heezen, who was barred from the data, would have had no difficulty in offering a fifth hypothesis. Ironically, Maurice Ewing seemed to be proving that the earth was expanding.

The second paper in this series by Ewing and his collaborators included large-scale strip-photographs of profiler records and detailed interpretations.[24] For the first time scientists anywhere could evaluate the evidence, and it was complex. There were three profiles across the entire Atlantic, although one crossed the ridge where it intersected the equatorial fracture zones, so it was difficult to understand. The northern profile was from Dakar to Halifax, perpendicular to the ridge. The crest was bare rock, but so was the lower east flank. The upper west flank had extensive pockets of sediment, and the lower flank was buried under a sedimentary blanket. The southern profile, from Buenos Aires to Cape Town, was also perpendicular to the ridge. The ridge crest, for 35 miles on each side of the rift, was again bare rock, but the flanks were a seemingly random mixture of bare rock, patchy blanket, and filled pockets.

In addition to profiling, *Vema* and *Conrad* were conducting other programs to test tectonic hypotheses. Denny Hayes, then a graduate student, was involved in a deliberate effort to core outcrops of older sediment to see if the data were "inconsistent" with sea-floor spreading. The oldest sedimentary rocks taken by coring or dredging the Mid-Atlantic Ridge were Miocene, although the aseismic Walvis Ridge had Cretaceous sediments. Photographs confirmed that the ridge crests were, indeed, bare volcanic rock. They also showed sparse occurrences of ripple marks and current scour that indicated transportation of sediment.

The available data hardly supported any simple hypothesis of sea-floor spreading, but much was being learned about the distribution of sediment. The coring indicated that the proportion of red clay and carbonate ooze did not vary in Cenozoic time. The calcareous sediment accumulates only in relatively shallow water such as a ridge crest. Red clay is typical of the deep water of the flanks. In the North Atlantic one deep flank was bare rock. What did it mean? The great field test was not working out.

In the third paper the Ewings summarized the global distribution of oceanic sediment based on 200,000 km of sailing.[25] There were a few crossings of the mid-ocean ridge in each ocean, and they were interpreted as indicating that the sediment was transported into pockets where relief was high, and "draped in a blanket of uniform thickness" where relief was low;

> the lack of thickening of the sediment layer outward from the crest . . . does not indicate a progressively older crust outward . . . as would be appropriate in the hypothesis of crustal spreading.[26]

After so much effort, however, in profiling, sampling, and photography,

> It is very difficult to state positively how these data bear on the various hypotheses about the history of ocean basins. It had seemed possible at one time to test the convection hypothesis of continental drift. . . . Unfortunately there are so many variables in foraminiferal productivity, carbonate solution, sediment transport, etc., that the validity of such a test is open to question.[27]

In due course, it would be one of the triumphs of plate tectonics that the thermal model of subsidence would allow detailed, quantitative predictions of sedimentary sections in ocean basins. Computing such sections of red clay over carbonate oozes now is a standard problem for beginning students.

The eclipse of Bruce Heezen revealed a whole galaxy of new stars at Lamont who included Charles Drake and Lynn Sykes as well as John Ewing and Manik Talwani. Chuck Drake (1924–) was born in Ridgewood, New Jersey, in the same year as Bruce, and graduated from nearby Princeton in 1948. He received a doctorate from Columbia in 1958 after the usual lengthy period of graduate research at Lamont. He was appointed Instructor and advanced to Assistant Professor in 1959. He was to be a full Professor in 1967, 10 years ahead of Bruce.

Drake's work spanned a wide range of problems in global tectonics in these years. He had already made a name for himself by his work with Ewing on mapping the two parallel geosynclines under the continental shelf and rise of eastern North America. This work provided the examples for Dietz's modern miogeosynclines and eugeosynclines. In 1963 and 1964, Chuck—more than any other—seemed to replace Heezen as Ewing's alter ego in tectonics. While Bruce published extensively on marine geomorphology and sedimentation Chuck published on elaborations and tests of sea-floor spreading. He showed, as I had done on the West Coast, that an offshore lineation appeared to extend as a series of tectonic structures deep into eastern North America.[28] His offshore lineation was only a line of seamounts rather than a fracture zone as I had defined them, but the fact did not seem important. We did not know the origin of fracture zones, so they might have different morphologies in different places. Anyway, the basement around the seamounts was buried under an abyssal plain, so a fracture zone might be concealed there. A particularly intriguing aspect of Drake's work was that the lineation trended northwesterly offshore but curved to east-west onshore at about the fortieth parallel. I had proposed that the Mendocino fracture zone extended on shore; Gilliland had proposed that it crossed North America; and Drake now proposed that it extended, with a curve, into the Atlantic. Like Gilliland, Drake believed that the fracture zone displaced blocks of Precambrian basement more than younger rocks, and it was active in Precambrian time. If this sequence of interpretations was correct, there was no possibility of differential movement between North America and the adjacent ocean basins in the past 500 million years, and no sea-floor spreading.

Drake also made a direct test of a prediction by Tuzo Wilson. If Labrador had drifted away from North America, there should be a spur of a mid-ocean ridge in the Labrador Sea. Drake showed that the floor of the sea was an abyssal plain, but the magical seismic profiler revealed a buried ridge whose crest was in the middle of the Labrador Sea.[29] It was aseismic, lacked a normal central magnetic anomaly, and was old enough to be buried, so it "developed at an early geological time and . . . its development ceased before it reached the mid-ocean ridge stage." Here was another indication that ridges were ephemeral, as Hess and I proposed, but it was also a reasonable confirmation of Wilson's prediction.

Drake also published two papers on the world-girdling rift, which had become a major object of study in the International Upper Mantle Project—the first successor of the International Geophysical Year. The first paper was a committee report about the rift and the proposed project. This was indeed ironical, that a discovery by a man as possessive as Bruce Heezen was now turned over to one of his colleagues. In the final paper in this series, Drake and Girdler expanded a detailed study of the Red Sea already started by the latter. A notable addition was the preparation of computer-generated curves of magnetic intensity that would result from various models of the distribution of magnetic rocks in the basement under the rift. It was assumed that the basement rocks were magnetized either parallel to the present field or reversed. There was, however, no reference to the Vine-Matthews hypothesis.[30] The rift in the Red Sea floor was tensional, and "this adds more evidence to the view that the whole of the world rift system is of tensional origin," but that did not signify that there was sea-floor spreading.

The work of Lynn Sykes was just beginning and its direction was still obscure, but within a few years it would very strongly support sea-floor spreading. Lynn R. Sykes (1937–; Fig. 43) was born in Pittsburgh and received both a B.S. and M.S. from MIT in 1960. He was a summer research fellow at Woods Hole in 1959 and 1960 and went cruising in the Atlantic and Mediterranean. He then worked with Jack Oliver at Lamont, where he received a doctorate in 1964. His thesis was on the propagation of short-period seismic surface waves across ocean basins. In the course of his thesis work, he recognized that the locations of earthquakes delineated a number of unrecognized major tectonic features in the oceans.[31] We began to correspond, and I sent him a copy of my unpublished map of the bathymetry of the South Pacific. His epicenters fell along the ridge crest in the region but also on the (Eltanin) fracture zone that I had mapped but not named. Later he recalled,

> The startling thing was that seismicity along these fracture zones was restricted almost exclusively to the region between the displaced crests of the oceanic ridge.[32]

Alan Coode and Tuzo Wilson were to concentrate on the fact that most of the earthquakes were so distributed. I was to focus on the fact that not all of them were. In any event, Sykes had demonstrated that the new accuracy with which earthquakes could be located had opened up a promising new field in tectonics. He had also confirmed my hypothesis that ridges end in offsets by fracture zones. He soon showed that the seismicity of the Carlsberg Ridge and the Gulf of Aden Ridge was offset on the Owen fracture zone just as Heezen and Tharp showed on their physiographic diagram. In this region, however, the evidence was difficult to interpret relative to global tectonics. Two earthquakes on the Owen fracture zone were far to the northeast of the offset of the ridge crests.[33] Here, the most that could be said was that ''much'' of the seismicity was on fracture zones between ridge offsets, a far cry from the seismicity ''almost exclusively'' between offsets in the South Pacific.

At Scripps, like our distant colleagues at Lamont, we continued in our endless collection of data. Our small navy crisscrossed the Pacific and began to venture into the Indian Ocean, where Bob Fisher would devote much of his time. Before the Indian Ocean work began, however, he participated in an expedition in the Gulf of California. The results were submitted for publication in June 1963 by G. A. Rusnak and Fisher.[34] There were no references to Dietz, Hess, or Wilson, and no mention of sea-floor spreading, but the American geologists believed that the gulf had opened by drifting. Wegener and Carey had said as much, but Rusnak and Fisher had a mechanism more like the one proposed by Gutenberg. The gulf opened by ''gravitational sliding, on extremely gentle slopes from the region of western Mexico uplifted by batholithic intrusions.'' They accepted my interpretation that the East Pacific Rise intersected North America under the gulf. Thus the source of the uplift and westward tilting was the development of the rise.

A.E.J. Engel (1916–) had been lured from the faculty at Caltech in 1958 by that matchless recruiter Roger Revelle. Al's notoriously tart nature was nicely neutralized by that of his wife, Celeste, who was also one of the world's leading analysts of silicate rocks. Born in St. Louis, he graduated from the University of Missouri in 1938 and received a doctorate from Princeton in 1941.[35] He worked for the U.S. Geological Survey until 1948 when he went to Caltech. At Scripps, Al became intrigued by the volcanic rocks that we had dredged from seamounts and the deep-sea floor. Until he arrived, marine geologists had just called all the rocks "basalt," meaning a dark volcanic rock. Hess, who could have done more, was really interested only in serpentine and peridotite, which might give clues about the nature of the mantle. I regularly sent him basalt samples from my dredge hauls, but nothing came of it. Continental petrologists had not shown much interest in our dull grey rocks until Al and Celeste appeared. The next thing we knew, Al was feverish with excitement and talking constantly about "low-K tholeiites." Petrologists are eternally using the most exotic jargon because they name each kind of rock after a type locality where it crops out. Usually one can ignore these rocks because they are rare, but Al was in the process of proving that the sea floor consists of tholeiitic basalt, which is thus the most common rock on the surface of the earth. So everyone had to learn how to pronounce "tholeiite." Fortunately, Al is the unusual petrologist who talks chemistry and physics when not among his peers. The oceanic tholeiites had extremely low concentrations of many elements, notably potassium and uranium, compared with continental basalts.[36] In time, so many petrologists would follow the Engels in devoting themselves to these rocks that they would feel the need for an acronym instead of a name. The mid-ocean-ridge basalts are now familiarly called "MORB" even by those who still see them as dull grey rocks.

The Engels made one more knowledgeable, pertinent observation. The lavas dredged at three widely spaced localities near the crest of the East Pacific Rise were surfaced with unweathered glass and probably were less than a century old. This was good news for a former Professor of Experimental Crystallography at Princeton.

I had some news for Harry Hess myself. I wrote to him on 27 March 1963, "Certainly in my own mind a possible clean sweep of the ocean bottoms every few hundred million years is an alternative to the development I have proposed." I was writing Hess, however, because what I had proposed was something very different from sea-floor spreading. About that time I was just finishing my first book, *Marine Geology of the Pacific*, which would update my views on the sequential development of mid-ocean ridges.

The book was written when I was on a sabbatical leave with Teddy Bullard at Cambridge University. I had a Guggenheim Fellowship, and we rented P.A.M. Dirac's house while he himself was on a sabbatical leave at Princeton.

Most of the book was written either in Dirac's study with a fireplace and a garden on two sides or in Teddy's Department of Geodesy and Geophysics in a former manor at Madingley Rise. Teddy last appeared as head of the National Physical Laboratory. He left it in 1955 to become a Fellow of Gonville and Caius College for 1955-1956 and then a Fellow of his old college, Clare, from 1957-1960. He transferred to Churchill College in 1960, when it was established, and at the same time became a Reader in Geophysics. He was to become Professor of Geophysics in Cambridge in 1964. Since 1955 he had been elected a Foreign Associate of the National Academy of Sciences and won the Chree Medal of the Physical Society and the Day Medal of the Geological Society of America. He had also accepted a collateral appointment as Professor of Geophysics at Scripps.

Teddy had done many remarkable, non-British things to Madingley Rise, such as heating the building. This was actually a disaster because all the wooden doors and window frames, which had never been heated, shrank and let in more air. Everyone loved the place, and Teddy made it a magnet. John Sclater was a first-year graduate student in 1963, and Dan McKenzie and Robert Parker were undergraduates. Walter Munk, who was on a similar fellowship and sabbatical leave, and I shared an office. Before leaving Teddy and turning to my book, I should comment on the status of the hypotheses of continental drift and sea-floor spreading at Madingley Rise. Certainly, continental drift was a respectable, even standard, idea in England. For example, my son was attending Cambridgeshire High School and his geography teacher taught that continents drift. But what of Teddy and the other geophysicists? This was the place where Vine and Matthews would pen their hypothesis soon after Walter and I left separately for home. Soon Teddy would speak in favor of drift, but his own personal account says he was converted by Blackett's presentation of paleomagnetic data, which was in 1960.[37] That statement actually does not say when Teddy was converted, only when Blackett published. Everyone met and talked twice daily at tea and coffee, so anything noteworthy was known widely at once at Madingley. I don't remember anyone, let alone Teddy, making any interpretation in which continental drift was important. I was not, however, working directly with Teddy as Walter was. They were joined in the composition of a computer program with the acronym "BOMM" of which the "B" was Bullard and the first "M" was Munk. The program was at the forefront of the use of computers in geophysics and was supposed to be capable of solving the important calculations in the subject. Surely the motion of continents would have been important to anyone who believed in it. In fact, Teddy was to help program a computer to determine just such motions in less than a year. Did Teddy talk about continental drift to his collaborator? "Not in my memory," says Walter. There is a vast difference between neutral acceptance and active advocacy, and it appears that Teddy Bullard made that shift no earlier than late spring in 1963.

Now to the marine geology of the Pacific—the first description and attempt at analysis of the geology of half the surface of the earth. I integrated all our Scripps data from a score of expeditions and described the characteristics and distribution of all the kinds of features we had discovered. I still considered abyssal hills to be universal features of a volcanic basement. I identified and extended great fracture zones throughout the eastern and central Pacific and showed how they offset magnetics and bathymetry. On island arcs I noted that,

> Almost everyone who sees an echogram of the side benches and bottom troughs of trenches believes that they are produced by normal faulting. Thus the topography of the trenches suggests tension and supports the hypothesis that trenches owe their existence to tension rather than compression.[38]

Bob Dietz and I were struck by the appearance of normal faulting displayed in the first echogram that we ever saw across the Mariana Trench. I would stand by my observations even now, but the faulting can be caused by flexing over the outer arch before the lithosphere rolls down into the trench. First-motion studies show that there is normal faulting on the arch and compressional faulting in the trench. At the time, my observations and conclusion could provide comfort to those who believed that the earth was expanding. They still provided ammunition for Carey in 1977. He quoted the same passage of my book that I have just quoted. For myself, however, the gravity observations were as important as the topographic ones. Raitt had shown the crust was not thinned as proposed by Worzell. There was some kind of root although not a tectogene.

> Accepting the root, we conclude that it must be held down by a downward stress, and this is taken to be the drag of a plunging convection cell as conceived by Vening Meinesz, Kuenen, Hess, and Griggs. By virtue of the drag, the root is in tension downward rather than horizontally away from the trench.[39]

The last sentence is correct although not exactly for the reasons I had in mind. Still, the topographic evidence for tension did not require an expanding earth.

One chapter was on the sequential development of mid-ocean ridges, which I called "rises" because of the proper name of the East Pacific Rise and because the Pacific ones were not necessarily in mid-ocean. This was a transitional period in the standardization of terminology, so I shall now call them by current terms except for proper names.

Hess and I had been proposing for some years that the deep central Pacific had been much shallower roughly 100 million years ago, but the idea attracted few supporters. Perhaps, therefore, I may be excused for the dramatic tone of my letter to him in March 1962 just as I was packing my books and data for shipment to Cambridge. "Enclosed is a paleo-bathymetric map of the central Pacific 10^8 years ago." I went on to explain how the map was constructed by assuming all the drowned ancient islands were contemporaneous and by measuring the paleo-depth by the relief of the guyots. "In short the datum is the height of the guyot platforms, not the present depth." The map showed an enormously broad mid-plate swell with a relief of about 2 km. I called it the Darwin Rise, after Charles Darwin.[40] As I conceived it, this was not a mid-ocean ridge in the sense that Hess and others would propose. It was not originally a site of sea-floor spreading but rather a subsided mid-plate swell. In my interpretation, which now seems correct, the swell had had a remarkable history of volcanism that resulted in a great number of high volcanoes and the archipelagic aprons around them. I calculated the volume of the aprons from new data and decided that the collapse of the Darwin Rise had been accompanied by the most intense volcanic outpouring in the geological record.

In my hypothesis, mid-ocean ridges and the Darwin Rise were the same kinds of features but in different stages of development. The evidence of the Darwin Rise seemed to prove that ridges are transient and that in their later stages they produce distinctive volcanic traces that formed an almost continuous layer. Continuity did not seem compatible with sea-floor spreading, and as I wrote to Hess (15 March 1962), there were other problems for his hypothesis:

> it seems to me that this poses serious difficulties for some of your recent ideas on the history of ocean basins. For example, why are the guyots on the flanks of the Darwin Rise higher than the ones on the crest if all of them were formed at the crest and then moved to the flanks?

The point I made in this letter I would make again in print but not in *Marine Geology of the Pacific*. Indeed, I cannot find a word on sea-floor spreading in this book written about two years after the hypothesis appeared but outlined and begun before Hess was in print. I wrote at length, however, on the origin of mid-ocean ridges. Hess's concept of serpentinization-deserpentinization had become untenable. The expansion of the earth had many virtues as a hypothesis. Among others "it permits continental drift of the Americas away from the Mid-Atlantic Ridge without motion toward the East Pacific Rise," but the idea had many flaws. It did not account for the down-bowing of trenches, and, as proposed by Carey and Heezen, it involved doubling the diameter of the earth in the last fraction of geological time. The density of the earth before expansion was thus 44 gm/cc, "which is impossible in the present state of the universe for a body with the mass of the earth."[41]

The convection hypothesis remained as the only one not discredited. Where the flanks of ridges pushed against continents, compression produced trenches, and there were no trenches in the Atlantic because convection cells carried the continental crust aside. Still, there was a basic difference between Hess's and

Dietz's ideas and mine for the Pacific if not the Atlantic. I thought the Pacific was old. If so, what of the thin sediment? New measurements gave much slower rates of sedimentation even than those only five years earlier. Carbonate ooze was deposited at 1-10 m per million years and the widespread red clay at only 0.3-1.6 m per million years. If the consolidation calculated by Ed Hamilton is taken into account, the sedimentary layer of the seismologists could require 1.4 billion years to accumulate. My analysis of the nature of the seismic second layer was that it was largely volcanic but could include layers of consolidated sediment as Hamilton had proposed.

In any event there is no compelling reason to believe that the Pacific Basin has not been receiving sediment from the continents for most of geological time.[42]

I had no data indicating regional variations in thickness of sediment except that it was thicker because of high organic productivity near the equator. Instead of bare rock, I knew we had cored and Agassiz had dredged sediment at the crest of the East Pacific Rise. Thus the available evidence in 1962 in no way supported sea-floor spreading in the Pacific. The book was not published, however, until 1964, and by that time the evidence was beginning to change. By then the Engels were reporting fresh volcanic rocks on the crest of the East Pacific Rise.

Chapter 17

1963-1964
MAGNETIC REVERSALS AND THE VINE-MATTHEWS
HYPOTHESIS

In order to advance their fruitful hypothesis, Fred Vine and Drummond Matthews had to believe three hypotheses simultaneously at a time when few scientists believed any one of them. They also had to disbelieve the only observation directly related to the hypothesis. Even so, they were not the only ones to produce such a hypothesis. The year 1963 clearly was a time when the stage was set for a conceptual leap forward. Perhaps it was because Dietz had come so close in his original paper on sea-floor spreading in 1961. His idea that the magnetic anomaly stripes off California are destroyed by subduction of the spreading sea floor had been dangling enticingly for two years. In any event, the spectacular complementary idea that the anomalies are created at spreading centers was about to be discovered. It would be almost completely ignored for another two years.

THE EVOLUTION OF THE REVERSAL CHRONOLOGY

When it came in 1965, the overwhelming power of the proof of sea-floor spreading was based on the demonstration of two facts: the magnetic stripes are bilaterally symmetrical around ridge crests, and the geographical spacing of the anomalies is proportional to the temporal spacing of dated magnetic reversals on land. On the simple assumption of a constant speed of spreading, distance could be equated with time. Thus the evolution of the reversal chronology becomes important in this account of the revolution in tectonics. The evolution has been chronicled in great detail as befits a major achievement recognized by many prizes and awards.[1] To general astonishment, magnetic reversals provide the long-sought global stratigraphic markers that are revolutionizing most of geology. At sea, as though by a miracle, magnetic anomalies give the age of the sea floor without even collecting a sample of rock. Kirk Bryan, my Professor of Geomorphology at Harvard, used to say that field geologists some day would merely aim a black box at an outcrop and know its age. The box is bigger than he had in mind, but in marine geology we don't even need to aim it.

An isotope of potassium decays to certain isotopes of calcium and argon, and if everything is favorable the argon gas is retained and the potassium-argon method can be used to date rocks. Potassium decays at an extremely slow rate, and the atmosphere contains argon that can contaminate the measurements. By the early fifties, however, the method was being used to date ancient rocks because of the great interest in Precambrian stratigraphy and the age of the earth. In the mid-fifties the method was improved by John Reynolds at Berkeley so that it could be used to date very young rocks. As fascinating as the early history of the earth may be, the early history of man is more so, and the human and prehuman fossils of Olduvai Gorge and its surroundings in East Africa were interlayered with volcanic rocks that could be dated by the potassium-argon method. Garniss Curtis and Jack Evernden quickly applied Reynolds' mass spectrometer and showed that man is much more ancient than could previously be determined.

The machine that could date human history was rapidly put to use on more geological problems, and one of these was dating magnetic reversals. The uncertainty of the method was at least 3%, and, as it later developed, the reversals commonly occur at intervals of a few hundred thousand years. Thus reversals could be dated no further back than 4 million years before it became impossible to correlate from one region to another.[2] Two groups of scientists undertook to date magnetic reversals, first to prove that they were caused by field reversals instead of self-reversal phenomena, and second to investigate the time and nature of the reversals. Each group required a Reynolds-type mass spectrometer, a paleomagnetic laboratory, and the different kinds of scientists required to use the instruments and collect field samples. The future nuclei of the two groups assembled at Berkeley in the late fifties and early sixties and learned the necessary skills as graduate students or visiting fellows.

The effort at the Australian National University at Canberra arose through a rare combination of inspiration and common sense in the conception of a new department of geophysics as established by J. C. Jaeger. He reasoned that he could not hope to compete with geophysics departments elsewhere in the full range of established specialties and focused his resources on making the department a world leader in a few subjects that were not yet but would become very important. Of course every chairman says this, but among Jaeger's few selections were paleomagnetism and isotopic dating. Ted Irving, one of the pioneers in paleomagnetism, came to Canberra, and Ian McDougall was dispatched to Berkeley in September 1960, primarily to learn the techniques of potassium-argon dating from Curtis, Evernden, and Reynolds.[3]

McDougall was interested in dating the Hawaiian Islands to see if they were in an age sequence as proposed by Dana a century earlier. This was before the proposal of sea-floor spreading and Wilson's origin of the islands by drifting. He sampled the islands on his way home but then began to work with Don Tarling, a graduate student of Irving who was interested in the paleomagnetism of young rocks. Tarling had his own collection from the Hawaiian Islands. In combination they worked on the chronology of reversals from 1961 until Tarling completed his doctoral work and returned to England in 1963.

Meanwhile, the second group had formed under circumstances that were administratively different from Canberra but intellectually similar. The Jaeger of the U.S. Geological Survey was James Balsley, Chief of the Geophysics Branch from 1953 to 1960.[4] Jim graduated from Caltech in 1938, the year I entered, and received a doctorate in geophysics from Harvard. He went on to pioneering research with an airborne magnetometer as well as more conventional laboratory studies of rock magnetism. He was a natural to nourish innovative research when he was in an administrative position to do so.

Soon to join his group were three students at Berkeley. Two of them, Allan Verne Cox and Richard Rayman Doell, would one day have their names linked together like Vine and Matthews or Gilbert and Sullivan. Together they were to be elected to the National Academy and win the Vetlesen Prize. They seem curiously matched. Both were born in California. Both obtained bachelor's degrees late in life, Allan at 31, Dick at 29, and doctorates quickly, in two and three years respectively. Allan was already in the Geological Survey in Menlo Park, and Dick joined him in the Geophysics Branch after a few years of teaching at the University of Toronto and MIT.

The Survey's policy was, and perhaps still is, to hire as many Ph.D.s as possible and give them as much opportunity as possible to do research in their specialties. As a consequence, the Survey has a matchless pool of independent experts. When the Carter administration wanted to know the characteristics of the bed of an Iranian dry lake, hoping to rescue the embassy hostages, it found in the Survey a man who had published on that very subject.

The Geological Survey was the organization with the world's largest staff of geologists with doctorates, and the largest budget for geology exclusive of drilling for petroleum, but it was locked into an antique style of individualistic research. Meanwhile, the style of much scientific research was changing. Advances in particle physics, radio astronomy, and oceanography required expensive centralized equipment and cooperative use. Revelle and Ewing had seized the opportunity and molded Scripps and Lamont into new types of organizations. Jim Balsley had the vision to create a little group in the modern style to work on paleomagnetism for the Geological Survey. It included Cox and Doell and Gary Brent Dalrymple who had learned how to do potassium-argon dating as a graduate student at Berkeley. It had technicians to support the principals, instead of more Ph.D.s and no staff. It had an old building of the sort that Parkinson established as the most suitable for creative work. It also had funding adequate for the objectives. How could it help but prosper if Balsley had picked the right people? Clearly he had.

In 1963 both the Australian and the Menlo Park groups began to publish geomagnetic-reversal time scales. The first paper appeared in *Nature* in June, the next two in October.[5] The reversals were global; therefore they were caused by reversals of the magnetic field. The race was on to refine and expand the time scale. From 1963 through 1969, 17 ever-more-elaborate time scales would appear. They do not concern us at this time. Vine and Matthews were able to read and cite only the June paper because they themselves published in September. The June paper was based on only nine dated polarity observations: three in Europe, six in California.

MAGNETICS AT SEA

Research on marine magnetic anomalies was continuing at Lamont and Scripps, and some of it was closely related to the Vine-Matthews hypothesis. At Lamont, Drake and others mapped anomalies parallel to the continental slope of North America and found them offset 100 miles at 40°N latitude. The trans-America extension of the Mendocino fracture zone seemed to be confirmed. Another paper on magnetics at sea described the use of a near-bottom magnetometer in the search for the lost American submarine *Thresher*. The paper was principally notable, for present purposes, because it was by a new man at Lamont who would be generally responsible for the underway magnetic program right through the proof of sea-floor spreading. James Ransom Heirtzler (1925–) was born in Baton Rouge and graduated from Louisiana State in 1947. He received a doctorate in physics from New York University in 1953, taught at the American University in Beirut from 1953 to 1956, and was then a physicist at General Dynamics Corporation until 1960 when he joined Lamont. In the early sixties he was making every effort to obtain magnetic profiles across the mid-ocean ridges because of Ewing's interest in them. This interest manifested itself in another way. It had become widely accepted that the Mid-Atlantic Ridge surfaces in Iceland. The shallow Reykjanes Ridge is the connection between the submarine structure at normal depths and the similar structure on the island. The island was spreading slowly in the center along narrow rifts, and it appeared that the Reykjanes Ridge might be a good place to determine whether the sea floor was doing the same. In 1963, Maurice Ewing dispatched Manik Talwani to make what would become a classic survey of the ridge.

At Scripps a direct test of the Vine-Matthews hypothesis was conducted even before the hypothesis was conceived. Dale Krause was a graduate student working with me in the late fifties, and he did his thesis on the marine geology of the region off northwestern Baja California. That became the region where a site was needed for the Mohole drilling test, so Dale made a bathymetric and magnetic survey of the deep-sea floor east of Guadalupe Island. There were low abyssal hills trending north-south and magnetic anomalies parallel to them.[6] Although in a new region, the anomalies were just like those discovered by Mason and Raff. Bill Bascom and his merry band were going to drill into the second layer; why not find out what was causing the anomalies? The topography and sedimentary layers were about the same everywhere, so the hole was drilled on a magnetic high. The basalt from the bottom of the hole was sent to Menlo Park, and Cox and Doell duly reported that the rocks were reversely magnetized.[7] They made no comment about the relation between their observations and the marine magnetic anomalies. The reason for this was simple. "Unfortunately in the distribution of the survey data, Cox and Doell failed to receive a copy," wrote Art Raff later.[8] Thus it happened that Raff, who helped Krause in the survey, made the first correlation between sea-floor rocks and magnetics based on oriented samples. The magnetic high itself could be explained simply by the presence of a body magnetized parallel to the present field or by a gap in a body reversely magnetized. This being so, the fact that the cored sample was reversed required a more complex explanation. Possibly the sample was not representative of the average rock below. Possibly it was reversely magnetized in a narrow band that was drilled, but the adjacent positive anomalies overlapped and concealed a narrow negative anomaly from the magnetometer that was towed two miles above the rocks. Possibly the whole sea floor was reversely magnetized and the "positive" anomalies were merely less intensely magnetized than the "negative" ones. All that the magnetometer measured was the total intensity of the earth's magnetic field. It was observable that the field was greater in some stripes and less in others, but it was only a convenience, at that time, to call the greater field "positive." When Vine and Matthews came to think about the origin of magnetic anomalies, the only available datum indicated that a positive magnetic stripe is reversely magnetized.

Drummond Hoyle Matthews (1931–) attended Kings College, Cambridge, where he received a B.A. in 1954. He was a geologist in the Falkland Island Dependencies Survey in 1955-1957 and returned to Kings for an M.A. in 1959 and a Ph.D. in 1962. His thesis was on basalts that had been dredged from the North Atlantic and were thought to be a "unique" sample of the second layer.[9] By December 1961, however, he was already in Mombasa and writing to tell me that he had shipped the granites from the Seychelles Islands that I had requested for isotopic dating. He had already made a survey of the Carlsberg Ridge and the Owen fracture zone and would continue this work as part of the International Indian Ocean expedition.[10] When he returned to Madingley Rise in 1962, a young graduate student, Fred Vine, was assigned to work with him.

L. W. MORLEY

The hypothesis that the earth is expanding was independently conceived and published repeatedly over many years. From the very nature of scientific research, most ideas that are published by one scientist are conceived but not developed to the point of publication by several, and conceived in passing but speedily dismissed by many more. Ideas are a basic, common commodity in science. They are offered for discussion in hallways and at lunch and are analyzed heatedly at cocktail parties. Ideas, in short, are cheap. Credit, however, is dear. Credit for ideas, moreover, is difficult to assign if an "idea" includes everything from an unpublished flash of inspiration to a brief published note (Wallace) to an integrated hypothesis documented by decades of meticulous research (Darwin). Consequently, credit and prestige in science usually are allotted arbitrarily on the basis of priority in publication. However, the rule may be bent. Hess received most of the credit for the idea of sea-floor spreading even though Dietz published first. It is particularly difficult to be just in assigning credit when one person gives the first formal talk before another publishes the first paper. That happened with Jason Morgan and Dan McKenzie/Bob Parker regarding plate tectonics, but there was ample credit for all. It also happened for Morley and Vine/Matthews but with a less fortunate outcome.

Lawrence W. Morley received a doctorate from Tuzo Wilson's department at Toronto in 1952 with a thesis correlating the magnetic properties and petrology of Precambrian rocks in Ontario. He became involved in an extensive airborne geophysical survey of the Canadian arctic. By the early sixties he was familiar with all aspects of rock magnetism and magnetic field reversals. In 1962 he read the 1961 papers by Mason and Raff about magnetic stripes and by Dietz about sea-floor spreading. Dietz had said that the stripes were destroyed by heating at subduction zones. Morley conceived in a flash that they were first created by cooling at spreading centers.

Morley wrote a manuscript incorporating his ideas in December 1962,[11] and it would finally reach print in 1964. He reasoned that rising lava at a spreading ridge crest would be magnetized parallel to the earth's field. Continual spreading and repeated reversals of the earth's field would inevitably generate a pattern of magnetic anomalies like that found by Mason and Raff.[12] He argued for the acceptance of his hypothesis on the grounds that it was much more probable than the alternative of varying susceptibility in the stripes. He went on to discuss how geomagnetic field reversals could be studied if the rate of sea-floor spreading could be determined. Finally, he explained the observation that some seamounts have the opposite polarity from the underlying stripes. The volcanoes are younger than the stripes, and they cooled during a period of opposite polarity.

Morley, at a blow, had explained the origin of the mysterious magnetic stripes discovered almost a decade earlier, but the mere reality of the stripes had been difficult to accept at first, and his explanation proved more so.

What subsequently happened to Morley's manuscript is obscure at present because his correspondence was lost in a fire in July 1978. Whatever happened it reflected "regrettable errors of editorial judgment."[13] The events have been widely recounted, but there is no agreement about when they occurred even though the source cited is usually Morley himself. Presumably, the confusion merely demonstrates that most of the chronology is unimportant. In any event, the paper was written in December 1962 and submitted to *Nature* either in the same month[14] or in February 1963.[15] It was rejected in January, April, or late June 1963.[16] The editor of *Nature* indicated that there was no space available. The paper was submitted to the *Journal of Geophysical Research* soon after it was rejected by *Nature*. About the same time Morley gave a talk on the subject at the June 1963 annual meeting of the Royal Society of Canada.[17] By August Morley was "confident enough" that he was in press in the *Journal of Geophysical Research* to discuss his idea with Hess and Runcorn at a meeting in Berkeley. The paper was rejected either in late August or September.[18] The letter from the editor indicated that a reviewer had stated that the ideas were suitable for discussion at a cocktail party but not for publication. Variations on this statement are heard not infrequently in conversations among scientists. In fact, serious science may be done at cocktail parties by people who do only what counts, regardless of the circumstances. The statement usually implies more about the state of development or documentation of an idea than about its correctness. It was at a cocktail party at Roger Revelle's house that Dave Griggs told Harry Hess to "stop beating a dead horse." He thought that Harry's hypothesis that the Moho was a boundary between serpentine and peridotite was no longer tenable. I thought so, too, but Harry did not accept this excellent counsel. None of this was any comfort to Morley, who had expected that everyone would immediately agree with his ideas.

VINE AND MATTHEWS

Frederick John Vine was only 23 years old when he conceived of the spreading sea floor as a magnetic tape, but he had already been thinking about continental drift for a decade. When he was no more than 14 he opened his new textbook in geography and found a discussion of the fit of Africa and South America on the first page.[19] The book went on to say that geologists did not know whether the fit was significant.[20] Fred was unimpressed by this indecision among geologists, and he believed almost immediately in the reality of continental drift. "I guess I was always looking in studying geology for evidence that confirmed my prejudice."[21] He entered St. John's College, Cambridge, as a physics major but transferred to geology before graduating. His "prejudice" conflicted with that of most of his professors, so his belief in continental drift was strengthened by the necessity of defending it. In January 1962, Harry Hess visited Cambridge and presented his still unpublished ideas on sea-floor spreading to the Sedgwick Club of undergraduate geology majors. Fred reviewed Hess's ideas in a later talk to the club under the title "HypotHESSes."

Fred began to work with Drum Matthews in the fall of 1962 just when I began my sabbatical year at Madingley Rise. By the time I left in May of 1963 he was at the point of formulating his magnetic tape hypothesis. There was a general awareness of mutual activities because of the intimacy of the small, isolated group, so I knew of Drum's work and saw him and Fred in their quarters above the stables. Drum had asked Fred to remove the effects of topography from the magnetic survey. Other things being equal, the effect of a magnetized rock on the total field will vary with the distance. I was vaguely aware of the procedures for correcting for topography because Vacquier was teaching them to our students at Scripps. Vine was mathematically trained, and Teddy Bullard, who was busy programming with Walter Munk, strongly encouraged him to learn how to make the corrections by computer. Vine scouted about and obtained a suitable program from V. Kunaratnam, a student of Ron Mason at Imperial College. He also obtained Vacquier's unpublished procedures. It was soon common knowledge at Madingley that the central anomaly of the Carlsberg Ridge was normally magnetized—that is, in the same direction as the present earth's field. That result was reassuring for anyone like Vine who was attracted to sea-floor spreading. Moreover, the seamounts on the flank just southwest of the central anomaly were reversely magnetized.

Drummond Hoyle Matthews and Elizabeth Rachel McMullen were married and departed on a honeymoon. When Drum returned, Fred gave him a manuscript containing an explanation of the magnetic stripes by a combination of sea-floor spreading and magnetic reversals. The manuscript and the explanation circulated around Madingley without causing much reaction from Maurice Hill or Teddy Bullard. Maurice and Drum prevailed upon Fred to expand the paper to include some original, unpublished data from the Carlsberg Ridge. This was ordinary advice. The data presumably would have value for others working in the region or planning to, so even if the ideas were wrong the paper might be worth publishing. Moreover, as Charles Darwin observed, the only reason for a scientist to write a book (or a paper) is that his colleagues will judge that he has done a reasonable amount of work before coming to his opinions. Reviewers are apt to reject wild ideas in which an author apparently has invested little time or effort. They are more tolerant of such ideas if based on field or laboratory observations. Vine, however, had not collected the data. Matthews had spent the months at sea, and he became, in accordance with common custom, a coauthor of a manuscript that was submitted to *Nature* no later than late July 1963. Hill, Matthews, and Vine spent the next four months at sea in the Indian Ocean.

"Magnetic anomalies over oceanic ridges" was published on 7 September, five or six weeks after it was submitted. That was fast but not extraordinarily so; Wilson's paper on transform faults was published at least as fast. Moreover, the two authors were not around to quibble about any errors in the proofs.

The paper offers a hypothesis based on three assumptions:

1. The sea floor is spreading at ridge crests.

2. The seismic second layer is basalt rather than consolidated sediment and has a high remnant magnetism.

3. The earth's magnetic field reverses itself.[22]

In 1963 all these assumptions were highly questionable, although the last one was generally accepted by the residents, if not the visitors, at Madingley Rise. In any event, the paper did not begin with an explicit list of assumptions and merely deduce the logical consequences. Such a format would generally lose the reader if utilized in the geological sciences because the list of assumptions would be forbidding.

Vine and Matthews began by listing the chief magnetic features of mid-ocean ridges and illustrating them from their own data. The three features are a pronounced central anomaly, short period anomalies on the upper flanks, and longer period anomalies on the lower flanks. The increase in the period on the lower flanks was "almost certainly associated with the increase in depth to the magnetic material." This was not a good beginning. At Scripps we had entertained the same idea regarding the change in anomalies across the Mendocino escarpment for several years. It was wrong. The large central anomaly was correctly explained by a block of material magnetized very strongly in the direction of the present field. All that remained to be explained was the cause of the flank anomalies. "Recent work in this Department has suggested a new mechanism." There is the statement that the authors had done enough work to justify presenting their wild ideas. Charles Darwin was a Cambridge man.

They went on to show that the relation between topography and magnetics required rocks with a high intensity of magnetism. The continuity of magnetic anomalies suggested that whole blocks of the volcanic layer are reversely or normally magnetized.

The new hypothesis was "virtually a corollary of current ideas on ocean floor spreading and periodic reversals in the Earth's magnetic field." The cooling crust is magnetized in the current field and spreads. The field reverses, and the new cooling crust is reversely magnetized. It was Morley all over again, but Vine and Matthews also showed that the anomalies would vary with latitude and orientation and could be simulated by computer. This ability would be essential in proving that the hypothesis was correct a few years later. Perversely, the simulations that they illustrated focused on the amplification of the central anomaly by the reversals on each side. The flank anomalies were drawn in the same shape rather than bilaterally symmetrical.

That was the hypothesis, but what of the observation? What of the reversely magnetized basalt that was cored by the Mohole test drilling on a positive magnetic anomaly? The "only reasonable" explanation was that the drill had sampled a lava younger than the crustal block. So there was no problem if the new paradigm was already accepted. There was, in fact, no problem if the sample was from a sill, as those who examined it generally believed. The sill had to be younger than the crust.

THE IMMEDIATE REACTION

A contemporary at Madingley and future member of the Royal Society—John Sclater—said, "I never took the Vine-Matthews hypothesis seriously during the first year and a half."[23]

From the Department of Geodesy and Geophysics, Annual Report 1963-1964: The publication of the Vine-Matthews hypothesis is not mentioned among the year's achievements.

As one of the authors said to the other in the summer of 1964: "That was a very good idea you had over a year ago."[24]

From a letter to me by Matthews in September 1964: "We collected . . . combined bathymetric and magnetic profiles across mid-ocean ridges . . . to see if the central magnetic anomaly has any reality. . . ."

In an evaluation published by Cox, Doell, and Dalrymple in June 1964 of the importance of various contributions with regard to understanding reversals of the magnetic field: no mention of sea-floor magnetics.

A comment by a distinguished American geologist: "rather ridiculous."[25]

At least one person, however, took the idea seriously, and what he did provides an example of the great leaps that can be made if a new idea is fairly accepted as a firm base for speculation. Only five months after the Vine and Matthews paper, George Backus published a note in *Nature* with the same title they had used. His purpose was to propose a field test of the Vine-Matthews hypothesis and of continental drift by spreading convection. He noted that symmetrical magnetic stripes should occur on the flanks of the Mid-Atlantic Ridge in the South Atlantic. The width of the stripes would vary with latitude. "Of course, any rigid displacement of a continent on a spherical globe is a rotation about some point. . . ."[26] Bullard and others had already found the point by their computerized fit of the two sides of the Atlantic (see the next chapter). The fit, soon to be famous, was unpublished, but Backus knew about it. Given the point, or Euler pole as it would be called (see Chapter 21), and assuming a constant rate of spreading, he predicted the variation of the width of magnetic stripes in the South Atlantic. Thus a few magnetic profiles would discover magnetic symmetry, and hundreds of matching anomalies, each of which would demonstrate sea-floor spreading as convincingly as the fit of Africa and

South America demonstrated continental drift. Moreover, the profiles should demonstrate that the sea floor was as rigid as the continents.

It seems pertinent at this point to consider who was proposing this test, for, according to folklore in some circles, there is a correlation between scientific prestige and success in receiving research grants. George Backus, born in 1930, was full Professor at the Institute of Geophysics and Planetary Physics (at Scripps) at age 32 after receiving a doctorate at Chicago. In 1964, when he published his note in *Nature*, he was well known to leading geophysicists and would be elected to the National Academy of Sciences within a few years. It was not an obscure outsider who applied to the National Science Foundation, nor was he associated with an organization incapable of the proposed work on ships. Nonetheless, the NSF rejected his proposal as "too speculative." It would be another year and a half before Vine and Wilson showed that data already existed in the northeastern Pacific to demonstrate magnetic symmetry, and three years before Morgan showed that the spreading rate in the Atlantic varies with the Euler latitude as Backus predicted.

Even though the Vine-Matthews hypothesis was discovered repeatedly in 1963, the world was not ready for it. Fred Vine would receive the Day Medal five years later at the age of 29, but it was not just because he conceived of this hypothesis. His contribution was much greater than that. He developed appropriate methods and used them in 1965-1966 to prove that he was right and with him Hess and Dietz and Wilson.

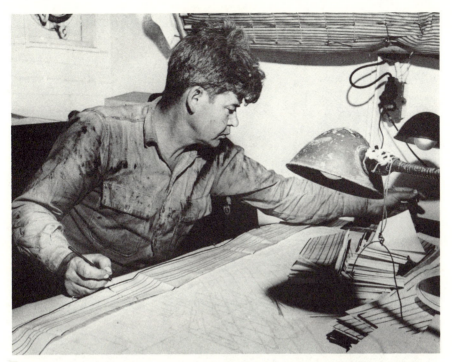

Fig. 25 Maurice Ewing working at sea studying seismic recordings that are spread out on a plotting sheet of soundings of water depth. Probably 1950s. (Lamont-Doherty archives)

Fig. 26 Maurice Ewing resting at sea. (Lamont-Doherty archives)

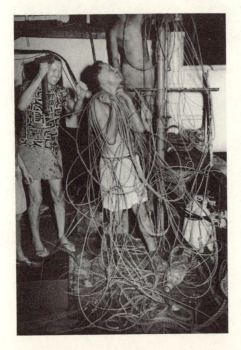

FIG. 27 Roger Revelle at sea on the Capricorn
expedition. The youthful friend is Gustaf
Arrhenius. (Scripps archives)

FIG. 28 The man who showed
in 1950-1952 that oceanic
sediment is thin and the crust
thin and uniform. Russell Raitt
in 1984.

FIG. 29 (*Above*) Some marine geologists from the
Midpac expedition at ease on a barely radioactive
Bikini Atoll in 1950. Bottom row: Frautschy,
Dietz, Menard, Emery, Shipek. Top row:
Garrison, Dell, Revelle, Brinton, Hamilton.

FIG. 30 (*Right*) Dredging in the Gulf of Alaska
after surveying the Mendocino escarpment in
1951. Left, Menard; right, Stewart.

FIG. 31 (*Above left*) Arthur Raff two decades after he and Mason discovered the magnetic stripes off California.

FIG. 32 (*Above right*) Bruce Heezen in his office. The physiographic diagram behind him is the South Atlantic in an early version before adding fracture zones. (Lamont-Doherty archives)

FIG. 33 (*Left*) Beno Gutenberg (right) and F. A. Vening Meinesz (in the 1950s?). Photo courtesy of Walter Munk.

FIG. 34 R. L. Fisher in 1958 on the Downwind expedition.

FIG. 35 J. L. (Joe) Worzel. (Lamont-Doherty archives)

FIG. 36 Dietz (right) in the early 1950s; center, Stewart; left, Menard. We were moonlighting on weekends as diving geologists consulting for oil companies. Probably late 1954 because when we started diving Hugh Bradner had not yet invented the wet suit.

FIG. 37 Bob Dietz in 1984 in La Jolla.

FIG. 38 Tuzo Wilson in 1984 in La Jolla.

FIG. 39 V. V. Beloussov presiding at a press conference at the Second International Oceanographic Congress in Moscow in 1966. On his left, Menard, and farther left Laughton. It was in exactly such a conference that Bruce Heezen chose to flout his agreement not to talk about magnetic reversals in sediment.

FIG. 40 (*Above left*) E. C. Bullard making geophysical observations in East Africa in the 1930s. Courtesy of Walter Munk.

FIG. 41 (*Above right*) Roger Revelle (on the left) and Bullard when one of them was receiving the Albatross Award about 1973 (Scripps archives). Vacquier also received the bird.

FIG. 42 (*Right*) Walter Pitman. (Lamont-Doherty archives)

FIG. 43 (*Above*) Lynn Sykes. (Lamont-Doherty archives)

FIG. 44 (*Left*) Dan McKenzie.

Chapter 18

1963-1964
CONTINENTAL DRIFT

Research and publication on marine tectonics and paleomagnetics as well as on the older subject of continental drift increased in 1963-1964. There were more conversions to the concept of drifting as well as the first stirrings of a new vocal opposition. The proportion of geologists who accepted drift, who were merely disengaged, or who actively opposed it varied markedly by country. Thus this chapter is organized in three parts dealing with views in the United States, then with those in Britain, and then with the confrontation of views at the Royal Society Symposium on Continental Drift in March 1964.

CONTINENTAL DRIFT IN THE UNITED STATES

Of the hundreds of geology papers published in 1963-1964, less than a score were even remotely associated with global tectonics, and most of them were on marine geology. Perhaps these years marked the nadir for Wegener's hypothesis. Paleomagnetism had had its try and had failed to inspire mass conversions to drift. Marine geology and particularly magnetics still had promise but were unconvincing. Only a few geologists in America were bold enough or thought it timely to publish on global tectonics or continental drift.

G. P. Eaton, future Associate Chief Geologist of the USGS, identified new interest in polar wandering and drift but believed that ancient patterns of sand dunes were not very useful for paleoclimatology. Volcanic ash distribution perhaps was more useful because it was determined by the relatively simple circulation of the upper air.[1]

W. N. Gilliland developed a hypothesis of zonal rotation of the mantle, resembling that of the sun, as an explanation of the east-west fracture zones shown on Heezen's maps of global tectonics.[2] Unfortunately, Heezen was in a phase of drawing fracture zones with an east-west trend whenever he lacked data to define a trend or even whenever the data did not forbid such a trend. For the southwestern Indian Ocean his drawings of fracture zones were very trendy indeed, and once snapped through a fully illustrated 90° change in a matter of days, to the general amazement of marine geologists.

The formidable Jim Gilluly, former President of the Geological Society of America, et cetera, et cetera, continued his support of drift in the Seventeenth William Smith Lecture to the Geological Society of London.[3] He observed "the complete independence of the large strike-slip faults" in the western states compared with the adjacent Pacific sea floor. Either the continent was "drifting westward over the ocean floor" or the sea floor was doing the opposite. "It is probable that the continent as a whole is moving away from a widening Atlantic." The last former President of the GSA and Penrose Medalist to make such a statement had been Reginald Daly; now it was Gilluly. Harry Hess was President in 1963 and would be Medalist in 1966. If part of the establishment clamored against Wegener's continental drift, there was little or no such opposition to Hess's sea-floor spreading. Gilluly was at the peak of the establishment. Judging by his papers and comments at the time, the older, more geological establishment was quite receptive to global tectonics. The younger, more geophysical establishment was on the average rather more neutral.

Warren Bell Hamilton (1925–) had been a student with Gilluly, and he also published in support of continental drift in 1963. The student, however, had been ahead of the professor with regard to drift. Warren was born and educated in Los Angeles, and in 1949 he was a graduate student at UCLA. He read DuToit's book on wandering continents and was convinced it was correct, but Gilluly was then perceived by the students as an outspoken opponent of drift; unlike Fred Vine at Cambridge, Warren remained mute.

In 1958 Warren went to the Antarctic to do geological mapping. On the way down he spent 12 hours side by side in a propeller plane with Tuzo Wilson, and they argued not about continental drift but about Wilson's ideas on continental growth.[4] Tuzo was on a VIP trip and soon departed. Warren and his collaborators began to find that conventional ideas about the geology of the Antarctic needed revision. It had been accepted that a coastal mountain belt of Mesozoic-Cenozoic age lay on one side of a Precambrian shield. DuToit, however, had predicted a Paleozoic fold-belt between the young mountains and the shield. It was found. Moreover, Warren quickly recognized the similarity between the rocks in Australia and those in Antarctica in regions that DuToit said were once joined. He talked about these discoveries in 1960 and published them in 1963:

> the major geologic elements of Antarctica, although mostly unknown to DuToit, are in positions consistent with drift origin. Antarctica provides yet another group of similarities between neighboring continents that seem better explained in terms of drift than in terms of coincidence.[5]

DuToit could not have put it better himself. Hamilton published on the origin of the Gulf of California by drift in 1961, and on the occurrence of island-arc rocks embedded in continents in 1963.[6] He was fairly launched as a new American champion of drift who had a background that skeptics should accept—a field geologist-petrologist with the Geological Survey in Denver. The opposition, however, was tougher than that. Even several years later, when he sought Survey approval, he found that a

> manuscript was blocked for a year by a Branch Chief and some senior geologists in Menlo, who regarded the analysis as so absurd that the Survey should not be associated with its publication.[7]

The International Union of Geological Sciences held a symposium on the Upper Mantle in New Delhi in 1964. Beloussov opened the meeting, and there were many talks on paleobotany, paleomagnetism, geology, and drift. I submitted a paper on tectonic effects of upper mantle motion, although I was unable to attend. The paper updated my views since the writing of *Marine Geology of the Pacific*. Block faulting occurred on the flanks of the East Pacific Rise rather than on the crest. An ancient fossil feature, the Galapagos Rise, existed between the East Pacific Rise and South America, so it was not necessary to have a trench between ridges, and the hypothetical Darwin Rise was not the only evidence that mid-ocean ridges are ephemeral. My emphasis, as usual, was on vertical motion of ridges, but I discussed the standard hypotheses of horizontal movement by convection and expansion. Sea-floor spreading appeared to explain much, but I was troubled by the crustal sections that it did not seem to explain. I had also come to realize that the broad upward bulge that I had advocated for some years produced gravitational potential energy. I introduced the possibility that the oceanic crust spreads away from a ridge crest because gravity pulls it down slope on a surface of minimum friction. This effect of gravity is, in fact, one of the driving forces of plate tectonics, but I merely suggested it, and suggestions are cheap.

The overt opposition to continental drift in 1963-1964 was limited. The distinguished paleobotanist D. I. Axelrod attempted to show that the persistence of climatic zoning in the geological past suggested stable rather than drifting continents.[8] Warren Hamilton prepared a lengthy refutation from the geological literature, but the editor requested that it be extracted to a three-page letter suitable for a geophysical journal.[9] The essence of the argument is the same as offered by DuToit decades earlier. Without drift the data are difficult to understand; with drift they are simple. Hamilton concludes with the prescient statement, reminiscent of Termier's language in the twenties but with diametrically opposed intent:

> the continents appear to be but temporary aggregates of randomly wandering and complexly deforming sialic flotsam.[10]

The second attack on continental drift and polar wandering, based on an analysis of published paleomagnetic studies, was by John W. Northrup III and

A. A. Meyerhoff.[11] As of 1984 it is accepted that Runcorn and colleagues in the 1950s were correct. The pole wandered in pre-Tertiary time, and its path relative to any given continent can be plotted as a line that approaches the present pole by moving into higher latitudes through geological time. Continents drift, so the lines for different continents can be considered to follow different longitudes. Northrup and Meyerhoff plotted these paleomagnetic poles on polar coordinates—the earth seen from above—and naturally found that the poles for any geologic age plotted at about the same distance from the present pole. They interpreted this plot as indicating that "prior to Tertiary or late Mesozoic time the present axial dipole field did not exist or was developing." As Warren Hamilton might have remarked, there was a simpler solution. John Northrup, although an active geologist,[12] does not appear further in this history, but Art Meyerhoff would launch a massive assault on all aspects of plate tectonics and the evidence purporting to document it in the early seventies.

The major arguments opposing drift in the early sixties were geophysical and based on new types of analysis of totally new types of data collected by artificial satellites. The "new moons" that were first launched during the International Geophysical Year when Tuzo Wilson was President of the sponsoring agency were beginning to produce data. To understand and interpret the observations would require a new generation of mathematical geophysicists.

Gordon James Fraser MacDonald (1929–) was born in Mexico City and moved to the United States when he was 12 years old. He was captain of the football team in his senior year at school, but an attack of polio ended his football career before he could enter college. He initially majored in mining engineering at Harvard but dropped "that nonsense" and graduated summa cum laude in geophysics. He was a Junior Fellow and received his doctorate in 1954. He advanced from Assistant to Associate Professor at MIT from 1954 to 1958. He then transferred at the urging of most of the other former Junior Fellows in geophysics, including Griggs and Kennedy, to their department at UCLA. He was a full Professor and elected to the American Academy of Arts and Sciences at age 29. Soon he was scheduled to speak at Scripps. Francis Shepard was an old friend of Gordon MacDonald, the premier vulcanologist of Hawaii; so when he heard MacDonald was coming, Shepard invited his many friends at Scripps to a cocktail party. The host was surprised when Dave Griggs walked in accompanied by an unknown youth. As the confusion was resolved, Dave spoke of the scientific abilities of "this" Gordon MacDonald in terms that I heard him use only one other time, in referring to the youthful Dan McKenzie. In 1962, Gordon was elected to the National Academy of Sciences and soon was a member of President Johnson's Scientific Advisory Committee, later Vice President of the Institute of Defense Analysis under Maxwell Taylor, and a member of the Council of Environmental Quality under President Nixon. Gordon and Walter Munk were among the leaders in a new field, geophysical fluid dynamics, that emphasized the unity of the basic equations for fluid flow whether air, ocean, or mantle, and utilized the new computers to obtain previously unobtainable numerical solutions for the equations.

In 1963, Gordon MacDonald was like Harold Jeffreys in 1923: young, productive, justifiably confident, and officially proclaimed as brilliant. Like Jeffreys, he was to use his analytical powers to contest continental drift. This time, however, the contest would not be an indecisive side attraction for decades. It would be the main event, brief, and won by a knockout.

In 1964, less than a year after being elected to the National Academy, MacDonald published a lead paper in *Science* on the deep structure of continents. It was based on a convincing array of published and original observations, and the logical structure of the paper was appealingly simple:

1. Isostasy prevails.
2. The surface rocks of continents are less dense than those of ocean basins; therefore the deeper rocks of continents are denser in order to obtain isostatic balance.
3. The surface heat flow of continents and ocean basins is equal, but the content of radioactive, heat-generating elements is higher at the surface of continents; therefore the content of such elements is compensatingly lower deep under continents.

The problem is thus reduced to determining how deep the differences in density and composition extend. Seismology provides a clue. Fragmentary results suggest that seismic surface waves behave differently to depths of 500 km under continents and ocean basins. Moreover, Benioff had proposed that earthquakes extending to 720 km lie along the initially vertical boundary between the deep structures of continents and ocean basins. MacDonald utilized data on the chemical and physical properties of rocks to calculate that the global heat flow and gravity observations also implied that differences between continents and ocean basins extend to 500 km. If the crust above that depth, for example a layer 100 km thick, should move horizontally for any significant distance, the observed balance of heat flow and isostasy would be destroyed. It inevitably followed that continents are formed dominantly by vertical segregation, and large-scale horizontal transport played only a minor role. This was a most welcome conclusion to those of us who attached importance to the vertical motion of mid-ocean ridges. There was a further conclusion:

> The deep structure of continents places heavy restrictions on any theory of continental drift. A relative motion of the continents must involve the mantle to depths of several hundred kilometers; it is no longer possible to imagine thin continental blocks sailing over a fluid mantle.[13]

Jeffreys had said that Wegener's continents could not sail through the mantle. MacDonald said that Hess's thin crust could not sail over the mantle. The battle was fairly joined. On one point MacDonald was certainly wrong. Within a few years, scores and then hundreds of people would be quantitatively calculating what he viewed as unimaginable. And yet the continents still appear to have deep structure. The paper was viewed as "News" by *Geotimes* in its April issue. The theory of continental drift, "which has been gaining support in the last few years," had been challenged by G.J.F. MacDonald. It appeared that the challenge was welcome.

CONTINENTAL DRIFT IN ENGLAND

Continental drift was much more acceptable in Britain than in the United States in 1963-1964, just as it had been from the beginning. Nonetheless, acceptance was by no means unamimous among geologists or geophysicists. Because of Jeffreys' influence the latter were once outspoken disbelievers, but the younger geophysicists included many paleomagneticians and among them drift found favor. Keith Runcorn, for example, was as much in favor of drift as Jeffreys was against it. For some time Runcorn had devoted himself to the geological evolution of convection and how it would affect drift. In 1963, however, he felt called upon to refute a new kind of evidence that appeared to eliminate the possibility of drift. Jeffreys had interpreted the nonhydrostatic bulge of the moon toward the earth as a fossil tidal bulge left from a period of a billion years ago when the moon was closer to the earth. Munk and MacDonald had proposed that broad gravity anomalies, discovered by tracking low satellites, were also fossil features of great antiquity. It had even been proposed by others that the gravity lows were the scars left by the separation of the moon. If true, these interpretations required a mantle with enough rigidity or strength to prevent convection. Materials at the temperature of the mantle creep slowly, so Runcorn calculated the maximum rate of creep implied by the interpretations. It was very slow indeed:

> It is remarkable that our geophysical colleagues have not made widely known to engineers their technological breakthrough in discovering a material which, at elevated temperatures, has a creep rate only one millionth of those found in the laboratory.[14]

In direct opposition to MacDonald's paper in the same year, Runcorn concluded that, lacking strength, the mantle convects to produce the broad gravity anomalies. Inasmuch as the anomalies did not correlate with the positions of continents, he inverted MacDonald's analysis and calculated the direction of convective flow from my map of the distribution of ridge crests.

The British preserve the custom, prevalent when science was more intimate, of recording the discussion after some talks. They also print written comments and responses with the original paper. This practice can make each controversial talk a small symposium and provide insight about the character of opinion at the time. When I left England in the late spring of 1963, I had no impression that my host, Teddy Bullard (Figs. 40 and 41), was preparing to be an active convert to continental drift. Nonetheless, on 19 June 1963 he delivered a paper on the subject before the Geological Society of London in Burlington House. The comments of the audience were preserved.

In his straightforward, sometimes blunt, style he dismissed the geographical, geological, and paleontological evidence as manifestly inconclusive regarding drift. The discovery of offsets of 100-1,000 km on the Great Glen, San Andreas, and particularly the great fracture zones of the northeastern Pacific was another matter. In the latter region, large-scale horizontal movements of crustal blocks constituting 2% of the earth's surface were demonstrated. Likewise, the observations in paleomagnetism were highly significant:

> Either we must accept the mobility of continents or suppose, without any supporting evidence, that the magnetic field in the past was quite different from what it is today and demonstrably has been since the Eocene.[15]

On the basis of uniformitarianism, the geologist's creed, he accepted mobility. The geologists Northrup and Meyerhoff, considering the same evidence, had rejected the creed.

Bullard followed many others in noting that seismology at sea "makes it difficult to believe" in former land bridges. Although in themselves unconvincing, however, the older geological data supporting drift were remarkably consistent with the new discoveries of paleomagnetism and the new fitting of Atlantic continental margins by computer. Convection was the only plausible cause of drift, despite the doubts raised by his fellow geophysicists—Jeffreys, Munk, and MacDonald—regarding the possibility of convection in a strong earth. Bullard's analysis was different from Runcorn's. Convection can produce the broad gravity anomalies, and therefore the case for a strong mantle vanishes. As to the the lack of a relation between continents and satellite gravity, convection should correlate more directly with heat flow than with the distribution of continents, and such a correlation was indicated by the recently published analysis by Willie Lee and Gordon MacDonald.[16] However, a new and potentially more damaging argument was emerging regarding the strength of the earth. Surface observations of gravity had indicated that the ellipticity of the earth was as it should be for a rotating fluid. Satellite orbits gave a more accurate determination, and they were showing a deviation of 1 part in 200 from hydrostatic equilibrium.

> This evidence . . . for a finite strength must be taken seriously and may be fatal to any theory requiring motions within the earth.[17]

Not even in a paper espousing drift, not even in the first enthusiasm of conversion, was a proper scientist prepared to eliminate the possibility that geophysics would prove drift was impossible. The earth, as the spherical harmonics were then interpreted, had an excess bulge around the equator. It was a shape appropriate for a faster spin. The rate of rotation is slowing. What could be more reasonable than that the bulge is preserved because the earth was exceedingly rigid? This would be a troublesome question until 1969 when Goldreich and Toomre offered a truly elegant answer.

Judging by the printed discussion, the reaction of the distinguished geologists was rather restrained—both to the hypothesis of continental drift and to its capture by geophysicists. Professor O. T. Jones had opposed continental drift three decades earlier on the grounds of insufficient evidence. Now there was more evidence that it required further consideration. It was curious that the offsets of the Pacific magnetic anomalies did not extend ashore. Moreover, Axelrod had recently claimed that the distribution of fossil floras proved that paleomagnetic positions were "grossly in error." The evidence opposed drifting continents.

Professor S. E. Hollingsworth recalled that a great physicist, Lord Kelvin, had persuaded geologists that the earth was younger than they thought. He noted that the lecturer's "admirable and balanced" statements that continental drift was a dead issue in the thirties and forties was at variance with his own experience. Many textbooks advocated drift. Moreover, "before the geophysical dawn" many geologists present had made important contributions to demonstrating the probability of drift.

Sir Edward Bailey recounted how he rather accidentally became the first person to lecture on drifting continents in 1910. G. W. Lamplugh, an outstanding glaciologist, was newly returned from South Africa where he had been impressed by the evidence for Permo-Carboniferous glaciation. Bailey had just returned from viewing the great thrust faults in the Alps. They spent a weekend in discussions during which Lamplugh "unfolded his theory of drifting continents." Bailey gave his nongeological lecture at the Old Vic, and Lamplugh never published his idea. Returning after the First World War, Bailey was assured by Lamplugh that Wegener's discovery was independent.

Professor W. D. Gill "regretted the lecturer's assertion that geophysical theories like continental drift were not obliged to consider the facts of structural geology." Taking the disagreements among geophysicists into account, geologists might well pin their faith on such facts instead of the meaningless concept of drift or "ludicrous" paleogeographic reconstructions based on paleomagnetism. It was remarkable that drift had to be linked to convection, which was no more than an "attractive speculation."

"The Lecturer, in reply" gracefully backed down on his remarks on structural geology.

Dr. E. R. Oxburgh pointed out that old-style paleoclimatological evidence might well be a more reliable indicator of latitude than the more fashionable paleomagnetic evidence. The relation between the axis of rotation and the magnetic pole was not as well known as that between the axis and climatic zones.

The Lecturer stressed the importance of more observations. He noted that Axelrod's evidence was difficult to credit. It required plants in the Antarctic where it was dark for months in the winter. "Forty years of discussion had shown that arguments of this kind did not lead to agreement; he did, however, find it a comfort not to have to try to believe such very improbable things."[18]

Mr. B. Webster-Smith questioned the significance of the Permo-Carboniferous glaciations to the drift hypothesis.

A written communication from Mr. J.E.G.W. Greenwood noted that Vening Meinesz's well-established concept of a global shear net might not be reconcilable with the concept of drift.

Dr. J. V. Hepworth disputed the geological evidence that had been cited relative to the history of the African Rifts.

Dr. N. Rast felt "that the lecturer was only right in deploring wild speculations . . . by geologists. Nevertheless geophysicists . . . had also to be careful . . ." in considering the role of structural data. Such data indicated problems for convection.

The Lecturer hoped that structural evidence combined with heat flow would show "what had really happened."

Dr. R.W.R. Rutland posed problems in relating convection to mid-ocean ridges and fracture zones. Continental drift "must now be accepted as a working hypothesis," but the currently proposed mechanisms had limitations.

A subsequent written contribution by Dr. R. B. McConnell noted that the Vening Meinesz net can already be reconciled with drift. He also remarked that the African rifts are Precambrian. Perhaps convection currents plunged under these rifts.

Another written contribution by Dr. M. J. Rickard noted the importance of structural geology to the questions under discussion. The proposed patterns of convection were unscientific because they did not agree with structural geology. An expanding earth was a more probable cause of continental drift.

The Lecturer concluded the discussion by acknowledging that he had ignored expansion of the earth in the "interests of brevity and clarity."

I have included an extensive exposition of this discussion because (1) it gives a fair sample of the flavor of geological controversy, (2) it illustrates the plight of the structural geologist, and (3) it demonstrates that in mid-1963 continental drift was still controversial in Britain. Regarding the second point, Carey had long since bluntly told the structural geologists that they were working on second-order features. Structural geologists did not take this kindly.

Many of them were among the most famous of geologists, and they had climbed the great mountain ranges of the world and discovered fantastic over-thrusting with offsets by stunning amounts—kilometers or even tens of kilometers. In Scotland and California they had found comparable horizontal offsets on vertical wrench faults. These were the geologists who legitimately speculated on the origin and history of the largest scale features. These were the ones who were studying the deformation of materials in laboratories in order to apply scientific rigor to their interpretations of field observations. Now, outsiders with no credentials in structural problems were having the effrontery to say that their work had little bearing on the gross deformation of the earth. They didn't like it. However, the people who would develop plate tectonics in the next few years, would not be structural geologists nor need to become such. Teddy seems to have backed down on this issue in the discussion, but presumably he did so only in the interests of decorum. He would be proved right, and I believe that he already expected it.

As to the third point, the status of acceptance of drift in Britain, this appears to contrast with Teddy Bullard's statement regarding the organization of the Royal Society Symposium in March 1964, which is the subject of the next section:

> A number of people from the USA attended and there were some complaints that the meeting had been packed with believers and that the opponents had not been given a proportionate representation. This was not a subtle plot; the fact was that almost everyone working in the field in England had become convinced a year or two before a comparable near unanimity was reached in the USA.[19]

Perhaps it depends on the definition of "working in the field."

THE ROYAL SOCIETY SYMPOSIUM

Sir Patrick Blackett, Sir Edward Bullard, and Professor Keith Runcorn convened a symposium on continental drift on 19-20 March 1964, under the auspices of the Royal Society, of which Blackett would one day be President. The meeting was expected to be large, and therefore it assembled in the Lecture Theater of the Royal Institute—the organization where rich Sir Humphrey Davy once gave popular lectures and poor Michael Faraday prepared exhibits to educate merchants and marchionesses. The speakers or discussants included numerous people who have already appeared in this book. Among the expected but missing were Beloussov, Dietz, Ewing, Hess, and Jeffreys. The results of this meeting, like any meeting, were largely predetermined for the audience by the selection of participants. On the other hand, the participants generally had known one another and one another's views for some time, so the meeting merely offered them an opportunity to renew acquaintances and restate refurbished ideas. The opening shots of a scientific revolution are not heard at well-attended, carefully organized symposia: Neil Opdyke had to round up a few listeners to hear Walter Pitman present the magically symmetrical, Eltanin-19 magnetic profile at the annual meeting of the American Geophysical Union. A hint of the unsuspected intricacies of the relationships among scientists in an invisible college appeared during a casual luncheon in London, which I believe was held during this meeting. The subject of undergraduate days arose, and I said to Walter Munk, "Do you know that you were my German instructor at Caltech?" He didn't. Gordon MacDonald then said, "Bill, do you know that you were my introductory geology instructor at Harvard?" I didn't.

Blackett introduced the meeting. From 1920 to 1950 a few people had thought continental drift was obvious, but to most people it seemed "wrongheaded and impossible." He spoke of the differences in methodology in experimental sciences and in geology,

> where the observational facts are highly complex and difficult to reduce to quantitative terms. In such subjects, a highly simplified model which can explain a large number of observed facts is invaluable, especially when it suggests new observations. . . . However, highly simplified models which prove that some supposed phenomenon cannot have occurred must be treated with caution, for they may discourage new observations.[20]

So much for Harold Jeffreys and his ilk. I don't recall a burst of applause, but the many geologists in the audience probably felt so inclined.

The meeting began with the topic of continental reconstructions. With such a topic the speakers—Runcorn, T. S. Westoll, K. M. Creer (all from Newcastle), and Bullard—naturally did not view drift as a question. Their task was to organize the supporting details. Teddy's paper was soon to be famous because it "proved" at last that there was a precise fit between the two sides of the Atlantic. He later told me about the great care he had taken to sell this fact that had been obvious to Wegener and had been demonstrated conclusively with continents drawn on a large globe by Carey in the 1950s (Fig. 1 F), but which nonetheless was still generally disputed. First, his method:

> By the fixed point theorem, usually called Euler's theorem in this application, any displacement of a spherical surface over itself leaves one point fixed; that is, any displacement of a contour line or of a continent may be considered as a rigid rotation about a vertical axis through some point on the surface of the earth.[21]

The method had style. It was based not on mere empirical relation but on a named theorem by a famous mathematician. The procedure itself had style. All possibility of bias or innocent error in plotting was eliminated. The fitting of the continents was done with a dazzling, advanced technique by digital computer. (Those were the days when people commonly still said "digital" computer because the device was so unfamiliar.) That still left the problem of a convincing presentation of the fit; after all, Carey's global models effectively had been analog computers of Euler's theorem, and they had not compelled acceptance. Teddy insisted that the Royal Society present his maps of the fit of Atlantic continents on a large scale even though that required expensive foldouts. Moreover, he insisted that the maps show overlaps in one color and gaps in another, although color printing was both expensive and unusual for the Philosophical Transactions. The illustrations constituted overwhelming evidence that the continents were once united (Fig. 2 G). Having artfully and successfully, although of course honestly, merchandized the evidence, Teddy closed with the properly reserved statement that none of it was "inconsistent with the hypothesis" of drift. A scientific revolution requires more than a letter to *Nature*. Even Darwin had his Huxley. A new idea must be advertised and sold. Dietz had long since been on his Distinguished Lecture tour, and in the next few years Tuzo Wilson and the Lamont group would be active in what the latter called "roadshows." Teddy knew what he was about.

In the discussion period, R. M. Shackleton (Leeds) observed that the Bullard fit could not extend very far in the past because of the distortion that had occurred during crustal shortening in fold belts. This was the structural geology viewpoint again. Moreover, not every geophysicist was overwhelmed by mathematics and computers. Joe Worzel was unusually articulate:

> We have seen the continents authoritatively reconstructed by a computer. This seems most convincing except of course it seems necessary to discard Central America, Mexico, the Gulf of Mexico, the Caribbean Sea and the West Indies along with their pre-Mesozoic rocks![22]

The second part of the symposium was on horizontal displacements of the earth's crust, and Clarence Allen (Caltech), Girdler, Heezen, Vacquier, and I proceeded to document such displacements on land and sea in elaborations of our previous papers. There were some changes and surprises. The Vine-Matthews hypothesis had been out for six months. What did Vacquier think of it?[23] "The origin of the north-south lineated magnetic pattern is unknown." Regarding Vine-Matthews, "Unfortunately this attractive mechanism" was probably inadequate to explain all known observations. In the discussion, the famous Dutch structural geologist, L. V. DeSitter, thought that the offsets of the magnetic pattern were improbable because the faults could not be traced on land.

Bruce Heezen presented an interpretation of his physiographic diagram of the Indian Ocean.[24] The diagrams were marvelous displays of the general characteristics of the sea floor and merited and received great praise as such. They showed what the sea floor was like in detail along the lines where echograms were available. Two logical choices were then possible: leave the regions between lines blank, or fill in the regions with imaginary detail like the lines. Heezen and Tharp chose the latter course. The result was an impressive drawing with the same validity, as I told my students, as a detailed drawing of the number of angels on the head of a pin. Heezen, of course, knew this because it was his imagination that was filling in the blanks. I proposed to Bruce that he put labels in the blanks indicative of legitimate ignorance. ''Here be giants'' was one suggestion. But no, the blanks were painted with hypotheses. In his symposium paper, the risk inherent in the diagrams surfaced; it appeared that Heezen was beginning to believe them. His main point was that the ''aseismic ridges of the Indian Ocean appear to be micro-continents.'' It was true they so appeared on his diagram. Regarding convection, ''in the Indian Ocean the existence of such divergent trends and the scattered ancient micro-continents make such an explanation extremely difficult.'' But the fracture zones with divergent trends were creations of his imagination.

In contrast to his discussion of his own physiography, Heezen's analysis of Ewing's sediment isopachs was cogent and scientific. He plotted sediment thickness against productivity of microorganisms in the overlying waters. They seemed to correlate in the South Pacific and the Atlantic, and this might be because of the youth of the sea floor ''but on the other hand may simply reflect the very low productivity of these areas.'' There the research mind is displayed. He had a new correlation, and he was prepared to entertain all possible explanations as working hypotheses even though one of them undermined the belief in a youthful sea floor that he had advocated for several years. That advocacy had been based on all known facts deemed pertinent, and this was a new ''fact'' and must be included in the hypothesis. Heezen concluded with a single sentence that ''a general expansion of the earth better explains the physiographic diagram.''

Bob Fisher had to restrain himself for several talks until the discussion for this session began. Then, ''Before we spend too much time philosophizing and glibly talking about micro-continents . . .'' we should look at the observations. Raitt, Shor, Fisher, and others at Scripps had shot 76 seismic refraction stations in the Indian Ocean. The aseismic ridges were volcanic, only the Seychelles were a granitic micro-continent.

Directly after Bruce, I had talked about what I called the worldwide oceanic rise-ridge system, a title that displays the difficulties that we were having in standardizing terminology for what are now ''mid-ocean ridges.'' I stressed that volcanism and fracture zones were associated with ridges and that the fracture zones were roughly perpendicular to the ridges rather than east-west. I also noted that an epicenter occurred on the Eltanin fracture zone at some distance from the ridge crest, so it seemed that the fractures might be active along their entire length. I proposed convection for the origin of ridges as usual.

S. E. Hollingsworth (University College, London) noted that the subsidence of the Darwin Rise might have caused the Cretaceous eustatic lowering of sea level. I had had the same idea as had Anthony Hallam, and it was to be a fruitful one.

In the next talk Ron Girdler proposed a mechanism for producing normal oceanic crust from the crust observed in the Red Sea. Joe Worzel found this conversion "mysterious."

Tuzo Wilson spoke in the following session, which was on convection currents and continental drift. He recapitulated his papers about islands published in 1963. He had perfected his explanation for the uplifted islands beside trenches. The distribution suggested "that they may be on the crest of a standing wave in front of the trenches and that the crust is rigid." He thought, however, that the islands in the eastern Pacific must therefore form on the crest of such a wave and are uplifted. This was simply wrong. In the discussion I questioned his observation that island age increases with distance from a ridge crest. It was "doubtful in itself and, if correct, does not require migration of volcanoes from a hearth at the crest of the ridge." Island groups at the edge of the Atlantic are still active. Moreover, there was no evidence of deep submergence—such as guyots—of the Canary and Cape Verde islands as would occur if they had formed at the ridge crest. My voice was full of tension, I recall. Perhaps I had had enough of this hypothesis.

The last session was on the physics of mantle convection, and Gordon MacDonald restated his 1963 analysis of the subject:

> In considering the possibility of continental drift it is necessary to inquire whether the present geophysical knowledge of the Earth's interior permits recent large continental displacement.[25]

Blackett, in his introduction, had argued against such an approach, but MacDonald went on to show once again that the continents are too thick and the mantle too rigid to permit drift. Inasmuch as paleomagnetic poles were different from the present one, he proposed that the field was more complex in the past.

Teddy Bullard in the concluding remarks effectively responded to MacDonald:

> many precedents suggest the unwisdom of being too sure of conclusions based on the supposed properties of imperfectly understood materials in inaccessible regions of the earth.[26]

Silent cheers from the geologists, whether drifters or not.

After this session, Harold Jeffreys recorded that he had been asked to make a written contribution "but had nothing new to say." If widespread, such an attitude would virtually eliminate the admirable scientific custom of holding symposia.

Gordon MacDonald and I flew home to California by the polar route, and somewhere over the arctic wastes, as we admired the glacially rebounded shorelines, I remarked that it was odd that he and I, people from a state being torn in half by the San Andreas fault, should be less inclined toward drift than the residents of the earthquakeless England. Geology, once locally oriented, had indeed become global.

Chapter 19

1965
SEA-FLOOR SPREADING CONFIRMED

In 1915 Wegener published the first book on continental drift. Half a century later his brilliant hypothesis was confirmed. DuToit might say "It was already confirmed in 1937 by my geological maps of the southern hemisphere." "And by my precise plotting of the geology on a globe in 1956," Carey might echo. Quite so; it will be a characteristic of this scientific revolution that most specialists were only convinced by observations related to their specialties. It would still be some years before the headlines proclaimed "Continental Drift Confirmed" because of the discovery of some fossil in Antarctica. It would even be a few years before the Deep-Sea Drilling Program spent millions of dollars and then announced triumphantly that it had confirmed sea-floor spreading. The above statements do not include the repetitive use of "at last," but it should be understood that "confirmed" is normally preceded or followed by "at last." So, was 1965 merely the year when all was confirmed by *my* specialty of marine tectonics? Perhaps so, but it was the discovery of magnetic symmetry that tipped the balance. Every matching pair of anomalies on the two flanks of the Mid-Atlantic Ridge was the equivalent of the matching continental slopes of Africa and South America. What is questionable for a very small sample may be indisputable for a large one. The symmetry of the magnetics was capable of convincing almost all geologists, regardless of specialty, if only they would look, and the symmetry was first demonstrated by Fred Vine in this miraculous year of 1965. In the next two years, we marine geologists, plodding along under a heavy load of data, would merely display better pictures to make the obvious more so.

The year 1965 brought another inspiration, namely the transform fault, which apparently came independently and simultaneously to Alan M. Coode and J. Tuzo Wilson. Transforms would capture the imagination of geologists and lead to a prediction and confirmation of a type rarely seen before in geology. For the most part, however, it was a year of probing and testing. Ewing and his young colleagues had been conducting field tests of the hypothesis of sea-floor spreading ever since Dietz had published it in 1961. After four years they were prepared to present their findings and dispute the hypothesis.

I myself was rather isolated from the events of this year. Gordon Mac-

Donald, aged 35, had been appointed to the President's Science Advisory Committee, and I went off to Washington from July 1965 through June 1966 to work with him on the White House science staff. I, naturally, did not stop writing scientific papers during this year, but I was away from my data base at Scripps. Thus it was that I never really thought hard about what Vine and Wilson had done until I returned to La Jolla. Doubtless others had similar experiences. It is one thing to be published, another to be read, and quite another to be absorbed.

HESS AND THE COLSTON SYMPOSIUM

The Colston Research Society is an organization of some antiquity in Bristol that periodically sponsors symposia. In 1965 the subject was "Submarine Geology and Geophysics," and 75 invited guests assembled on 5-9 April on the grounds of Bristol University and settled into the dormitories emptied by the spring vacation. Bob Dietz reported on the meeting to *Science* a few months later. Many of the distinguished senior marine geologists were present including Andre Guilcher from France, and Emery, Kuenen, Shepard, and John Wiseman who had mapped the rift on the Carlsberg Ridge three decades earlier. There were also the familiar geophysicists: Bullard, Girdler, and Worzel, and many newer men including Matthews and Vine. Maurice Ewing was again absent, but Harry Hess was for once not too busy to come. Bruce Heezen and I joined the more geological group and eschewed tectonics. He spoke on the discovery in piston cores of the ash emitted by the volcano Thera when it exploded and supposedly destroyed the ancient Minoan culture on nearby Crete. I talked about the deep-sea fans I had mapped in the western Mediterranean in 1962 while on the way to Cambridge for my sabbatical year.

Regarding tectonics, Dietz reported that "drifty strike-slip solutions to sea-floor tectonics enjoyed favorable treatment" and drift was "endorsed by at least a vociferous minority." Teddy Bullard states that the Royal Society symposium the year before did not include British scientists who opposed drift because they were so few. For the Colston meeting they had miraculously proliferated to a majority.

Most of the papers on tectonics were familiar stuff, but the one by Harry Hess was an exception. The last few years had found him busy with his commissions and boards and especially the AMSOC project to drill to the mantle. The Colston meeting was to result in his first published ideas since he had produced the famous preprint in December 1960.

I do not actually recall his talk, but I must have enjoyed it immensely because he repeatedly noted that most of it was based on my data and much of the remainder was a comparision of our ideas. Mid-ocean ridges were produced by rising convection and are ephemeral. The East Pacific Rise, Mid-Atlantic Ridge, and Darwin Rise differed in ways that were related to their age, and each individual ridge went through comparable stages.

Hess introduced an extreme variation on the theme of the age sequence. The East Pacific Rise was not yet spreading; it was only in a nascent state (Fig. 20 G). Hess had at last produced the discussion to accompany the ignored figure in his 1960 preprint. Some time was required to elevate the rise, so it had been developing, as a topographic feature, for perhaps a million years. There was yet, however, no median rift, and despite the seismicity the sea floor was not spreading. I had reported a band of ridges and troughs on each flank of the East Pacific Rise. Hess proposed that rising convection was spreading under the crest but diverging so it did not touch the old crust inside the ridges and troughs. The crest was ancient crust covered with sediment. Allowing for the thinning of the median crust, Hess had come to agree with my model rather than Dietz's. Hess wrote,

> It is evident that our difference of opinion [regarding spreading] arises from our concepts relating to the evolutionary history of ridges. Menard's being based largely on the East Pacific Rise and mine on the Mid-Atlantic Ridge.[1]

But I had accepted the opening of the Atlantic by drift and possibly spreading, and he was now accepting my thesis that the East Pacific Rise was not spreading. Harry was unlucky in his timing; in less than six months, the young Fred Vine, who presumably heard him talk, would show that the Juan de Fuca section of the rise was spreading.

If the East Pacific Rise was "prenatal," where did the ancient crust above it come from? Hess proposed that the whole Pacific crust was generated by the now-dead Darwin Rise. Most of the guyots, even in the Gulf of Alaska, originated in shallow water on the crest of that rise. The great fracture zones of the northeastern Pacific originated on that rise rather than on the East Pacific Rise as I proposed. The subduction zone along the California coast in Mesozoic and Tertiary time was caused by motion from the Darwin Rise toward North America. If one had heard him then, it would have been hard to believe that within a decade the mere existence of the Darwin Rise would have been doubtful and the subduction of an unheard of, unseeable, feature—the Farallon plate—would be commonly accepted as the cause of the deformation of the California coast. It is a curiosity of Harry's style of tectonics research that he believed he had discovered that the bathymetry around the great fracture zones was, "strangely enough," offset by the same amount as the magnetic anomalies. He just plain forgot that he had seen that in my papers in 1960 and 1961. The displacement was the main reason for my claim that the fracture-zone offsets occurred after the rise had formed. I cite this lapse on the part of a man most generous to me as proof that there is no meaningful plagiarism without intent and to point out, if the question arises, intent is hard to prove.

Hess added another new theme that was bound to please two members of the audience. He had already told Vine at Madingley Ridge that his hypothesis was a "fantastic idea."[2] Here he said that the Vine-Matthews hypothesis was "most fruitful," and he quickly made use of it. Given that the average time between reversals is 1.5 million years, the spacing between the ones off California indicated a rate of spreading of 1-2 cm/yr, and

> As a result, this would mean that the relative movement between the two sides of the Mendocino Fracture-Zone took more than 10^8 years to operate and possibly an even longer time.[3]

It is impossible to know what Hess had in mind in this brief statement, but it appears that he visualized a uniform, unfractured crust spreading from the straight Darwin Rise crest and, later, differential faulting that produced the off-sets over a period of 100 million years. He had told me repeatedly and this paper says that he did not accept my idea that fracture zones are associated with ridges, and this apparently was his solution for the offsets. In the same vein I considered that the East Pacific Rise was initially straight and then the flanks were differentially offset. Tuzo Wilson, who was not at the meeting, conceived of transform faults within a few months. If there is anyone who does not appreciate the magnitude of that achievement, he should compare Hess's statement with Wilson's concept. The offset did not take 100 million years to develop, or even 100 years; the offset was an active transform fault.

DIETZ

In March of 1965, Bob Dietz wrote to *Geotimes* to propose Project Gondwana—a study of the land and sea geology of the southern hemisphere. "We can't lose," he said. If drift was discredited, at least the geological viewpoint would be more global. "If continental drift is sustained, our work for the next decade is defined." In May, H. H. Harrington wrote in support of Project Gondwana.

In general, however, Dietz was increasingly involved in controversy about astroblemes and the origin of geosynclines. It reminded me of Teddy and the structural geologists, or Hercules and the hydra. K. Jinghwa Hsu sent a Discussion of Dietz's 1963 paper to the *Journal of Geology*. Dietz replied. Hsu agreed with Dietz that the crust under eugeosynclines is thin but it is continental, not oceanic. Dietz rejected this as no true agreement. Hsu refuted the significance of some of Dietz's data on crustal structure. Dietz quoted as yet unpublished observations and went on to refute the significance of Hsu's data on

paleocurrent directions. Dietz began with mention of ''sesquipedalian nomen-
clature and semantic confusion'' and ended extolling simplicity and actualism.
This discussion provided a clear example of the reason for a young scientist to
work in a new field, such as planetary geology, when it has no literature to be
quoted.

COODE, WILSON, AND TRANSFORM FAULTS

If we judge by the citations, the acclaim, and the rewards, the transform fault
was a solitary discovery by Tuzo Wilson and the unique inspiration of his fertile mind. I certainly thought so for almost 20 years.[4] According to the published record, however, the discovery, like so many described by Merton, was
multiple. Apparently the idea of the transform fault was hanging in the air,
available for anyone who visualized a spreading ridge with an offset, or two
spreading ridges connected by an active offset. For those who believe in the
concept of simple priority, or attach importance to it, the circumstances will
provide the usual difficulties. The paper that was submitted first was published
second. Those who see in history inklings of conspiracy will note that the first
paper published was afforded special handling by *Nature* and would otherwise
have been second. All that is certain is that one paper was completely ignored
and the other immediately became famous. Considering the scope of the papers, the fame of the authors, and the circulation of the journals, it is by no
means obvious that the impact of the papers would have been changed if the
priority of publication had been reversed.

Alan M. Coode sent a brief manuscript from Keith Runcorn's Department
of Physics at Newscastle to the *Canadian Journal of Earth Sciences*, and it was
received on 30 March 1965.[5] It was one of the most remarkable documents in
all of geological literature. Coode assembled a few newly discovered facts and
drew from them some spectacular but compelling conclusions. First the facts.
Transcurrent faults have almost wholly or wholly horizontal offsets. They offset magnetic anomalies and ridge crests. Sykes had showed that earthquakes in
fracture zones ''are almost always contained within the limits of the separated
ridges.'' Furthermore, ''Oceanic magnetic anomalies suggest the production
of new crust takes place at the same rate everywhere along the ridge.'' These
transcurrent faults are not active along their whole length as indicated by great
fracture zones.

> Rather a model is proposed where the center line of the ridge is broken and
> displaced by a transcurrent fault. As the new crust is generated the older ma-
> terial with the fossil fault and fossil magnetism is pushed to either side.

When the cause of the [earthquakes] dies down, the linear structures, which were once parallel to the rift, and the transcurrent fault are all that remain.[6]

Coode's only figure in this two-page note shows a three-dimensional ridge off-set by a fault (Fig. 9 J). There are symmetrical magnetic anomalies on each side of the ridge, and they are offset on the fault by the same amount as the ridge crest.

In short, and without adumbration, Coode realized that the magnetic anomalies are symmetrical, that the flanks of ridges move rigidly, and that the offset of magnetic anomalies merely preserved the original offset of spreading centers. His diagram was far in advance of anything that would appear elsewhere for many years. Certainly, it was superior to the related illustrations that Wilson and Vine would publish this year and the next. He had shown that the earthquake epicenters confirmed the only logical explanation of what happened when a spreading center had an offset. He had invented what is now called a ridge-ridge transform fault, but he had not felt it was different enough from an ordinary transcurrent fault to give it a separate name. One way to call attention to a new idea, however, is to give it a special name, like "sea-floor spreading." Moreover, if the creator wants credit for an idea, it is desirable to put it in a journal with rapid publishing and wide circulation. Alan Coode did none of these things. He sent his revolutionary little note to a relatively local journal that published quarterly. It would not appear until August. Meanwhile, another man entered the field, a man who would instinctively make all the right moves.

"This is a splendid place to work isn't it?" wrote Tuzo Wilson to me in June 1965. He had arrived at Churchill College, a short walk through the woods to Madingley Rise, in January. Harry Hess had appeared not long after, and although he was associated with another college and resting, they met occasionally at Madingley.[7] Tuzo recalls that he got the idea for transform faults under highly unpromising circumstances for creative thought. Writing textbooks is not doing what counts, which is one reason why the older scientists in the United States never thought to read Holmes's text of 1944. Revising textbooks generally is an even drearier patching of new data on old ideas. That, however, was what Tuzo was doing when he thought of transform faults. He was in the modest one-room library at Madingley because it had a table for plotting a worldwide map of great faults that he was drawing. It was just the sort of map that Heezen and I had begun to publish or had in press. The astonishingly straight fracture zones with enormous offsets had captured the imagination of many a geologist, and we all hoped that a global pattern would be revealing. At that time the pattern was still obscure. For example, Heezen thought that most fracture zones trended east-west; I thought (correctly) that most were perpendicular to ridge crests; and Hess thought they had nothing to do with ridges. Tuzo plotted the northeastern Pacific fracture zones and the San Andreas fault.

I had been saying that the East Pacific Rise crest disappeared under North America at the Gulf of California and reappeared north of Cape Mendocino. Ewing and Heezen had said that their world-girdling rift extended along the San Andreas fault from the gulf to the cape. Tuzo had long since joined Wegener in attributing the formation of the gulf to drift. What if the East Pacific Rise crest did not extend under the elevated region in the western United States? What if it was offset from the gulf to Cape Mendocino along the San Andreas, not as part of a rift but as the strike-slip fault it appeared to be? Tuzo looked at the offsets of magnetic anomalies on his plot. Vine and Matthews had shown that anomalies might be created at ridge crests. "And that," he says, "was it."

In the paper that he would soon publish Wilson observed that there are three types of major tectonic features: mountains (or equivalent trenches), mid-ocean ridges, and "major faults with large horizontal movements."[8] He proposed that these are not isolated features but are connected and divide the earth's crust into "large rigid plates." "Any feature at its apparent termination may be transformed into another feature of the other two types." There were logical short names and abbreviations for two types of features, ridges (R) and trenches (T), but what of the third?

A separate class of horizontal shear faults exists which terminate abruptly at both ends, but which nevertheless may show great displacements.[9]

He called them "transform faults" (F). (See Fig. 9 G.)

Next in this elegant paper Tuzo exercised logic to identify and illustrate all possible types of dextral and sinestral connections between ridges and trenches. His successors would call these R-R (ridge-ridge), R-T, and T-T transforms. Then he provided examples of transform faults from well-known regions of the Atlantic and Indian oceans. Those offsetting the Mid-Atlantic Ridge are R-R transforms, and the earthquakes should be confined to the ridge crests and the transform between them that constitute the plate boundary (I am using modern terms). The seismicity on the Owen fracture zone extends beyond the R-R transform, and I had used this as an indication that fracture zones on ridges are active along their whole length. No problem existed anymore. If the Owen fracture zone was active beyond the R-R transform it must still be a plate boundary and connect to something else. Wilson extended the R-R transform to the Himalayas. My "anomaly" was merely an R-T transform connected to an R-R.

For the northeastern Pacific Wilson used a peculiar map projection rather like a transverse mercator with the San Andreas fault as a line of latitude. It is just a "sketch map," however, and not a true precursor of the map of plate tectonics that McKenzie and Parker would produce two years later. But did it plant a seed? Both were students at Madingley in 1965. The map shows the

East Pacific Rise transforming to the San Andreas fault to a nameless ridge (the Juan de Fuca Ridge), whose discovery Tuzo credits to me, to a submarine transform fault recognized by Benioff, which extends into the Denali fault mapped by St. Amand in Alaska, which in turn is a transform leading to the Aleutian trench! It was a remarkable synthesis of large but isolated phenomena.

In one sentence Tuzo mentions that the magnetic anomalies and fracture zones represent the "shape of a contemporary rift" and R-R transforms. In another he says that if Hess's postulate is correct and the northeastern Pacific crust drifted from the Darwin Rise, the crest of that rise had the same offsets as the anomalies. It is a pity the timing was not different so that Hess could say this very thing at the Colston Symposium.

That was it, indeed—a remarkably imaginative and fruitful paper like the ones by Dietz and Hess on sea-floor spreading, and like the Vine-Matthews hypothesis. But the reception of this paper was different from that of the others. Sea-floor spreading was just one explanation for the observations, and, as stated, it had flaws. The Vine-Matthews hypothesis had no real competitors to explain the observations, but it was built on the sea-floor spreading model. No wonder the hypotheses were ridiculed or ignored. Compare the status of the concept of transform faults relative to all currently vigorous hypotheses of global tectonics. The idea of ridge-trench transforms was not troublesome. They were merely large-scale versions of the tear faults that geologists had been accustomed to mapping at the ends of great thrust faults (Fig. 9 F). Wilson cited St. Amand for the suggestion that the Denali fault is a tear fault connected to the end of the Aleutian Trench. St. Amand thought it improbable, but it was a possibility. Wilson's idea of plates had been a commonplace since the International Geophysical Year.

The principal innovation was the introduction of the concept of the ridge-ridge transform fault and its implications. It was exactly the same concept that Alan Coode had generated when he also analyzed the motion of a spreading ridge with an offset (Fig. 9 I). The simple fact is that no one had ever published anything about two sections of ridge connected by a fracture zone before. I doubt that anyone had done any serious thinking about the subject. Those of us who thought in terms of convection, an overwhelming majority, visualized a vast overturn of the mantle and flow rising and spreading under ridge crests— or spreading *as* ridge crests in the sea-floor spreading model. It was hard to conceive of a flow pattern with giant convection cells rising under each section of ridge offset by a fracture zone. Perhaps that is why Hess could not believe that fracture zones were related to ridges. In my endless sketches of cross-sections of ridges, I always thought of them as straight. Perhaps because there were no known offsets along most of the East Pacific Rise. I thought of the Eltanin fracture zone as connecting the East Pacific Rise and Pacific Antarctic

Ridge, but the region was as yet hardly sounded. I knew about the Blanco fracture zone connecting the Juan de Fuca Ridge to the Gorda Ridge, but the data
were classified and the western end of the fracture zone was unmapped and
might continue beyond the Juan de Fuca Ridge. Bruce Heezen had long said
that the Atlantic was born with a kink in it and was spreading in the middle.
Moreover, his model was an expanding earth with no internal problems about
convection. He was in a perfect position to think about ridge-ridge transforms,
but if he did he didn't talk to me about it or publish anything (see Fig. 9 H for
the state of his ideas). Heezen had believed for many years that an expanding
rift existed, wherever the ridge was seismically active, before he discovered
the equatorial fracture zones about 1960. He did not publish a paper about the
fracture zones until 1964. Perhaps he never really adjusted his mind to the implications if the equatorial Atlantic was not a spreading rift.

In any event, the immediate reaction to Wilson's paper generally ranged
from positive to enthusiastic. It was elegantly organized and lucidly written. It
gave a simple logical explanation for, or useful way of viewing, a host of well-
known problems. Best of all there was no reason to oppose it. It fitted admi-
rably into any hypothesis of global tectonics. I had arrived in Washington about
the time this paper came out. I sent one of the papers I wrote while there to
George Shor in La Jolla for comments. He was disappointed in it and said it
didn't seem to get anywhere. He analyzed my usual style of writing papers,
which no one else has ever done. After presenting a few critical observations,
he wrote, I revised some definitions, made an order of magnitude calculation
or two, and "emerged triumphantly waving a conclusion so obvious that the
reader kicks himself for not thinking of it." Most people who read Tuzo's paper kicked themselves.

The accounts of what happened after Tuzo conceived of transform faults are
colorful, but the chronology is clouded by time. Vine recalls that Wilson described transform faults to him in February 1965.[10] Matthews recalls that Wilson went sailing off Turkey and returned with the concepts of transform faults
and magnetic symmetry[11]—the two ideas in Coode's paper as well. Bob Parker, then a student, recalls that Wilson never discussed transform faults at the
morning coffee or afternoon tea sessions as was the custom when an idea was
developing. It sprang forth full grown at a scheduled seminar. John Sclater,
also a student, remembers that most people at Madingley took transform faults
no more seriously than the Vine-Matthews hypothesis.[12] Vine and Matthews
took transform faults very seriously and had a marvelous time working with
Wilson for a few weeks.[13] Hess read Wilson's manuscript and suggested some
changes with the expectation that further discussions would follow a revision.
The next day Wilson told Hess that he had modified the manuscript overnight
and it was already on its way to *Nature*.

In early June I received a letter from Tuzo mailed on 1 June from Churchill

College. Hess had talked to him that night, and Tuzo had gotten the impression that I believed that he, Tuzo, had recommended against publication of one of my manuscripts and that he, Tuzo, in fact had not seen it. I have no idea what this was about, but it appears from the context that Tuzo felt that if I thought he had rejected my manuscript I might believe it was for some reason related to one of his manuscripts.

I have it is true sent a letter to *Nature* interpreting many large features in terms of a new class of faults and Harry has seen the paper and appears to agree. It is of course based on the observations of many others. . . .

If more scientists were as sentitive as Tuzo, there would be far fewer nebulous allegations of plagiarism in the air.

Tuzo sent the manuscript before 1 June. To his surprise, an editor phoned instead of writing him to say it was accepted. It was published in three weeks, on 24 July, in the most prestigious scientific journal in the world. Alan Coode's paper appeared a few weeks later in Canada—five months after it was submitted. This appears to be a classic example of multiple discovery. With regard to the working of science, it is profitable to consider what might have happened if only the Coode paper had been published. It is difficult to conceive that the advances that followed the Wilson paper would have occurred so quickly. Thus the introduction of the concept of the transform fault may also be a classic example of the importance of merchandizing to the success of scientific ideas.

From the foregoing it might appear that Tuzo had been wholly engaged with transform faults in Cambridge that spring. But in addition to the revision of his book, he wrote two other papers and one jointly with Vine to develop the concept of transform faults. The first of these appeared in August, the other two in October, by which time he and Vine were in America.

The August paper identified itself as the fourth in the series on transforms, but it is unrelated to the second and third.[14] It was concerned first with the question of why aseismic ridges and fracture zones are not parallel if they both reflect motion of plates. It is necessary at this point to jump ahead a year or two and explain the relation between transform faults and fracture zones. A ridge-ridge transform fault is seismically active and produces characteristic linear topography. This topography is preserved as a fracture zone when a rigid plate drifts away from the ridge crest. Thus a fracture zone is a band of linear topography that is generated in a central segment called a transform fault.

Wilson even then understood that fracture zones indicate the *relative* motion between two spreading plates. Aseismic ridges and lines of volcanoes show the motion of a plate relative to a magma source or hot spot in the mantle. The significance of this difference escaped many people in subsequent years. In his paper Wilson deduced possible interactions of transform faults and aseismic ridges or lines of volcanoes. A straight volcanic line indicates that there is no internal motion within the plate that crosses the line. If a transform fault drifts across a hot spot, part of the volcanic line is on one plate and part on another. As movement on the fault continues, the two segments of the line drift apart. Consequently, a line that ends at a fracture zone probably had active volcanoes on it when the transform was active. Wilson provided examples and developed the geological history of oceanic regions by application of his discoveries. He continued in an unfortunate attempt to define the western boundary of the East Pacific Rise, which merely shows that he did not quite understand everything yet. He believed that such a boundary ''should surely be discernible,'' presumably because there would be a discontinuity at the boundary of the crust generated at the Darwin Rise. In the same vein he entered the discussion about whether the great fracture zones were related to the East Pacific or the Darwin rise. He supported me rather than Hess; the fracture zones were related to the younger rise. Who would have thought that there was no distinction between the rises with regard to generation of crust?

The paper was yet another illustration of Wilson's extraordinary certainty in his beliefs. It was one of his most significant characteristics as a research scientists. Most creative scientists seem capable of only one leap forward at a time, and they have reservations about the solidity of their own discoveries. This is one reason why young scientists can make great creative leaps. They jump from a solid belief in what has gone before. Einstein and Wegener had this characteristic of certainty, so they could work out the corollaries and implications of their own ideas, which are often rather obvious once the ideas are truly accepted. Tuzo Wilson emerged unscathed from believing successively in each of the major tectonic theories. Now, with the advent of sea-floor spreading, magnetic anomalies, and transform faults, he was in his element. He really believed them all. Like Wegener he was a very trusting man.

The two October papers came out side by side in *Science*.[15] Tuzo had no time for journals that published more slowly. Furthermore, publishing several papers in rapid order is like firing a salvo instead of a single shot. It is a much more effective way of hitting the target of the potential reader. Tuzo had done it before. The problem for most people is to have enough new ideas to fill a salvo, even allowing for extensive overlap in the targeting. The paper began with such overlap as was necessary to define transform faults and explain the same transverse sketch map of the northeastern Pacific, which was the focus of this paper.

When he came to new details, Tuzo, who is characteristically generous in acknowledging sources, made a curious error that was to crop up repeatedly thereafter.[16] In his first paper on transform faults he had noted that I had shown a section of ridge north of Cape Mendocino. He did so again but with a twist.

Menard has already [identified] a young mid-ocean ridge in this area. The new interpretation does not change its location, but differs in two respects.

Menard regarded the ridge as the northern end of the East Pacific Ridge with a northwest strike, but the Juan de Fuca Ridge is here considered to be an isolated ridge with a strike of N 20°E separate from the East Pacific Ridge off Southern California, although linked to it by the San Andreas transform fault.[17]

Near the end of this brief paper he ventured another geographical discovery.

R. G. Mason from a study of magnetic anomalies, and H. H. Hess, from a consideration of measured heat flow, have both independently proposed to me in discussion that there may be another young and growing ridge between the Mendocino fracture zone and the Juan de Fuca Ridge. This neat solution [is] named the Gorda Ridge.[18]

Talk about the blind leading the blind. The Coast and Geodetic Survey soundings as of 1949 were contoured by me for the first paper on the Mendocino escarpment in 1952.[19] Figure 1 in that paper shows, using Wilson's names, the Gorda Ridge beginning and ending as a northeasterly trending feature with a median rift. The map also shows the southern part of the Juan de Fuca Ridge as a group of low ridges. Figure 4 is a diagram of bathymetric lineations and earthquake epicenters. The San Andreas epicenters seem to overlap the northeasterly trending Gorda Ridge and reach the southern end of the Juan de Fuca Ridge. This diagram extends farther north than the map in Figure 1. The Juan de Fuca Ridge is shown extending northeasterly for its full length and terminating at a line of epicenters that extend off to the northwest along the base of the continental slope.

I never had any reason to renounce that information. Quite the contrary, I used a Navy Electronics Laboratory ship to survey both ridges early in 1952 and traced the median rift of the Gorda Ridge for its full length. All that was unclassified. In addition, I had the classified maps that I had contoured from the extremely detailed *Pioneer* survey. I had spent some time unscrambling the soundings in the Blanco fracture zone because no one had believed the topography could be so rough, but I knew it extended at least from one ridge to the other. The topography of the Juan de Fuca and Gorda ridges was better known than almost any other region of the sea floor anywhere in the world, and as of 1965 it has been known for more than a decade. It was unfortunate that the east flank of the Juan de Fuca Ridge was buried under turbidites, but the trend and lineations of the crest and west flank were known, and the Gorda Ridge was wholly exposed. Even though the detailed maps were classified, the physiography was not. The large physiographic diagram included as a foldout in my *Marine Geology of the Pacific* shows the entire ridge-transform-ridge-transform pattern very clearly. It even shows the small offset of the northern end of the Juan de Fuca Ridge that Wilson had not yet recognized.

Apart from geography, Wilson had a few points of interest in this paper.

Without really knowing the geography he explained it—which I, knowing it for so long, had not. He drew a distinction between the magnetic anomalies trending parallel to the ridge crest and those farther out on the flanks that had a slightly different trend. The parallel anomalies he took as those created at the ridge crest and referred to their "rough bilateral symmetry." He noted that paired aseismic ridges in the Atlantic demonstrated a similar symmetry of topography. The width of his ridge is compared with the offset on the San Andreas fracture zone, which he expected would be equal. The principle that he here applied for the first time would be fruitful, but the actual example was meaningless because of the existence of a third plate between the ridge and the continental slope.

The final paper in this series was written primarily by Vine who had moved to Princeton in September.[20] It was concerned with the magnetic anomalies on the "newly described Juan de Fuca Ridge." The introduction noted that

> the Vine and Matthews hypothesis was particularly speculative in that no large-scale magnetic survey was thought to be available for an oceanic ridge. . . .[21]

Considering that Ewing and Heezen had declared that their rift extended off Oregon in 1956 and I had published several papers saying that the crest of the East Pacific Rise could be seen in that region, this statement seems surprising. It may indicate something about the general attitude of disbelief for the hypothesis that no one who knew about the ridges thought to check them for symmetry. Perhaps, even more important, few if any people other than Vine had the vaguest idea of what the shape of an anomaly should be. The original paper with the V-M hypothesis certainly did not help because the magnetics illustrated are not bilaterally symmetrical (Fig. 21).

In this paper Vine showed profiles across the Juan de Fuca Ridge, and they were essentially symmetrical just as Backus, Wilson, and Coode had thought they should be. Vine also used the current Cox, Doell, and Dalrymple reversal time scale to compute synthetic magnetic profiles. They were in surprisingly good agreement with the observed profiles, and the power of the modeling technique that Vine had developed was revealed. The tectonic features mapped by field geologists at sea can be explained by exact physical theory. That fact had the same kind of impact as matching continental boundaries by Euler's theorem.

The models were not quite in agreement with observation, and Vine had to make an untidy assumption that the spreading rate varied with time. He presented this paper, however, at the annual meeting of the Geological Society of America in Kansas City in November 1965. There he met Brent Dalrymple who told him about the recent discovery of the Jaramillo event. Vine "realized immediately that with that new time scale the Juan de Fuca Ridge could be interpreted in terms of a constant spreading rate.''[22]

Thus, with another elegant, powerful paper, this series ends. It was not supposed to end so quickly. Tuzo submitted yet another paper to *Nature* describing the evolution of triple junctions—a kind of feature he would have had to invent in order to explain. *Nature* rejected it.

SCRIPPS

Not much happened at Scripps that is pertinent to this history. Al Engel and colleagues continued to emphasize the distinctive characteristics of sea-floor basalts and their relation to the mantle. Stuart Smith and I described the Molokai fracture zone, the one that was needed in 1955 to provide uniform spacing between five great fracture zones. Since then the discovery of the Pioneer fracture zone had already destroyed the uniformity of the spacing. The Molokai trend was interesting, however, because it appeared to affect tectonic trends in the Hawaiian Islands. My major effort before going to Washington had been the completion of a summary paper on sea-floor relief and mantle convection. It had been requested by Keith Runcorn, but it was not published until 1966, and by that time Keith certainly would have been happy if I had withdrawn it. It was my last gasp before converting to sea-floor spreading.

LAMONT AND THE GREAT SYNTHESIS

Maurice Ewing had developed the salvo approach to publication long before Tuzo Wilson, and by 1965 Lamont was ready with the first two of a numbered series of papers on the crustal structure of mid-ocean ridges. These were not like the earlier papers in which Ewing, Heezen, or Worzel presented data in great detail and with a full discussion of techniques and limitations. These were summaries of enormous quantities of data and interpretations of their significance. Ewing was coauthor of a few of these and Heezen of none. The new generation had taken over this phase of Lamont's work. Heezen, in fact, was relatively inactive this year, being first author of only two papers although coauthor of five others. Ewing was going full blast as far as volume is concerned, first author of three, coauthor of 10 more. His range included the origin of ice ages, oceanic heat flow, suspended matter in the ocean, crustal structure of the Strait of Magellan and the Bering Sea, sediments of the Argentine Basin and Tyrrhenian Sea, and the Pleistocene stratigraphy of the North Atlantic.

In the first paper, in January, Le Pichon and others analyzed the ridge structure on the basis of 22 seismic refraction stations in the North Atlantic and compared it with Scripps data from the East Pacific Rise. They concluded:

> The data, therefore, suggest a sequence beginning with a structure of east Pacific rise type, which develops with uplift into a fractured but yet unaltered crust of the type found for the equatorial mid-Atlantic structure.[23]

Further maturity led to the North Atlantic type of crust. In short, they had joined Hess and me in believing in the sequential development of ridges but did not follow Hess about the role of sea-floor spreading in the Atlantic, nor did they agree with anyone about continental drift.

In the second paper, also in January, Ewing discussed gravity and seismic data together, and he was a coauthor but without Worzel.[24] The Lamont group changed their definition of "crust" and "Moho" to agree with my reasoning of 1960. They thereby inevitably concluded that instead of a crust-mantle mix, ridge crests have a crust of normal thickness and velocity but that the underlying mantle has low velocities.

The third paper, in August, was concerned with magnetic anomalies and was based on 58 profiles across ridges.[25] It dealt largely with the apparent properties of the rocks that caused the anomalies and how the anomalies were affected by such factors as local heat flow and the local intensity of the earth's magnetic field. The Vine and Matthews paper had drawn specific conclusions about the relation of anomalies to water depth and to the depth of the Curie-point isotherm.

> It is clear from this study that most of the profiles do not follow the pattern assumed by Vine and Matthews.[26]

This conclusion did not apply to the central point of the Vine-Matthews hypothesis, namely, that the anomalies were generated by reversals of the magnetic field. Heirtzler and LePichon, however, had found "no basis for including" reversals in their modeling of the source rocks, and they did not note that they had the data to test Backus's predictions about the width of anomalies.

A fourth Lamont paper in November was not part of this numbered series, but it was on tectonics. Talwani, LePichon, and Heirtzler wrote a critical analysis of the tectonics of the northeastern Pacific. They began by presenting some of the characteristics of magnetic anomalies. They are linear, parallel to ridge crests, and bilaterally symmetrical about the crests. The last characteristic is based on observations on the Reykjanes Ridge southwest of Iceland. Figure 1 shows eight perpendicular crossings of the ridge crest and identifies seven linear anomalies on one side and their symmetrical twins on the other side. The profiles seem convincingly symmetrical now, but at the time the scale was too small and the diagram was just not up to the standard required to overwhelm the viewer. It was prepared about the same time as Vine and Wilson were working on Juan de Fuca symmetry, so it appears to be an independent discovery of symmetry—by scientists who were not immediately convinced that the sea floor spreads. That fact was clear from two statements:

the symmetry and linearity, as well as the trend of the anomalies parallel to the ridge strike, indicate that these anomalies are genetically related to the formation of the ridge:

but,

The flank anomalies are *not* axial anomalies at greater depth.[27]

The flank anomalies were produced in parallel tension cracks on the flanks.

Given this assumption, the paper turned to the tectonics of the northeastern Pacific—the very region discussed by Wilson the month before. The Lamont scientists were superior scholars. Instead of rediscovering or speculating about the geography, they reproduced my physiographic diagram with earthquake epicenters added. They also reduced this to a diagram showing, with perfect clarity, the crests of the Gorda and Juan de Fuca ridges, the Blanco fracture zone, and the small fracture zone offsetting the northern end of the Juan de Fuca Ridge. They went on to accept and elaborate on my theses that the East Pacific Rise extended under western North America and that the fracture zones and magnetic anomalies were produced on the rise. The authors then took a surprising tack:

it is possible that the various segments of the rise crest were never displaced at all but developed at their present positions. In that case one would have to ascribe the match in the magnetic anomaly pattern not to enormous strike-slip displacements but to the great similarity of magnetic patterns at constant distance from the rise axis, even for distant segments of the rise.[28]

They went on to state "that the long fracture zones . . . existed before the rise" and "that the major faulting on the fracture zones is normal faulting" to explain the vertical relief.

A footnote mentions Wilson's recent paper on transform faults but draws no distinction between the probability of creating identical anomalies at ridge crests, splitting and spreading them, and the probability of creating them symmetrically at a uniform distance from the crest. As Teddy Bullard said in the discussion after his talk at the Geological Society of London meeting, I did "find it a comfort not to have to try to believe such very improbable things."

In retrospect, it seems clear that marine geology was at a turning point in 1965. For the decade of the fifties, the deep-sea floor had abounded with opportunities. There was enough ocean to avoid overlap. There was no old literature to read. There was no basis for controversy, replies, and rejoinders. It was then my objective to write only one paper on each kind of discovery. In the early sixties we ran out of new kinds of discoveries. The paths of ships from different institutions began to overlap. I wrote a book. Heezen published physiographic diagrams—endless expanses of mere examples of a few kinds of things. Maps of global distributions appeared. I wrote three papers on topography and mantle convection. The latest Lamont series showed the trend. Marine geology had passed beyond the discovery phase to one where other people's data and ideas would not be accepted but rather reinterpreted. We had come to the same phase as had land geology. Better rocks and shoals than replies and rejoinders. From them we were saved by sea floor spreading, which takes 200 million years to sweep the sea floor clean but took only six years to sweep the cobwebs out of marine geology.

UPPER MANTLE SYMPOSIA

The International Upper Mantle Committee continued its tectonic symposia with one on East African Rifts at Nairobi in April. Participants included Beloussov, Bullard, Drake, Girdler, and Heezen. In Ottawa there were two symposia in September, and the list of attendees was much more extensive. The first symposium, on "The World Rift System," was presided over by the Chairman of the convening commission—Bruce Heezen—who, although little honored at home at Lamont, was still a major international figure.

The second symposium was presided over by V. V. Beloussov, who was in good humor. In his opening remarks he referred to believers in continental accretion as optimistic and to believers in oceanization as pessimistic for the future of the human race. "The boldest of us move the continents or inflate the globe." He appealed for collection of more data. The Russian presence was unusually strong and offered an opportunity to see the direction of their work. It seemed unfortunate. Gleb Udintsev, a marine geologist with whom I exchanged reprints and visits, described dredging on their research ship *Vitiaz* in the Indian Ocean. He divided the sea floor into tectonic provinces using 21 complicated symbols and identified ancient "monocrations" or stable areas. The mid-ocean ridges were "geotaphrogens." R. M. Demeneskaya used the same complex tectonic maps to show that the Baffin Bay rift once connected across the Atlantic to the Red Sea and was not offset on the Mid-Atlantic Ridge. These complex tectonic maps can be useful on the equally complex continents, but it was hard to believe that the approach would be worth the effort at sea.

Most of the other speakers talked about variations on more familiar themes: linear magnetic anomalies, rifts, and so on. Tuzo Wilson spoke on patterns of growth of ocean basins and continents. Continents grow, break up, and reassemble as a consequence of convection in a single giant cell. He demonstrated transform faults, and the discussion focused on them. H. P. Laubscher (Switzerland) pointed out that transform faults are merely the familiar tear faults of Alpine geology. Adumbration appears early. Wilson "didn't take the point of view that transform faults are wholly new." Bruce Heezen concurred. "The idea of these transform faults, I think, is clearly a fresh one, but if you examine the rift graben in the South Atlantic you can see that the idea wasn't exactly unknown to us." It wasn't unknown to Alan Coode either.

Harry Hess presented a paper titled "Comments on the Pacific Basin." He restated his ideas on the evolution of ridges, the history of the Darwin and East Pacific rises, and the origin of the northeastern Pacific fracture zones on the Darwin Rise. Tuzo Wilson proposed in the discussion that the East Pacific Rise (EPR) was not extremely young and that the fracture zones were related to it. Manik Talwani concurred. How did Hess, he asked, counter my "very impressive arguments" about the slope and the offsets in the slope of the west flank of the EPR? Hess had not thought about that but said he would do so. Wilson cited a paper by William Riedel and Brian Funnell showing fossils as old as Eocene on the EPR flanks. Hess had not seen the paper. K. L. Cook thought that the EPR extends under western North America and cited the slope from California to Hawaii. Elevation of the rise probably produced the early Tertiary (Basin and Range) structures of Nevada and the surrounding area. If so, the EPR could not be extremely young. Hess concurred that the EPR might extend inland but the fracture zones did not cross the San Andreas fault and therefore were older. The discussion continued on similar topics.

K. L. Cook himself gave a talk on the world rift system in the Basin and Range and cited many suggestive lines of evidence in support. G. A. Thompson raised the same issue in the (printed) discussion following Wilson's talk. He thought that there was overwhelming evidence that the East Pacific Rise extended into western North America. After minor confusion Wilson explained that he thought that the EPR *once* went into the western states but now was transformed through the San Andreas fault. The discussion did not stop at that point, but it should have because he had just solved everybody's problems—as usual in that year.

Chapter 20

1966
CONVERSION OF THE INVOLVED

Sea-floor spreading, magnetic recording, transform faults, and magnetic symmetry were all presented by the beginning of 1966. They were already broadly correct ideas, but no one seemed sure that they were, except possibly Tuzo Wilson and Bob Dietz. During 1966 Fred Vine and Drummond Matthews became convinced that their hypothesis was really correct. This was the year of the Eltanin-19 magnetic profile that was a miracle of symmetry. That profile alone was capable of converting to sea-floor spreading almost everyone who was involved enough to appreciate the issues. By the end of 1966 the hypothesis would be accepted not only by the pioneering oceanographers but by the American establishment. It would be accepted by Maurice Ewing and by me.

To those who were not already involved, however, it seemed an ordinary year. Research continued unabated on a broad front by innocent specialists in area X who little dreamed that they would soon be giving talks on "Sea-floor spreading and X." The great ferment was at Lamont where Walter Pitman first spread out Eltanin-19, but no one who was not involved could tell it from Lamont publications until December. It might seem that scientists who are converted to a new theory would immediately withdraw all work in press that invokes the old paradigm. They do not always do so. Thus the ferment at Lamont was covert. The overt record for the year had geologists trying to understand the cryptogram of earth history with no more success than usual. Meanwhile, the covert operations in the code room at Lamont were using the key of Eltanin-19 to obtain a message *en clair*.

The Lamont work would be exposed to public view in November and published in December. Consequently, in this chapter the overt activities perceived by almost all geologists will be presented for the whole year first. The covert activities will follow, leading to the exposure at the end of the year. I hope this approach gives the feel of the times. It is only by viewing the peaceful scene before that one can appreciate the violence of the revolution and the change after.

THE PEACEFUL SCENE

Hundreds of papers were published in geological journals in 1965 and 1966 with only two that mention sea-floor spreading in the journals that I have examined. One of those refuted the Vine-Matthews hypothesis. Only a very small fraction even mentioned continental drift or any other concept of global tectonics. What were geologists writing about as the revolutionists were routing the old order? It does not do to list titles of papers because, as Senator Proxmire has shown, they are specialized to the point of appearing ridiculous. The paleontologists, mineralogists, sedimentologists, petrologists, geophysicists, and stable isotope geochemists, among others, were doing normal science of the same quality as at any other time. Geologists with the same specialties were just as eager to read their research papers as before and after. There was no perception of doom or euphoria; all was normal. *Geotimes* reported in September 1966, that Academician Vinogradov, of international fame, had "dealt a flat 'no' to the theory of continental drift." I'll risk one title, though, and mention Keith Runcorn's "Corals as Paleontological Clocks," a fascinating account of the integration of John Well's careful observations of Devonian corals into the history of the rotation of the earth. It was excellent science and may ultimately lead to important conclusions about global tectonics, but it had nothing to do with the ongoing revolution. Runcorn was long since on the side of the angels and was waiting for the rest of us to join him.

Waiting was a facet of the old geology. The ideal had become the patient, thorough, even leisurely study—for decades—of a problem or area. There was no haste to publish. Why hurry when a geologist had exclusive claim to a field area? Important papers could be submitted to the International Geological Congress with deadlines years before publication without fear of losing priority. No one else was working on the problem. The *Bulletin of the Geological Society of America* delayed publication by a few years and the U.S. Geological Survey by many.[1] Persnickety editing was more important than the rapid dissemination of science. Science would come out fast enough for the need. Warren Hamilton captures the times in a letter to me in 1984.

I have a Permian manuscript of about 400 pages, untouched since ca. 1968, integrating paleoclimatic, paleobiogeographic, paleomagnetic, and tectonic data into a global synthesis. When I wrote it in the mid 1960's, few geologists were drifters, and it seemed that I had all the time in the world to do the job leisurely and properly. Then plate tectonics erupted and I was off on other things.

"Leisurely" means by Warren's standards; in 1966 he was one of the few Americans actively appealing to continental drift to explain his data. Regarding the Caribbean and Scotia arcs,

If the continents are regarded as having drifted westward into the Pacific Ocean basin, then the two arcs . . . lagged behind.[2]

The statement was prepared for a symposium in Canada, and the recorded discussion was sharp.

Prof. Runcorn: Your last statement, which pictured a raft moving independently of what's underneath, of course, is nonsense.

Dr. Hamilton: Perhaps. But. . . .[3]

Before responding to such statements, geologists should always recall (but of course not mention) that Kelvin and Jeffreys were superb physicists. Hamilton also explained the tectonics of the western states in terms of some form of continental drift or deformation on the scale of drift.[4]

As far as the journals were concerned, even 1967 was more of the same geology at the same pace. The more "geological" journals contained three papers that mentioned sea-floor spreading, and one of those was opposed. Only in the *Journal of Geophysical Research* did the hypothesis become significant. It was not until 1968 that the overwhelming proof of sea-floor spreading was published, and by then the hypothesis was obsolete and had been superseded by published versions of plate tectonics.

MARINE GEOLOGY AS PUBLISHED

The number of marine geologists was doubling every four to five years, and new people began to publish important results.[5] One of them was George Peter (1934–), who was born and received an undergraduate degree in Budapest. From 1957 to 1963 he was at Lamont and then took employment as a geophysicist with the U.S. Coast and Geodetic Survey. He worked with Harris Stewart on the Seamap Project, which was a detailed survey of the marine geophysics between the Aleutian and Hawaiian islands. He began to talk about their results, and on 18 March he received an enthusiastic letter from Harry Hess.

That 15-minute presentation of your latest Pacific results is so exciting that I have spent at least fifty hours during the evenings puzzling over what it might mean and do I have to revise some cherished hypotheses about the Pacific.[6]

The rest of the letter shows that Hess was in fact confused about what Peter had described, but he would have been just as excited if he had remembered it correctly. The data were published in November. Peter had traced some of the magnetic anomalies on the outer flank of the East Pacific Rise across several fracture zones. The trend was almost exactly the same for 1,500 km. The great length was interpreted as suggesting that the anomalies followed ancient lines of weakness, and that the anomaly bands now located over the flank of the rise were older than both the rise and the fracture zones.[7] He was supporting Hess's published view, rather than Wilson's or mine, about when the anomalies were created but not about how. What Harry believed at any given time, however, was what the data then indicated. He must have been wondering during that 50 hours how the anomalies could have been created on the Darwin Rise if they formed a continuous series from the longitude of Hawaii to the active crest of the East Pacific Rise. Peter seemed to be destroying the logic of Hess's argument while agreeing with his conclusion. The Seamap survey showed that something strange happened to the flank anomalies just south of the Aleutian Trench. The happening was interpreted as an offset on a fracture zone, but it would prove to be much more novel. The Coast and Geodetic Survey had discovered the "Great Magnetic Bight" where the anomalies curved around in a vast, puzzling bend. Its existence would cause a great stir in the next year.

The flank anomalies were being traced outside the northeastern Pacific. Christoffel continued his work on the magnetic anomalies south of New Zealand—"the magnetic anomaly pattern cannot be completely explained by any current theories on ocean floor formation"[8]—but Vine-Matthews could explain everything not affected by the topography of the ridges.

In 1966, Peter Vogt and Ned Ostenso published one of their first detailed geophysical surveys of the Mid-Atlantic Ridge that would ultimately lead to highly detailed maps of the whole North Atlantic like those of the Seamap Project in the Pacific.

At Scripps we continued to publish our largely descriptive papers about the deep sea. I, for example, took advantage of our newly acquired computer to redetermine the hypsometry, or frequency of depths, of ocean basins. I also found myself in the odd position in May of refuting J. B. Carr's refutation of continental drift. He had proposed that the existence of Vening Meinesz's ancient shear net on the continents ruled out the possibility that the continents had drifted. I pointed out that his shear net did not agree with the fracture zones on the sea floor and "should be allowed to rest in peace." I was not yet accepting continental drift, except in the Atlantic, but science is science and fair is fair.

At Lamont, Jim Heirtzler managed to publish two more papers on marine magnetic anomalies without even referring to Vine and Matthews. John Ewing produced a status report on marine seismic studies showing the vast network of reflection lines obtained by Lamont up to 1966. The chief work of the early part of the year, however, was part four of the crustal structure of the mid-ocean ridges, and the first author was Maurice Ewing himself. Five years earlier he had undertaken to test sea-floor spreading, and at last he was prepared to discuss the Cenozoic history of the Mid-Atlantic Ridge. The circumstances merit some review. The model that Ewing was testing was simple. As Teddy Bullard said, he had learned the style of physics while a student at Rice. Pelagic sediment falls onto the sea floor; other things being equal, the thickness of sediment is in direct proportion to the age. If the sea floor spreads at a constant rate, the thickness is in direct proportion to the distance from the ridge crest. He had not found such a simple relation in the North Atlantic. The sediment was thinner on the average on the crest than on the flanks, but locally it was not. He proposed that sediment was removed from the steep slopes of the crest by slumps and turbidity currents.

Now Ewing, and his colleagues, pointed out that the South Atlantic is the type example for the hypothesis of continental drift and thus the logical place for a test. What did they find? The sediment distribution was as complex as in the North Atlantic. The sediment of the crest was always thin, but in two crossings it formed a continuous cover. The sediment thickness on the flanks increased with distance from the crest in some profiles but was uniform for long distances in others. Ewing drew on his hard-won data bank. Cores indicated Tertiary sediment on the Mid-Atlantic Ridge and on the Walvis Ridge and Rio Grande Rise. In other places Pleistocene sedimentation appeared adequate to produce the entire sedimentary layer. Given the rates of carbonate sedimentation,

> the indication of very recent age or destruction of sediments over the crestal area is compelling.[9]

This time the destruction was not by erosion. Ewing harked back to the samples collected on his Mid-Atlantic expedition of 1947. He had dredged basalt with glassy surfaces in which was embedded baked Globigerina ooze. This suggested that

> over the central part of the ridge, sediments older than Miocene were buried under lava or consolidated by contact with lava, thus explaining the thinness of unconsolidated sediments recorded by the profiler.[10]

The existence of a continuous cover or patchy cover was associated with smooth or rough topography respectively. The roughness was caused by the same volcanism that engulfed the sediment.

Considering the data on sediment ages,

> The hypothesis that best fits the facts is that some major event, probably in early Miocene time in the north . . . possibly later . . . in the South Atlantic buried or consolidated most of the sedimentary cover north of 29°S.[11]

Certainly for the sediment isopachs in hand, and ignoring all else, the hypothesis was superior to the hypotheses of continental drift or sea-floor spreading that were being tested. Moreover, Ewing et al. offered additional tests as well. The South Atlantic was supposed to have started opening when marine sediments were first deposited on its borders about 120 million years ago. If spreading was at a constant rate, "500 km of new crust should have been added on each side of the crest" in the last 20 million years. Yet sediment younger than that age had been obtained in six locations, and

It is becoming increasingly clear that if continental drift occurred at all, it must have terminated in a few million years. Or if it took as much as a few tens of millions of years, the process of continental drift was one of adding new crust at the axis without disturbing the sediments in the basins or at the margins of the continents. In any case, all movement had ceased 20 m.y. ago at the latest.[12]

I accepted the data as valid and was impressed by these arguments. I wrote to Hess in March 1966:

the Ewing et al. papers on Atlantic stratigraphy really eliminate significant Tertiary Atlantic sea-floor spreading.

Hess replied,

I don't agree at all. Ewing et al. have one misplaced sample (the first one in their table) and one bad K-Ar age. If these are disregarded everything fits as it should.[13]

The paper was submitted in September 1965 and was received after revision on 7 December—almost exactly at the time that Walter Pitman realized that Eltanin-19 was symmetrical. It was in press only four months, but those were the months of the greatest ferment at Lamont.

During that same period, Ewing was involved in another controversy. He and two colleagues had prepared a manuscript, describing the old fossils embedded in basalt, that was submitted to *Science* in January and published in March. They said,

The data rule out the possibility of large-scale continental drifting or spreading of the ocean floor since the Lower Miocene.[14]

Egon Orowan, Professor of Mechanical Engineering at MIT, had been interested in mid-ocean ridges since their discovery, and he responded to the paper.[15] *Science* sent his response to me for review, and I wrote to him on 28 April,

You have illuminated a very important point just when I thought that Ewing's data had struck a mortal blow at sea-floor spreading.

He had tried to determine just how the Lamont group visualized the flow of mantle material as it changed from rising to spreading. In this he had difficulty, as he indicated to me in a letter of 9 May:

There seems to be some mystery behind the Lamont picture of ridges and the crust. . . .

Orowan had obtained a preprint of an illustration from Lamont that showed vertical flow bending through a right angle to horizontal flow, but "this flow pattern is not mechanically possible."[16] He proposed instead that the rising convection diverged under the ridge crest in a pattern "essentially identical with that proposed by Menard." If so, the ridge crest was a stagnant zone, and the occurrence of old fossils there did not bear on the question of sea-floor spreading or continental drift of the ridge flanks. The spreading zones might be on the ridge flanks where I had found the mountainous relief on the East Pacific Rise. There was an alternative possibility in which the crust was created at a ridge crest by the injection of dikes. If so, "the whole oceanic crust must consist of basaltic dikes."[17] It soon developed that there was no stagnant zone, but it was an attractive possibility until the origin of magnetic anomalies showed that Orowan's alternative possibility was essentially correct.

DIETZ

Harris B. Stewart, Jr., who long before was tied to the echo sounder as we surveyed the Mendocino escarpment, had become an important official in the Environmental Sciences Services Administration, which had been formed by amalgamating the venerable Weather Bureau and Coast and Geodetic Survey. One of his wise acts was to hire Bob Dietz as a senior scientist free to do research of his own choosing. As far as I could tell, this did not change Bob, but he was free at last. He published two papers on continental drift in 1966. One on ancient supercontinents was straightforward, normal science, but the other—on "Continental Drift and Earthquakes"—may have been one of the first attempts to apply drift to a practical problem like predicting earthquakes. The application is now commonplace.

Dietz's major effort relating to drift, however, was the elaboration of his model of the formation of geosynclines that he had proposed in 1963 and that had been attacked by Ken Hsu. In one paper he turned to the evidence that Precambrian continental crust is largely undeformed. The belief common then— that the most ancient rocks were deformed—was based on circular reasoning that could be demolished by isotopic dating. A second paper was concerned with "mioclines," Dietz's new name for "miogeosynclines," in space and time.[18] He elaborated on the theme of "thickening out," providing many convincing examples of the structure of continental terraces. He then showed that several classical stratigraphic sections in the interior of North America and Africa can be interpreted as having similar structures. A third paper, by Dietz and

John Holden, elaborated on the eugeosynclinal section of Dietz's 1963 paper.[19] It focused on the ''Doctrine of Vertical Permanency''—that continents and ocean basins are not interchangeable—and how the idea was influenced by the apparent absence of deep-sea sediments on land. Dietz and Holden compiled the numerous worldwide occurrences of radiolarian cherts and shales, which they accepted as evidence of deep water. The occurrences are largely in eugeosynclinal prisms. Thus Dietz and Holden deduced that deep-sea deposits are in but not on the continents. It is difficult to overestimate the number of regional specialists who might have been offended by having ''their'' regions and ''their'' sections reinterpreted by a marine and cosmic geologist. Some would certainly reply, but most of the land geologists probably took the approach of Gilluly,

> I tentatively suggest that the eugeosyncline developed in large part on an oceanic crust[20]

or that of Warren Hamilton,

> The modern analogue of the oceanic tholeiitic eugeosyncline is the floor of the ocean along [a] continental margin[21]

without any reference to the works of Robert S. Dietz.

FERMENT AT LAMONT

Ewing was looking to sediment thickness to provide a test for the sea-floor spreading hypothesis. He and his close associates poured over each record as it arrived at Lamont. The ironic fact is that his ships were collecting another kind of information that gave a better test, and no one was looking at it. In the fall of 1965 two graduate students, Walter Pitman (Fig. 42) and Ellen Herron, were on the twentieth cruise of *Eltanin*, a large ship suitable for the South Pacific. They were compiling data in the normal routine way, annotating records and so on. Their job was to obtain data that could be used at home, and indeed the format of the data at that time did not lend itself to interpretation on board. The magnetic records were long rolls of paper, and they showed not anomalies but the total field. Moreover, the anomalies had a larger amplitude than the width of the paper and did not appear very simple. The main thing that was obvious on a magnetometer record was the latitudinal variation in the earth's field.[22] Ellen and Walter were on Eltanin-20 from September to November collecting data for Lamont.[23] They might ultimately use the data themselves or it might be used by someone else in pursuit of a problem; meanwhile, the prime concern was simply to acquire data, and after three months they would have had many boxes.

Back home in December, Walter was working on the Eltanin-20 results and made a computer printout of the magnetic profile. Soon after he returned, he had read the Vine and Wilson paper that had been published in *Science* while he was at sea. Sea-floor spreading and the Vine-Matthews hypothesis had hardly captured his attention before, but the very short magnetic profiles across the Juan de Fuca Ridge closely resembled those he had just plotted for Eltanin-20. Walter wandered down the hall to display his puzzling find but was greeted with skepticism and ridicule. A scientist, as a scientist, however, cannot help but follow where the data lead. Every effort had been made to direct the ship tracks perpendicular to the ridge crest on the reasonable assumption that most phenomena would vary with distance. Moreover, such tracks would avoid off-sets on fracture zones. Circumstances and logistics, however, sometimes required oblique courses (Fig. 3). The computer made it possible to generate synthetic profiles to show what would have been recorded if the ship had gone perpendicular to the crest and not crossed any fracture zones. This made it easy to compare profiles from different cruises. Luxuriating in the new ability to plot, Walter summoned up a profile of Eltanin-19 (Fig. 22). "It hit me like a hammer";[24] the magnetic anomalies were symmetrical on the two flanks of the East Pacific Rise.

Lamont was more compact than now, and most of the geologists and field geophysicists and related graduate students were in a few adjacent rooms. One of the scientists was Neil Opdyke, who was working on magnetic stratigraphy. He had graduated from Columbia, where drift bordered on the unspeakable, and received a doctorate from Newcastle on paleoclimatological research that supported drift. He had worked closely with Keith Runcorn, and his acceptance of drift was complete and widely known when Ewing brought him to Lamont. He was the resident heretic. One night in January 1966, Walter was working late after acquiring profiles of Eltanin-19, -20, and -21, and the Juan de Fuca Ridge. They were all symmetrical, and they all showed the same anomalies allowing for different rates of spreading. Like Martin Luther at Wittenberg, Walter pinned them to Opdyke's door and went home.

The next morning a conversion more like the Protestant Reformation than a scientific revolution began at Lamont. The activities were feverish like those at Madingley Rise the previous spring, but the revision of basic assumptions was enormously more painful. As at Madingley there is no conflict or confusion regarding the unfolding of the facts, and they have been well documented in interviews by Wertenbaker, Frankel, and Glen. By February, Pitman had generated a reversal time scale for the South Pacific. The reversal time scale on land so painfully acquired by the groups at Menlo Park and Canberra extended back only 4 million years. Assuming constant spreading, the marine time scale spanned about 80 million years (Fig. 22). Just beyond the oldest anomaly dated from land was a distinctive anomaly called "four-fingers Brown" (Fig. 23) by

Pitman after a gangster in Newark where he was brought up. Due to a misunderstanding, we later called it "four-fingers Jack" at Scripps after a western desperado. In June, Opdyke found the four-fingers-Brown anomaly in a sedimentary core. Pitman, Heirtzler, and others had been working at top speed but quietly. Also in June, they called in Lynn Sykes and Jack Oliver among others and had a show. The conversion snowballed.

Meanwhile, there was the question of publication and priority. Fred Vine visited in February, and Heirtzler gave him a copy of the Eltanin-19 data. In April, Heirtzler displayed the Eltanin-19 profile at the American Geophysical Union meeting in Washington. Vine attended the meeting bearing a draft of a manuscript that, among other things, discussed Eltanin-19. Perhaps shocked out of their euphoria, Pitman and Heirtzler in May and June began to write a brief report on Eltanin-19 for quick publication in *Science*. It would appear in December 1966. They were joined by a growing group of colleagues, and after April they also began the massive documentation of sea-floor spreading that would be published in early 1968.

So much for the facts; how and when various people reacted during this conversion is rather more difficult to establish with confidence. The basic problem is that Lamont was so fragmented at some level or in some way that different people had perceived what had been happening in totally different ways. And what they had perceived influenced what they were perceiving.

I shall restrict this discussion to the views and actions of the most conspicuous person at Lamont, Maurice Ewing. At the time in 1965, I was aware of only two groups at Lamont: Bruce Heezen plus associates and students, and Maurice Ewing plus everyone else. Apparently there was a third group. The Ewing group actually consisted of Ewing's older associates, and confidants other than Heezen, most of whom had known him as a young, approachable scientist. They were apprised of his methods and goals. A third group consisted of graduate students and newly hired scientists who knew Ewing only as a man in his late fifties and as their famous director. Naturally, one could get a very different idea of Maurice Ewing by talking to members of the different groups. Joe Worzel had worked with Ewing for almost 30 years. To him, Ewing was a brilliant scientist of great versatility with a continuing record of prizewinning solutions to difficult problems. Bruce Heezen saw Ewing as a flawed hero who was not keeping up with changing ideas. The younger group, presumably, saw or heard of Ewing as a great man with strong, unchanging views.

The virture of the above speculation is that it may help account for the very different statements that have appeared about Maurice Ewing. It may account for the statement of one historian about another:

> Wertenbaker's *The Floor of the Sea* (1974), the most complete treatment of Ewing, fails to convey his adamantly fixist views.[25]

In any event, Talwani and Heirtzler have both described how Ewing's immediate reaction to Dietz's sea-floor spreading was that it was worth testing, and he focused the Lamont effort on that objective. That was one of the reasons why Ellen Herron and Walter Pitman were recording magnetics on profiles that Jim Heirtzler had laid out perpendicular to the crest of the East Pacific Rise. Yet Ellen understood:

> Doc Ewing's philosophy that the oceans were permanent features was the party line at Lamont; sea-floor spreading was anathema.[26]

Likewise, Eric Schneider, a student of Heezen, reacted with amazement bordering on incredulity in 1984 when I described Ewing's reaction and actions to him. All he knew was that Ewing, Worzel, and Heirtzler strongly and openly opposed continental drift.

As to Heirtzler's opposition to drift, and the objectives of the ship program, Jim Heirtzler himself remembers only that he and Ewing were seeking the correct answer before committing themselves. He was not a ''fixist''; he was ''not anything.'' His emphasis was on ''collecting data to test ideas rather than speculating on a finite set.''

Neil Opdyke was hired by Ewing, who knew he supported continental drift. Ewing did not direct Opdyke to do anything, but he suggested that the magnetic stratigraphy of deep-sea cores might merit study. Opdyke later made the following characterization:

> Ewing was an avid anti-drifter and the leader against mobility, both intellectually and emotionally.[27]

Dennis Hayes, who had been Ewing's student, says he ''never found anyone more open.'' That was my feeling. I observed Ewing to be a passionately committed scientist, and he avidly wanted answers. I talked to him off and on for decades, and I felt his opinions were strong but based on the facts as he knew them at the time. Bullard felt the same way:

> It is remarkable that Ewing not only allowed but encouraged Heirtzler, Opdyke, Lynn Sykes and others to pursue this investigation and to publish views that were basic to the subject on which he had spent his whole life but were contrary to his own beliefs. His open-mindedness led to what was, perhaps, Lamont's greatest success.[28]

There remains the question of Ewing's reaction to Eltanin-19 and the mass conversion going on around him. A scientist, as a scientist, has to accept facts, unless they can be disputed, and inferences drawn from facts, unless they are unwarranted or there are alternative explanations. The symmetry of magnetic anomalies that hit its discoverer like a hammer was a fact. It might be expected that scientists would accept the reality of this fact and its implications roughly inversely to their knowledge of competing facts and inferences. Teddy Bullard put it this way regarding Ewing and the implications of Eltanin-19:

> I think his initial difficulties were due to knowing too much. If you have in your mind an enormous data bank, there is sure to be some fact that appears to contradict any general theory. You then become very wary of all general theories.[29]

It was to be expected that Neil Opdyke would see and accept the symmetry immediately, because it was a magnificent confirmation of what he already believed. Nor is it surprising that Jim Heirtzler, confronted with Eltanin-19 after years of thinking about possible origins of marine magnetic anomalies,

> did not come to accept its true significance until three weeks later.[30]

What of Ewing who had thought so long and collected so much? William Glen writes,

> Whereas most others who had shared his fixist views were dislodged by the powerful magnetics evidence in 1966, his dedication to the fixist cause precluded his conversion to sea-floor spreading for three more years.[31]

That would be 1969.

There are three lines of evidence that support a faster adjustment to the fact of symmetry and its implications. First, there is a witness of Ewing's reaction upon seeing Eltanin-19, which, it seems probable, was as soon as he was available at Lamont to see it. Walter Pitman said,

> Ewing swallowed—and said he wished it would go away—but he believed it. [Walter nodded his head slowly and repeatedly] . . . he believed it.[32]

Second, Teddy Bullard talked to Ewing in November 1966, just before a meeting that will shortly be discussed:

> Just before the meeting started, Ewing came up to me, looking, I thought, a little worried, and said: ''You don't believe all this rubbish do you? He admitted that he did and I fancy that the following two days of systematic exposition, largely by his own students, convinced him (he did not contribute to the published proceedings of the meeting). He still found the ideas too simple and too uniformitarian.[33]

Third, the Ewings and others published a paper in December 1966 in which they explain that their data are compatible with sea-floor spreading, just not with spreading at a constant rate. As Teddy said, the ideas at the time were too simple to be compatible with all the data as Ewing understood them. In fact, the Lamont group became so committed to uniform spreading before the older reversals were dated that I later appealed to Jim Heirtzler not to risk discrediting his work.

Finally, Manik Talwani says that everyone at Lamont accepted the magnetic evidence of drift, but there was a total disagreement about the time of drifting. If there had been a simple relation between sediment thickness and distance from the ridge crest Ewing would have discovered it, and the magnetics would only have provided an additional proof of sea-floor spreading. The sedimentary pattern was complex. Thus the simplest hypothesis for Ewing, when he accepted the magnetic evidence of spreading, was that the timing and rate of the spreading were also complex.

BRUCE HEEZEN AND THE IOC MOSCOW

The Second International Oceanographic Congress was scheduled for 30 May–9 June in Moscow. Like all such great international spectacles, the preparations had begun long before. The abstracts of talks were already handsomely bound and waiting when we arrived in Moscow, so they could not possibly have reflected the ferment at Lamont or even Vine and Wilson's already published symmetry of magnetic anomalies. The abstracts suggested that the talks would not be the most interesting part of the Congress. Most participants would be content to adhere to their abstracts because they really had nothing new to add, but that was not true of the geologists from Lamont. They were able to dazzle if they wanted to. One of them did, and the consequences were tragic for Bruce Heezen.

Early in 1965, Ewing had urged Neil Opdyke to pursue the possibility of developing a reversal stratigraphy in deep-sea sediment. There had been intermittent efforts to do so ever since Ron Mason had tried it in the early 1950s. Opdyke, not unreasonably, was more interested in using his paleomagnetics laboratory to study the hard rocks around him for evidence of continental drift. He thought Ewing's idea was a waste of time. Ultimately, he adopted the standard solution of middle management: when confronted with some stupidity from on high, he dumped it to the next level below. By June of 1965, graduate student John Foster had built a spinner magnetometer sensitive enough for studying red clays. Early attempts to detect reversals were unsuccessful. One night in the summer of 1965, however, Foster and Billy Glass sampled a few cores as a courtesy to Donald Corrigan, a graduate student of Dale Krause, who was visiting from the University of Rhode Island to learn about magnetics. They found the first reversals, and the unparalleled Lamont core collection lay waiting around them. Teddy Bullard had once asked Ewing why he was so obsessed with taking cores and was told,

I go on collecting because now I can get the money; in a few years it will not be there anymore, then I shall have the material to keep my people busy for years.[34]

The people who would benefit would not even be the ones who had been collecting the cores for 15 years.

What then happened has been documented in some detail by William Glen.[35] Foster phoned Opdyke, Heezen, and Bill Ryan, one of Heezen's students. Opdyke was "not favorably responsive at that point." Heezen immediately drove to the laboratory, and with characteristic enthusiasm and energy he supervised the preparation of many more samples.

Heezen grabbed all of his graduate students and put them to work; Eric Schneider was going to do the North Atlantic, Ruddiman another ocean, Ninkovich the North Pacific, and Ryan the eastern Mediterranean.[36]

Opdyke objected to having his paleomagnetics laboratory overrun and requested Heezen to curtail his work. Heezen appealed to Ewing to at least permit Billy Glass to continue his thesis work that was underway before the invasion. That was granted, and Glass studied the magnetic stratigraphy of a series of Antarctic cores for which a detailed paleontologic stratigraphy had been established by James Hays—an earlier student of Heezen.

Then came Pitman's discovery of symmetry and the profiles pinned to the door. In February, Opdyke "decided to pursue the deep-sea core problem."[37] Consequently, for the first time there were two scientists and their students investigating the same problems and the same samples. Cooperation was one possibility, but Opdyke "told Bruce Heezen that he could not be an author on the first paper." This was a prescient evaluation of the power of the Matthew effect. Bruce would have had all the credit. Heezen had refused Ewing as a collaborator once he had tenure and had been refused as a collaborator by a former student who had received his doctorate. The multiauthorship and common-data-pool approach at Lamont almost inevitably led to these confrontations over authorship.

Any prospect of collaboration ended, but the spoils still needed to be divided. A meeting was held to discuss priorities and publicity. Heezen was expressly forbidden to talk on the subject of magnetism in sediments at the International Oceanographic Congress in Moscow.

We met in Moscow. I was in my last month on the White House science staff and had been allowed to accept an invitation from the Academiia Nauk, the Russian Academy of Sciences, to be its guest for two weeks. I had already been in Leningrad and the Crimea visiting oceanographic laboratories before I moved into the Leningradskaya Hotel in Moscow. Gleb Udintsev took Bruce and me to see "Swan Lake" at the Bolshoi Ballet, but they did not deign to join me in the frivolity of a musical play. As for the meetings, I talked about hypsometry, and Bruce was on the published program merely as the third author of a talk by Bob Fisher. Nonetheless, he did talk (see Fig. 39 for a comparable press conference).

The *New York Times* on 2 June (1966) had a story on an interview in Moscow. "Evolution Linked to Magnetic Field," said the headline.

> The mutation and extinction of species in the evolution of life on earth may be attributable in part to disappearances and reversals of earth's magnetic field. . . .

> The theory was advanced by Dr. Bruce Heezen in a report on the study of rock cores. . . .

> Dr. Heezen said that reversals . . . could be determined in dated cores of sedimentary rock. . . .

> The professor submitted his findings at the second International Oceanographic Congress. . . .

> Other members of the research team with Dr. Heezen were Dr. Neil Opdyke, Dr. Dragoslav Ninkovic, Dr. James Hays. . . .

The field reversals involved a period with no field and therefore no protection from cosmic rays.

> The result of this cosmic ray bombardment is the complete killing off of some species.

> the magnetic field is decreasing in intensity and will reach zero in 2000 years. . . .

> "I don't want to be an alarmist, but we may be next," he said with a smile.

A follow-up story about the Moscow talk appeared in *Time* magazine on 17 June.

> Working with . . . sedimentary cores taken from the bottom of the North Pacific, Geologist Bruce Heezen and his associates . . . carefully examine each one . . . for traces of residual magnetism. . . .

It was flagrant defiance of the restrictions placed on him at Lamont before the meeting.

Just what happened later at Lamont came to me in fragments, but they seemed consistent. One of the "other members of the research team" filed a formal complaint. Hearings were held at various levels and Heezen was censured, but his tenure protected his basic employment. Ewing said that granting the tenure was the worst mistake he ever made, and he adopted the only measures available to him.

The events that followed are partially revealed in a letter to President Kirk of Columbia, drafted in pencil by Bruce. Internal evidence indicates that the date was late 1966-1967, and the draft was found by Marie Tharp in 1984. There is no indication of whether or not it was sent, but it is a poignant statement of his feelings at the time.

Regarding administration attempts to separate him from his students,

> I regret to say that I have received numerous reports that border on blackmail. . . .

The students were quoted as being told,

> You may have the official cooperation of the University in your project if you exclude your professor from the work.

Regarding publications,

> I do strongly believe however that a Professor in the University should not find it necessary to submit his writings for approval by any officer of the University.

Concerning harassment,

> Proposals must not be unduly delayed [or] transferred to other people, and utmost care must be taken that they are not lost or vetoed without explanation.

> When a faculty member in the University presents a proposal for consideration the administration owes him an explanation if an adverse decision is reached. It is, I believe, not ethical for the administration to refuse to sign without giving cause, to submit the same or similar proposal under his own name or that of another staff member, to propose to a granting agency the funding of an existing project not conducted under another professor and representing it as his own work or that of another. It is not in my mind ethical to enter informal negotiations in an attempt either to stop funding of an existing contract or to discourage funding of a new project without the knowledge of the professor involved.

And what of that precious tenure?

> It is abundantly clear that the administration wishes by such tactics to circumvent academic tenure and force tenured faculty to leave. It is obvious that academic tenure to these men is not something to be respected but something to be circumvented. If such tactics are successful it would be clear to all that academic freedom cannot be exercised at Lamont.

At age 42 the internationally famous oceanographer Bruce Charles Heezen was forbidden access to the data of his home institution, denied the right to have Lamont as a sponsoring agent for his requests for grants, and denied access to Lamont ships. Under the circumstances he might have been expected to accept a position somewhere else, but apparently none was offered thereafter and he did not solicit one. Instead he chose to stay at Lamont and live by his wits.

BACK AT LA JOLLA

There were two things on my mind when I returned to La Jolla in July 1966: the Nova expedition and the reality of magnetic symmetry. The expedition has a side that is pertinent to this account, so I shall mention it first. The point of concern is the vulnerability of research scientists to charges of conflict of interest that may arise because they are thinking only about what counts. The book I wrote about this expedition[38] tells about its conception and funding. The idea for the expedition began to develop in the fall of 1965 when I was in Washington. A proposal for funding was submitted to the National Science Foundation by a group of us at Scripps on 13 May 1966. I was still in Washington on the White House science staff. After almost a year I was aware of some of the wholly unstated but fully understood power that the merest underling derives from being on the staff of the President—at least while in Washington. Much earlier I had found to my amazement that when I had a meeting scheduled with the Director of the National Science Foundation he expected to come to my office. I had learned that the Mohole project could be killed on a suspicion of a conflict of interest. Yet it never occurred to me that I might have such a conflict in asking the NSF for $1,690,000, and no one else raised the issue. Now, after three years of daily concern with minuscule conflicts of interest in the Geological Survey, I can only look back in wonder.

As to what counts, all I had on magnetic symmetry, other than hearsay, was the profiles published by Vine and Wilson. I had found them too small, too short, and too few to be convincing. I hauled out my copies of the drafts of the Mason and Raff maps. The drafts were three times as large as the published versions. I made numerous profiles across the full width of the anomalies at this large scale. The profiles covered a much wider area than Vine and Wilson had discussed. As I did the work, it became obvious what was coming. Even so, when I laid out all the profiles I could only shake my head and think, "Well, I'll be damned."

Tuzo Wilson wrote to me on 11 August, "I find submarine geology in a very exciting state." I could not have agreed more.

I wrote to him on 27 October,

I have been looking further into the problems of sea-floor spreading and magnetic anomalies and I must say I am getting increasingly convinced. The symmetry of the magnetic anomalies in a number of places is really overwhelming.

My expression "am getting" may have been the wrong form of the verb. I was already doing normal science within the framework of the new theory. I wrote that I believed I could show that the crest of the East Pacific Rise must have moved while spreading. The manuscript already existed, and I would talk about it at the NASA meeting in 10 days.

THE NASA MEETING AT COLUMBIA

In early August I received a phone call inviting me to a meeting, and I said I would come. Soon after, a letter arrived from Robert Jastrow. He and Paul Gast were planning a small conference on the "History of the Earth's Crust" on 10-12 November at the Institute for Space Studies at the edge of the Columbia campus. The Institute was part of the National Aeronautics and Space Administration, so this became known as the NASA conference. The meeting was fortunately timed because the scientists from nearby Lamont were ready to display their work, but nothing was published. The list of invited participants included only 39 people, but more than half of them were or would be members of the National Academy or Royal Society. This meeting fortuitously gave many of the young members of the establishment a preview of the mass revolution that was about to occur. It also gave them a little edge on the competition.

Hardly any of the titles in the tentative program that was first circulated are the same as those given at the meeting and eventually published.[39] Presumably, this means that people gave tentative titles when they were phoned to ask if they could come. Some other changes, however, seem significant. Maurice Ewing did not give his scheduled paper on "Sediment Cover of the Deep-Sea Floor," although he attended the meeting. Gordon MacDonald did not publish his portion of the Closing Review, although I don't recall whether or not he was mute at the time of the Review. The withdrawals give a clue about the way the meeting went. It began with a few talks on the upper mantle, including one by Dan McKenzie, whose reaction to the meeting may also show how it went. Dan had started graduate research at Madingley Rise just after the Vine and Matthews paper appeared, and he was there while Hess and Wilson visited and while transform faults and magnetic symmetry were being discovered. Yet,

At this conference various papers clearly showed me for the first time that sea-floor spreading was a global phenomenon and had probably generated the ocean floor. I think the papers by Vine and by Sykes were the ones which really convinced me. I returned to England for about six weeks and immediately started to work on the subject.[40]

Much more will be heard of the results of the labors triggered by this meeting.

The second part of the program included the papers that convinced McKenzie—and presumably almost everyone else. Certainly, many people learned a lot that day. After one of the excited talks on magnetic stratigraphy involving the Jaramillo and other events, Gordon MacDonald turned to me and said "My God! They've got names for these things now." That was only the beginning. Fred Vine showed that the youthful magnetic anomalies of the Juan de Fuca Ridge were symmetrical and could be modeled with the reversal time scale. The same was true of the Reykjanes Ridge and the East Pacific Rise. Assuming constant rates of spreading at each location, the anomalies from 3.5 to 11 million years old were the same on the Juan de Fuca, Gorda, and East Pacific ridges. Therefore the reversal time scale for that period was established. He presented a color map of the age of the crust of the Juan de Fuca Ridge, and it was manifestly symmetrical. He did the same for the Reykjanes Ridge with the same impressive result. He compared the older flank anomalies mapped by Mason and Raff and by Christoffel and Ross south of New Zealand with anomalies in the northeastern Atlantic and showed that they could all be modeled by the same reversal time scale.

My memory is that the audience was stunned, but certainly there was a discussion. Don Anderson asked why the crust was spreading. Because of convection. Jerry Wasserburg asked why, if spreading was symmetrical, the central anomaly was not. It was symmetrical, but it didn't always look that way because it was also a function of latitude and orientation. Frank Press asked if the symmetry had been tested statistically. Vine replied,

I never touch statistics. I just deal with the facts. [Laughter][41]

Walter Munk felt that the statistically determined orientation of anomalies might indicate whether spreading was influenced by the rotation of the earth. It was an old question, said Vine. He had easily fielded questions from what were normally some of the toughest hitters in geology. His future in his chosen career of geophysics appeared glorious, but a sensitive listener might have felt a tremor of foreboding. Fred Vine made several references to the Vine-Matthews hypothesis. Would he be a victim of eponymy?

Jim Heirtzler followed with a preview of the series of papers on all the oceans that would appear in 1968. He assigned numbers 1 to 32 to prominent anomalies and thereby established a classification that is still in use every day.[42] Some of his future coauthors were in the audience and must have wondered what the reaction would be to their remarkable and emotional achievement. Walter Pitman observed that

> One man who had been violently against continental drift just got up and walked out. But I remember that Menard from Scripps, who had opposed it, sat and looked at Eltanin-19, didn't say anything, just looked and looked and looked.[43]

There is no record in the published account of any discussion after this talk. The evidence was simply too good. Perhaps the violent man had been wholly convinced by Vine and saw no reason to be hit over the head again. I had no reason to say anything; I had already been convinced in La Jolla by the symmetry of the Juan de Fuca Ridge. No one had any reason to say anything to Heirtzler after the questions to Vine.

My paper was next, and it was a straightforward investigation of the terminations of fracture zones and apparent complications in sea-floor spreading. I had not succeeded, any more than Ewing, in starting with a clean slate. The East Pacific Rise still might have been straight initially, but if so it was before sea-floor spreading. If, as everyone was saying, the spreading was a consequence of mantle convection, was it not likely that the pattern was initially simple and the ridge straight and that when the convection broke into smaller cells, the offsets and the observed transform motions began? If so, the crest must have migrated. That the crest of the Gorda Ridge had migrated was demonstrated by analyzing the length of magnetic anomalies where the Blanco and Mendocino fracture zones converged. A drifting ridge, however, could produce symmetrical anomalies. It was simply a matter of defining the right frame of reference. In fact, one side of a ridge could remain fixed if the ridge migrated at the full spreading rate. This could explain how the Juan de Fuca Ridge could produce symmetrical magnetic anomalies without deforming the sediment on the east flank against the North American continent. I mention all these details because this shows what happens when a new theory is accepted. Tuzo Wilson and Bob Dietz had solved all kinds of large-scale problems immediately after accepting sea-floor spreading. I was merely solving the smaller problems closest to my expertise. It could be done only once for most problems left over from the old theory. The race was on, and the Lamont magnetics would not be published for a year and a half.

The published discussion illustrated the rudimentary state of thinking about the implications of sea-floor spreading—rudimentary except for Fred Vine. Heirtzler took exception to my proof of ridge migration. It might merely be a case of a ridge producing an anomaly longer than itself.

> Dr. Menard: What you are saying is that a section of ridge, of length X, is capable of producing a magnetic anomaly of length $X + Y$?

Dr. Heirtzler: Yes.

Dr. Gast: It has to be rotated.

Dr. Bullard: How can you elongate it?

(At that time Fred Vine graced the meeting with two lengthy but lucid comments.)

Dr. Menard: It certainly helps to think about these things at leisure. I agree completely that a feature is produced by a ridge of the same length.[44]

The final paper in the marine group was by Lynn Sykes (Fig. 43), who analyzed first motions on mid-ocean earthquakes as a test of the transform fault hypothesis. He had begun this work in February the morning after Pitman and Heirtzler showed him their profiles. The results were very impressive. All the earthquakes on ridge crests were tensional with motion in the direction of spreading, and all those on fracture zones were strike-slip in the direction predicted by Wilson's transform hypothesis; these facts were corollaries to seafloor spreading, if one had accepted it, or independent confirmation if not. Sykes concluded that if convection was involved, it must be much simpler at depth than the complex pattern of orthogonal offsets of a ridge crest.

The discussion was brief, but it offered Vine the occasion to refer to convection cells as "presumed" and "mythical." Certainly, the many problems related to convection that had been troubling the conference members would have been solved by eliminating convection entirely.

The general restraint in the discussion of marine data continued after the first talk on "Evidence from the Continents." P. M. Hurley showed that isotopically dated Precambrian provinces matched from West Africa into eastern Brazil if the Atlantic was closed. In the general silence, G. J. Wasserburg rose to say that he found this an impressive confirmation of drift. By the closing talk, however, the discussion expanded to the point of expressing a general euphoria about the outcome of the conference.[45] The paleontologist, Arthur Boucot, my former officemate at Harvard, had spoken on the paleogeography of early Paleozoic time.

Boucot responded to a question from Dr. Anderson, his colleague from Caltech, by saying:

I feel it would be much wiser to be conservative in these matters.

Dr. Bullard: Why is it more conservative not to believe in continental drift?

Rhodes Fairbridge, a shipmate from the Capricorn expedition, sought to enlighten the geophysicists on biological migration:

the fossil elephants of the Celebes have to have got there on foot because even a living elephant doesn't swim very far. Only the major organisms are involved in such rules. Most of the smaller organisms can go as passengers in some way—spores are blown in the wind, seeds are carried in mud and stick to branches and float. Seeds were found adhering in the mud of a duck's foot.

Dr. Bullard: They can take a ride on the backs of elephants, too. [Laughter]

Voice: I did read that an elephant was once seen swimming in the middle of the Bay of Bengal, 50 miles from land, which rather shook me. Maybe that will change some of these ideas. [Laughter]

Dr. Imbrie: Was it pregnant?

Dr. Fairbridge: That is a point. I understand in the rutting season the major animals do travel in very peculiar ways.

And so from this conference the happy new disciples of sea-floor spreading and continental drift sped across the country, pregnant with ideas.

REACTION AT SCRIPPS

The NASA meeting ended early enough on 12 November for me to take an early plane home. I immediately went to the office because I was due to leave the next morning for Australia to make arrangements for our now-funded Nova expedition. The memorandum that I wrote to the Director of Scripps may illustrate the response of the participants in the meeting:

Re: Magnetic Program

   I just returned from a remarkable meeting in New York and I cannot let a month pass before I return from the South Pacific without letting you know my views on the consequences for Scripps. Sea-floor creation and spreading centered on the mid-ocean ridge and rise system is now demonstrated although the mechanism is still in doubt. This means that the history of the ocean basins is capable of being unraveled through mapping and sampling of magnetic anomalies and the rocks of the second crustal layer. Lamont has a considerable jump in this work which will clearly revolutionize all thinking on the history of the earth. . . . The ideas have come from Hess and Vine at Princeton and Bullard and Matthews at Cambridge.
   The situation at Scripps is that almost everyone on the staff thinks that sea-floor spreading is either very questionable or funny. The routine collection of magnetic data has been neglected. Special magnetic surveys have been

few since Vacquier's splendid work in the northeastern Pacific. The sole exception is his work on seamounts which gives Scripps a lead in one aspect of the problem. From our recent discussions we are now faced with complete collapse of the underway magnetic program for lack of a few technicians and some equipment. This would have been regrettable in any event but under the present circumstances it would be absolutely disastrous.

Thanks to Spiess and his deep-tow group, Scripps has a commanding lead in the detailed study of magnetic anomalies. Thanks to Vacquier we are ahead in other fields. But the gross mapping of anomalies is neglected, even though it is the easiest to do. If none of the more qualified geophysicists is willing to divert his efforts to this problem then I would like to take responsibility for it. This may not be wholly unreasonable because the data collecting and analysis is relatively routine and the real interest lies in the geological and tectonic interpretation. Every marine geologist will shortly have to learn how to interpret magnetic anomalies just as everyone now learns how to interpret echograms.

It is my impression that marine geology and geology as a whole are at a turning point comparable to physics when radioactivity was discovered. It will be a very exciting time for participants but a sad time for onlookers. Walter Munk attended one day of the meeting and Stan Hart was at both. You might ask for their views.

I would appreciate it if you would send copies of this to Vacquier and Spiess. This is Saturday and the Xerox is locked up.

H. W. Menard

Spiess and Vacquier declined, and by default I inherited the underway magnetics program just when it became priceless. Bruce Heezen could have used the Lamont magnetics for a quantum leap in marine geology, but he merely became an onlooker. When I returned from the South Pacific I wrote to Heirtzler to tell him that we were beginning to digitize our magnetic data "following your practice." Although we merged the navigation in different ways, Scripps, Lamont, and Madingley Rise developed compatible data systems and could exchange soundings and magnetics by tape or punched cards. This greatly facilitated use of a much larger data bank and made it possible for a scientist to make short working visits to other laboratories. This additional advantage was one reason why so few poeple from other institutions would participate in the developments that followed in 1967-1970.

MORE MEETINGS IN NOVEMBER

The Geological Society of America met in San Francisco on 14-16 November. There were three sessions on marine geology, and the floodgates began to open. Vine and Morgan spoke on simulation of magnetic anomalies, and four papers had ''sea-floor spreading'' in the title. One of these was concerned with the lack of deformation in trench sediments and began a new phase of trying to understand all kinds of data if spreading was real.

The big event of the meeting was the awarding of the Penrose Medal to Harry Hammond Hess. The citation by W. W. Rubey said the award was for

> detailed studies of the pyroxine and plagioclase groups of minerals, of serpentine and ultrabasic rocks . . . widely recognized as classics.

It was also for

> His bold and sweeping hypotheses about the tectonic significance of peridotite belts, the role of island arcs and ocean deeps in mountain building, and the spreading floor origin of ocean basins—these hypotheses are perhaps not characterized by the same detail as the mineralogic studies but they are no less renowned—and, on the whole, favorably renowned.[46]

And so, after 50 years, the geological establishment in America formally, if unenthusiastically, accepted continental drift. How much longer would it have been if, long ago, Harry Hess had not been an undergraduate at Yale in that cluster of future academicians?

THE DECEMBER PAPERS

Three papers published in December spread the news officially and much more widely, but it was the same news.[47] The Ewings and their colleagues did not announce that they had accepted sea-floor spreading, but it is implicit throughout their paper. For example, the absence of sediment on the ridge crest

> may be evidence of recent spreading although it is possible that the sediments have been buried by lava. . . .[48]

In the paper earlier in the year, the statement was reversed.

The history of the publication of the papers on magnetics has some interest. Heirtzler gave Vine a copy of the Eltanin-19 profile, and Vine proceeded to write a manuscript incorporating it at the same time that Pitman and Heirtzler were doing so.[49] In Heirtzler's absence, Pitman suggested joint publication with Vine, who declined. They finally arranged that *Science* would give priority to Pitman and Heirtzler in a Short Note, and Vine would publish his interpretation in a major paper two weeks later. It was like the *Double Helix* story in which the hard-driving young scientist, Watson, obtains critical data prior to publication by Rosalind Franklin and succeeds in scooping the old master, Pauling, who had been thinking about the problem longer than anyone else. In the symmetrical magnetics story, however, Fred Vine occupied the roles of both Watson and Pauling, and priority fortunately went where it should have.

Teddy Bullard recalled that it was difficult to persuade *Science* to publish Vine's paper at all because of adverse comments by critics. Despite initial triumphs, the revolution was not over yet.

Chapter 21

THE OCEAN OF TRUTH

Sea-floor spreading was proved for those who were aware of and could understand the data. They were few in number at the start of 1967, and nearly all worked at the same institutions. A feverish search for enlightenment began, and anyone who was suspected of knowing anything was invited or importuned to inform eager audiences across the country. The year augured well; the ferment of Madingley Rise in 1965 and Lamont in 1966 would begin to intoxicate the geological world. The publication of Fred Vine's paper in December 1966, gives a convenient date for ending the period of uncertainty, speculation, and testing that began just six years before with Hess's geopoetry. There are no similarly obvious dates in 1967 or early 1968 because so many different activities were occurring simultaneously. These are difficult years to chronicle, so I shall focus on a relatively small number of major developments to simplify the story. These will be, first, the publishing of the proof of sea-floor spreading so that specialists in other fields could view their own work in a new light; second, the emergence of data and ideas contesting sea-floor spreading or mantle convection; third, mapping of magnetic anomalies and thus the age and history of the sea floor; fourth, investigating the cause and process of spreading; and fifth, inventing plate tectonics.

To visualize what 1967 and 1968 were going to be like for geologists one can hardly do better than contemplate the classroom that a new Scripps student, Tanya Atwater, entered as the winter quarter began in January 1967. At that time, Scripps required all students to take a series of introductory courses, so a mixed group of biologists, chemists, physicists, and geologists assembled for the first class in marine geology. They expected a professor to walk in and begin with the usual formalities of an introduction, an identification of the course number, and a schedule of the exams. Instead, the professor walked straight to the blackboard and said, "I've got to tell you about these magnetic anomalies" and started drawing pictures. Here and there around the United States and Great Britain a few students began to realize that they had had the incredible luck to enter graduate school at the right time. *Science* is read all over the world, and in due course professors and their students would absorb the importance of Vine's article. Absorption to the point of belief, however, might take years. Meanwhile, the students and colleagues of the few people who had visited Lamont or Princeton, or attended the NASA meeting, had a big head start. Speakers had to invent new techniques to spread the word. Tuzo Wilson had pioneered the way with his demonstration of the transform fault, which involved folding a sheet of paper along dotted lines and then making one cut. When the paper was unfolded to simulate sea-floor spreading, the cut was a transform fault. Without the paper, the concept that now seems so simple seemed almost beyond grasp. In mid-1966 Tuzo wrote to a well-known structural geologist,

> I feel that the fundamental problem is that you do not understand what is meant by a transform fault. I find that few people do until they have seen a demonstration of a folded paper model. . . .[1]

He enclosed such a model.

All of us constructed such models, and further dramatic business was contrived to emphasize the magnetic symmetry. One bit was to use a blackboard to draw anomalies away from a center line with both hands at once. Another was to call upon audience participation. Draw the right-hand anomalies, then ask if the audience notices anything about the left-hand ones before drawing left from the center until someone calls out "They're symmetrical." All that was necessary was to capture the eyes of the audience. The evidence itself was overwhelming.

CONFUSING OBSERVATIONS AND IDEAS

The acceptance of sea-floor spreading by those not directly involved with the marine data would have been faster were it not for the existence of apparently conflicting observations. In the past, the disarray of paleomagicians regarding their data had been sufficient cause to reject the paleomagnetic evidence of drift. The evidence related to spreading was less complicated than continental paleomagnetics but it was not all consistent, so the outsider had grounds for skepticism. In due course it would develop that the inconsistencies were caused by erroneous or misleading data and interpretations. Their existence in 1967 caused quite different reactions among geologists. Those who accepted sea-floor spreading assumed that the inconsistencies were only apparent, and they plunged ahead. Others found the inconsistencies grounds for caution and missed the opportunities that opened.

The most obviously inconsistent observations were that ancient rocks had been dredged from ocean basins. Most of these were from the North Atlantic and were commonly accepted as ice-rafted continental rocks. Consequently, Hess and Dietz, who were familiar with the evidence, had no qualms in saying that the ocean basins were young. Ewing, was however, also familiar with the evidence and knew that he should consider only dredge hauls that appeared to be freshly broken fragments of bedrock, or relatively undisturbed cores. Considering such, he did not find the North Atlantic old, but he did believe that it could not be spreading at a constant rate.

The same sort of apparently inconsistent age was identified in the Pacific in 1967, and it, too, caused some confusion. Cobb Seamount is 110 km from the crest of the Juan de Fuca Ridge, and in 1965 B. J. Enbysk used the potassium-argon method to date it at 27 ± 6 million years. Dean McManus discussed the significance of this seamount age.[2] He observed that Wilson believed that seamounts originate at ridge crests, and if so, the spreading rate was only 8 mm/yr, whereas Vine and Wilson had put it at 60 mm/yr. Moreover, Vine and Wilson had indicated that the central magnetic anomaly of the ridge required a rate of at least 10 mm/yr. Something was wrong. McManus proposed that the seamount had formed not on the ridge crest but on the flank and that no spreading had occurred. In his interpretation, the ridge was a purely vertical phenomenon with an age of 27 million years.

Something was indeed wrong, but it was the apparent age rather than the spreading rate. The actual age of Cobb Seamount is only 3 million years and it is consistent with sea-floor spreading. In 1967, however, the erroneous date seemed to put sea-floor spreading and magnetic stratigraphy in question.

After the apparently conflicting dates of ocean basin rocks, the most troublesome aspect of the sea-floor spreading hypothesis was the absence of direct evidence of convergence. There was no problem if the earth was expanding, but if it was not, enormous areas of old oceanic crust had to be plunging into the mantle along the line of oceanic trenches. It was generally expected that the sediment in trenches would show signs of this violent phenomenon, but none could be found. The lack of such signs would raise many questions in the next few years as more and more people acquired pertinent data. In 1967, however, Maurice Ewing was surely the one most troubled by this issue because he already had many seismic-profiler lines across deep-sea trenches. He visited my office at Scripps in the fall of 1967 to discuss this very issue. He had folders full of profiler records with him, as he often did, and we looked at critical crossings. One of the most likely places to look for disturbed sediment was where the east flank of the Juan de Fuca Ridge abutted the continental slope of Oregon and Washington. Turbidity currents from the Columbia River had built up a smooth plain, and the slightest folding or faulting would be obvious. We studied his profiler records in the area. The sediment was undisturbed. He had heard me talk at the NASA meeting, but I repeated my analysis about how a drifting ridge crest could produce symmetrical magnetic anomalies while the east flank was motionless. The magnetic anomalies extended under the continental slope, so at one time there had been convergence and subduction, but perhaps it had stopped before the turbidities were deposited. We looked at crossings of active trenches elsewhere but saw no signs of deformed sediment. There was no question in our discussion about the existence of sea-floor spreading, although neither of us was confident about it being at a constant rate. Neither of us believed for a moment in an expanding earth, so we were left with a puzzle. Ewing indicated that he would welcome some help in interpreting the profiler records and we discussed active cooperation, but nothing ever came of it.[3] He may have been missing Bruce Heezen.

DAWN OF PLATE TECTONICS

In January of 1967, I published a two-page note in *Science* that was nothing special in itself but was to be extraordinarily fruitful.[4] The paper merely presented the latest results of the sounding programs on Scripps expeditions in the central Pacific. The great fracture zones were traced from the eastern Pacific westward past the Hawaiian and Line islands, but they splayed at what I called the "branching line." The year before, I had shown that in addition to the great fractures there was another set of short fracture zones, with a northwesterly trend, that offset the crest of the East Pacific Rise. In the paper I did not accept the hypothesis that fracture zones are fossil transform faults. In fact, I just sounded discouraged:

> The origin and history of these great fracture zones are not becoming more obvious as additional facts accumulate.[5]

The fruitfulness of the paper clearly was not a consequence of a brilliant new hypothesis. Rather, it was a fluke of the style of illustration. I wanted to show that the fracture zones are evenly spaced in a vast region, so I used an equal area map for the purpose. I also wanted to show that the fracture zones are nearly straight for long distances, so I used a great-circle projection for a second map. On such a projection a straight line is also straight on a globe. A scientist can go for years casting papers into the presses and wondering what happens to them. For this paper I know.

William Jason Morgan (1935–) was born in Savannah and obtained a B.S. in physics from Georgia Tech in 1957. He went into the Navy for two years and then to Princeton where he received a doctorate in physics in 1964. His first paper in 1961, coauthored with J. O. Stoner and R. H. Dicke, was concerned with the gravitational constant. His fifth paper was to be the conerstone of plate tectonics.

Jason had already submitted his abstract for the April 1967 meeting of the American Geophysical Union when he saw my paper. He was due to talk on "Convection in a Viscous Mantle and Trenches," but he immediately stopped his project and spent two months generating a computer program. I had learned about the properties of great-circle sailing charts as a naval officer, and so had he. What struck him about my illustrations, however, was not that the fracture zones were almost straight but that they were not entirely straight. It appeared that they followed the arcs of enormous small circles and that the radii of the arcs increased from north to south. Like Teddy Bullard, he recalled Euler's theorem, and what he was programming was a means of determining an Euler pole from the geography of fracture zones (Fig. 24 C). An Euler pole is a motionless point around which a spherical cap rotates when it moves over a sphere.[6] When Jason could process the data on fracture zone trends he established that the zones did indeed lie on small circles with a common center—the Euler pole. As Backus had noted three years earlier, the angular velocity of separation of two rigid plates with a common Euler pole is constant, so the linear spreading rate of a ridge crest varies as the cosine of the Euler latitude (Fig. 24 B). Jason turned his attention to the Mid-Atlantic Ridge where he had the data on spreading rates that Backus was unable to collect for lack of funding. It would be more than a year before some of those rates would even be documented in print by Lamont, but meanwhile, anyone who had an urgent need could extract critical rates from published abstracts. So as the geological world tried to accommodate to the shock of Vine's paper in December, Jason Morgan was already developing the quantitative theory of plate tectonics that would subsume the qualitative miracle of sea-floor spreading.

ANNUAL MEETINGS IN APRIL AND NOVEMBER

Although Vine's paper in December 1966 had established sea-floor spreading, most of the Lamont evidence was still in press. As a consequence, the annual meetings of the major societies in 1967 became unusually important for hawking the news of what was happening. Two meetings, one in mid-year and one near the end, can serve as examples.

Magnetism was in the air as the American Geophysical Union assembled for the annual meeting in Washington on 17-20 April. Tuzo Wilson was scheduled to be among the select speakers at the All-Union session on Frontiers of Geophysics. His topic was "Transform Faults." There were three sessions of the Section on Tectonophysics that involved sea-floor spreading, and the halls were full for each of them.[7] The first was chaired by Peter Dehlinger, and it included all the studies of magnetic anomalies by the Lamont group in an ocean by ocean format. There was also a paper by Glass and Heezen about tektites in deep-sea sediments. The discovery of these fragments of debris from meteorite impacts was important, but the inclusion of this paper with the spectacular magnetics merely highlighted the opportunities from which Bruce was being excluded.

The second session was chaired jointly by Fred Vine and me. Heirtzler gave a summary of the implications of the regional papers on marine magnetics and proposed again his numerical chronology. Once again continental geologists had cause to muse on the simplicity of marine geology. On land, the stratigraphic column was established by the painstaking efforts of numerous paleontologists and stratigraphers to interpret complex and widely separated outcrops. It is difficult to conceive of a more demanding scholarly enterprise, and it had been going on for a century. For the two-thirds of the world that are not land, a few months of intensive effort by half a dozen people at one institution established a standard magnetic stratigraphy that remains unquestioned.

The next talk was by Lynn Sykes on first motions of earthquakes on transform faults; then the Ewing brothers on sediment distribution, trenches, and ridges; and finally me on the relation between spreading rate and sea-floor relief. All those talks would be published within the year. The session was on island arcs and trenches as well as spreading ridges. Consequently, my talk was followed by five on the regional geophysics of trenches. The last talk scheduled was the more theoretical one on trenches by Jason Morgan. It was clearly a long session, and at some point Dan McKenzie (Fig. 44) decided that he had had enough trenches—and the abstracts promised nothing else—so he walked out. Thus, Dan was not present when Jason Morgan ignored his abstract and gave the first presentation of plate tectonics.

That session incidentally provided me with an excellent example of how unreliable memory can be. A decade later Dick Hey, then a recent Princeton Ph.D., conceived of celebrating the tenth anniversary of the session at another Geophysical Union meeting. Fred Vine was unavailable, but Dick invited me to be co-chairman as at the original session. I not only did not remember hearing Jason's famous talk, I didn't remember presiding over the session. What I did remember was a conversation with Bruce Heezen at the end of that day in 1967. Bruce told me that someone had said that all the Lamont talks were excellent but could not say the same for mine because it presented no new data. Bruce had responded that mine was the only talk that he wished he had given himself. It was, in fact, just the kind of analysis that he would have done had he not been forbidden to use the Lamont data. In the third session on tectonophysics, Harry Hess spoke on the development of ideas on sea-floor spreading.

The Geological Society of America met in New Orleans in November, and in the six months since the Geophysical Union meeting much had happened in sea-floor spreading. The meeting, however, was almost isolated from the ongoing scientific revolution, and it did not reflect current events in tectonophysics. The technical sessions went on for most of a week, but only one, on marine geology and geophysics, had anything related to sea-floor spreading. Vine and Morgan discussed the simulation of magnetic anomalies. R. S. Yeats proposed that the Southern California Borderland was produced by spreading. Finally, Eric Schneider reported that Navy data confirmed the sediment distribution observed by the Ewings in the North Atlantic. He also supported their hypothesis of a hiatus in spreading.

The intrusion of sea-floor spreading into the meeting of the sedate Geological Society was neatly balanced by a talk by F. G. Stehli titled "Does paleomagnetism provide valid evidence of continental drift?" Like Wegener and like all other authentic field observers, Stehli believed in his own data—despite half a century of statements that paleontology was incapable of testing continental drift. He used variations in faunal diversity as a key to paleolatitude in Permian time. Paleomagnetic data did not give the same result, so he felt that use of such data was "unjustified."

COROLLARIES

Once sea-floor spreading was accepted as a fact, it naturally followed that almost every specialist reinterpreted his data. Hess, Dietz, and Wilson, the first believers, had reinterpreted the gross features of the ocean basins. Wegener, DuToit, and Carey had already done the continents and been confirmed by Blackett and Runcorn. Bruce Heezen was an exception, and perhaps this indicates something about how science works. He had believed since the late fifties that the sea-floor was spreading, but he thought that the cause was an expanding earth. This hypothesis seemed to put a rein on his creative abilities. He was right but for the wrong reason, and that was enough to keep him from finding corollaries of his hypothesis or, being right on the fact of spreading, from abandoning his hypothesis about the cause.

The published record indicates that the scientists who made the fastest reinterpretation of their data were Maurice and John Ewing. If Maurice was a "fixist" or dominated by any other antique philosophy, this record is astonishing. If he was a scientist who had been collecting data for six years to look for sea-floor spreading, it would hardly be surprising that he was able to correlate the two. The first paper incorporating sea-floor spreading in explaining sediment distribution had been in December 1966, and had the signs of a revision. The second paper by the Ewings was in June 1967, and it assumed spreading and focused, in the style of normal science, on the remaining questions of rate and variability of rate of spreading. Their data set had been enhanced by additional crossings of the East Pacific Rise. The logic of their interpretation was clear. The sediment on the ridge crest was very thin but increased slightly in thickness away from the crest on the upper flanks. On the lower flanks the thickness was several times greater and essentially constant. Spreading rates from magnetic anomalies indicated that the transition from thin to thick sediment occurred about 10 million years ago everywhere. Thus the sea floor had been spreading for the last 10 million years, and before that there had been a quiescent period of 30 million years or more. There was no record of events before 40 million years. The only alternative to a variation in rate of spreading was a variation of rate of sedimentation by orders of magnitude during a very brief interval. This appeared highly improbable.

Another paper came out of Lamont in July. Heirtzler and Hayes demonstrated that a magnetic quiet zone existed in a band along the base of the continental slope on each side of the North Atlantic. They correlated the band with a known period in late Paleozoic time when there were no magnetic reversals. Assuming a constant rate of spreading "(although we are aware that this is a hazardous assumption)," the quiet zone was at the right distance and was the right width to fit the magnetic stratigraphy on land. Once again the logic was clear, but the land geology indicated that Africa and South America had been joined long after the end of the Paleozoic era.

I had accommodated to sea-floor spreading in time to publish my first paper incorporating it in August.[8] Like Jason Morgan, I had collected information on rates of spreading, but I had plotted them on maps that showed other kinds of marine observations. I was aided by the fact that the Lamont group were very generous with their priceless data. Jim Heirtzler had sent me numerous unpublished magnetic profiles and maps. So assisted, I could show that a central rift might occur if spreading was slower than 40 mm/year but not if it was faster. Likewise, a transition occurred at the same speed in the low relief of ridge flanks. If slower, the flanks were mountainous like the Mid-Atlantic Ridge. If faster, the flanks had only low hills. I also showed that the thickness of the volcanic second layer varied inversely with spreading rate. These were the first correlations of any kind of data with spreading rate, but they would not be the last. Indeed, spreading and drifting rates would provide the foundation for the revolution that followed. Among other things, it was so easy to make correlations with the rates, and all of geology was ripe for reinterpretation.

One can contrast the ease of an instantaneous shift in thinking about marine geology, where hardly any frame of reference was as old as a decade, with the situation in the more traditional geology of the land. In 1967, Warren Hamilton, for example, was publishing bold speculations on continental drift in Antarctica and Melanesia; but without the constraints soon to be added by plate tectonics, who could tell which speculations were correct? Likewise, W. R. Dickinson and Trevor Hatherton published an important paper on the relation between magma type and depth of seismicity in island arcs, but "At that time, we had no idea we might be making a contribution to plate tectonics." The relation of their work to other facets of geology did not appear so quickly as in marine geology.

SEISMOLOGY AT LAMONT

Japanese seismologists are in the (scientifically) enviable position of living in the seismic belt of an island arc. Thus it was almost inevitable that Wadati, or a colleague, would be first to demonstrate the existence of the deep-focus earthquakes of the Benioff zone. It was equally likely that the superb Japanese seismographic network would first yield data on the material properties of the mantle of the Benioff zone. In 1960, M. Katsumata studied some earthquakes at 350 km beneath Japan and observed that

the seismic waves in the media in which earthquakes occur frequently have larger velocities and attenuate more slowly than the ones in other media.[9]

In a similar vein, T. Utsu demonstrated that attenuation was high on the inner side of the Japanese arc and low on the outer, Pacific, side.[10]

As a part of the Upper Mantle Project, Professor Jack Oliver of Lamont and students Lynn Sykes and Brian Isacks set out to study the seismicity of the remote and sparsely populated Tonga arc as the Japanese had studied their homeland. In 1964-1965 they established a network of five seismic stations in the exotic southwestern Pacific. By 1966 they were ready to pour out papers describing their results, which were spectacular. They independently rediscovered the variations in upper mantle properties found by the Japanese, but they interpreted them in a new way. They reasoned that

the zones of deep shocks should be anomalous if only because earthquakes occur there and not elsewhere.[11]

So they were

searching consciously for gross differences between the deep seismic zones and the surrounding mantle. . . .[12]

They were looking not merely for a boundary between two different materials along the Benioff zone. The zone was not so much a two-dimensional plane as a three-dimensional tabular volume. They anticipated that the zone was different from the material above and below.

The properties of the mantle they observed were indeed grossly different in and out of the seismic zone. In reading the conclusions they were triumphantly waving it is easy to imagine that more than one seismologist was kicking himself:

There exists in the mantle an anomalous zone whose thickness is of the order of 100 km and whose upper surface is approximately defined by the highly active seismic zone.

The new data can be interpreted . . . to show that the lithosphere has been thrust, or dragged, or has settled beneath the Tonga arc on a grand scale.[13]

Oliver and Isacks had found that seismic waves travel faster and more efficiently along the dipping seismic zone. Seismologists usually measure the efficiency of travel by the term ''Q,'' and Q[14] is a measure of the strength of a material. Thus the dipping zone was stronger than the surrounding mantle. It was, in brief, a strong slab—which corresponded to what Reginald Daly had discussed as the "lithosphere" (Fig. 16 I). Except in island arcs, the lithosphere was the 100 km thick, strong layer covering the surface of the earth, roughly the technical equivalent of the common term "crust," and

> it is apparent that theories of tectonics based on convection currents must take this layer into account, for its effects on surface geology could be large and varied. Such a layer, when involved in extension, thrusting, or flexing, could introduce great complexity into the effects of a driving force of relatively simple configuration.[15]

Oliver and Isacks, like Lynn Sykes, had been hearing about sea-floor spreading from Heirtzler and Pitman, and they knew of Sykes's confirmation of transform faults and Morgan's talk on plate tectonics. Now they had their own data to feed into the scientific revolution. At one blow they had published the first paper on the tectonics of large plates, showed where the crust created by sea-floor spreading was going, eliminated the need for an expanding earth, and determined how simple convection could generate sea-floor spreading offset by numerous transform faults. They had not, however, considered Euler's theorem and the motion of rigid plates on a sphere that was the essential element of plate tectonics.

DEVELOPMENTS AT SCRIPPS

The intimate linkage between Scripps and Madingley Rise led to the growth of a little colony of expatriate Cantabridgians by the La Jolla and Del Mar shores. In 1966, John Mudie and John Sclater joined the staff. Mudie had heard Jason Morgan talk on plate tectonics and returned full of enthusiasm to spread the word. He was in Fred Spiess's Deep-Tow Group, which included many students with Tanya Atwater among them. The basic ideas of plate tectonics became familiar to this group. It was also known to others on campus. When I had returned from Washington in the summer of 1966 a few students including Clem Chase and Dan Karig asked to work with me on marine tectonics. Clem received a desk in my laboratory, and I saw him constantly. Jason Morgan sent me a preprint of his manuscript in its early draft, probably in the late spring of 1967. I believe I also reviewed the paper for an editor. In any event the manuscript certainly circulated among my students, and we discussed it. The original draft, however, was difficult for me to fathom, and it did not have the impact of the final publication. Moreover, I had other things on my mind. I spent July through September on the Nova expedition in the southwestern Pacific.

Teddy Bullard was among the old hands to participate, and there were many students including those who knew about Morgan's paper.

John Sclater was at sea on Nova with me until 12 September, and he was out again from 29 October to 19 November. Meanwhile, two of the five geophysicists who received doctorates at Cambridge in 1967 had joined Scripps. Bob Parker was on the permanent staff, and Dan McKenzie was a visitor for six months beginning in June. Dan and John Sclater shared an apartment a few blocks from Scripps when John was ashore.

After the NASA meeting in December 1966, Dan had returned to Cambridge for about six weeks and immediately started to work on the theoretical implications of sea-floor spreading. Just before he arrived at Scripps he had sent a manuscript to the *Journal of Geophysical Research* on the relation of heat flow and gravity anomalies to sea-floor spreading.

> These remarks suggest that the outer perhaps 50-100 km of the earth must behave as a rigid layer and that the places where this is not true will be marked by shallow earthquakes. . . . As the sea floor spreads, the lithosphere must move as a solid layer and carry the magnetic anomalies with it.[16]

McKenzie was thinking in the same way and using the same conventional terminology as Oliver and Isacks at Lamont. He proceeded to analyze the flow of heat from a drifting slab that was created at a spreading center. This was the first analytical paper about the properties of such a slab, and it opened a whole new field. In sum, it is clear that the concept of plate rigidity was in the air regardless of the kind of data under consideration. Oliver and Isacks submitted their paper for publication on 13 April; Morgan talked about plates about a week later; and McKenzie submitted his paper on 30 June after starting it the previous December.

McKenzie had been thinking of the drifting slab in cross-section rather than in three dimensions, and

> it was not until I moved to Scripps in June that I had the idea of using rigid body rotations to describe the motions. This idea occurred to me while I was rereading the paper by Bullard, Everett and Smith on fitting continents together. This I was doing because I had certain doubts about the fitting condition they used. Their paper combined with Wilson's ideas suddenly made clear to me the importance of rotation poles and plate motions.[17]

He was also thinking about earthquakes and the flood of reliable new data on first motions. Hodgson and Benioff had analyzed a much more primitive data set to deduce the orientation of fault planes in island arcs with ambiguous results. Since that time the tectonics of island arcs had been analyzed by seismologists largely in terms of deductions about the direction of principal stress. The ambiguity had modified for the worse to become confusion—apparently because of the importance of local effects upon the stress field. Recently, however, Lynn Sykes had shown the advantages of plotting slip vectors in testing Wilson's idea of transform faults, so a different approach to first motions was open for consideration.

In November, John Sclater recalls that

> Dan had just received a tough review of one of his theoretical papers on the internal structure of the earth. As light relaxation he started thinking about plates and in two days produced the North Pacific plate tectonics paper.[18]

Bob Parker recalls that Dan was having difficulty presenting and analyzing the significance of spherical triangles, so he came to Bob, who was a mathematician.[19] Bob had recently generated a computer program for plotting data on any map projection. In the course of a week, they realized that the slip vectors should be plotted on a transverse Mercator projection, and the problem was solved. In such a projection, the lines of latitude are parallel and horizontal like the conventional lines of latitude on an ordinary Mercator map (Fig. 24 D). Lines of latitude depict small circles around the north pole of a spherical earth, so Parker's transverse Mercator map showed small circles around an Euler pole. All the slip vectors around the margins of the North Pacific could be plotted as parallel and horizontal on a properly fitted projection. This was true of earthquakes from spreading on the East Pacific Rise, from horizontal motion on the San Andreas fault, and from compression in island arcs. Thus it was demonstrated that the lithosphere of the North Pacific is a rigid plate rotating about an Euler pole near Greenland. It might appear that this was merely a trivial geographical extension of the work involved in the Bullard fit in the Atlantic. The fitting of the Atlantic, however, merely demonstrated that the continents are rigid plates as Wegener had proposed. McKenzie and Parker showed that the lithosphere of even the largest ocean basin is also rigid as the undistorted magnetic anomalies had implied. The whole world is composed of rigid plates that jostle each other. Their paper was published, very rapidly, by *Nature* on 30 December 1967.[20] This bit of "light relaxation," reminding one of Deitz's "potboiler," was subsequently generally cited as the first paper on plate tectonics.

THE END OF THE BEGINNING

January and February of 1968 passed uneventfully as far as the published record of the plate tectonics revolution goes. It was not until the 15 March issue of the *Journal of Geophysical Research* that the pioneering work that had been gestating for one or two years reached print. Jason Morgan published his seminal paper on plate tectonics, and Heirtzler, Pitman, and the Lamont group published four papers on marine magnetic anomalies. At the time the series of papers had a great impact on the uninvolved. The information was startling enough, and the conclusions were convincing and revolutionary. To the marine geologists and geophysicists who were involved, however, the main function of most of the papers was to provide something that could be cited to give proper credit instead of mere personal communications or abstracts from annual meetings.

Jason Morgan's paper provided the foundation for all subsequent work on ancient plate tectonics and may have been the most important paper ever written in geology, and certainly in tectonics. The concepts of continental drift and sea-floor spreading were bold and fruitful, but they left geology in the stage of qualitative naturalism. Morgan's plate tectonics permitted quantitative predictions of a wide range of phenomena. Predictions led to quantitative field tests, quantitative hypotheses, and geology as a normal physical science. Nonetheless, for those in the know, the paper merely repeated what Morgan had said in April. It was noteworthy, however, for the remarkable change in the style of presentation. The manuscript I originally received began,

> Consider the earth's surface to be made of a number of rigid crustal blocks.
> . . . Then, on a spherical surface, the motion of one block (over the mantle)
> relative to another block may be described by a rotation of one relative to
> another. . . .[21]

By the end of the first paragraph, he was already involved in a critical test of the hypothesis. I had found the paper difficult. The beginning of the published paper was,

> A geometrical framework with which to describe present day continental
> drift is presented here. This presentation is an extension of the transform
> fault concept . . . to a spherical surface.[22]

The content of the paper was the same as the manuscript, but the style had been transformed from geometry to geology.

In their summary paper the Lamont group observed that the three regional papers

> have shown that a magnetic anomaly pattern, parallel to and bilaterally symmetric about the mid-oceanic ridge system, exists over extensive regions of the North Pacific, South Pacific, South Atlantic, and Indian Oceans.[23]

The evidence for sea-floor spreading was overwhelming, and the mapping of the age of the sea floor was well underway. But what was the age of the magnetic anomalies? Beyond 5 million years none were dated. Heirtzler and colleagues investigated the possibility that the spreading rate in each region was constant. If so, the distance from the ridge crest to each anomaly in a region would be in constant proportion to the distance to the young dated anomalies. If that in turn was so, the distance to anomalies in different regions would also be in constant proportion. They were not; thus spreading was not at a constant rate everywhere. What region would be the standard? By brilliant science, black magic, or good luck they picked the South Atlantic (Fig. 22). With regard to matters other than the time scale, this presentation of the great discoveries in January 1966 at Lamont mainly cited Jason Morgan's plate tectonics, which had leap-frogged past it.

That was the end of what had begun with Wegener in 1915, the demonstration of the fact of continental drift. The cause and most of the effects of drift were not known, but the fact was. Moreover, the drift was in the form of rigid plates whose motion could be precisely defined by an Euler pole and an angular rotation. Fracture zones had a new importance, and a month after Morgan's paper was published, three years after the conception of transform faults, and 18 years after the discovery of fracture zones, Tuzo Wilson and I were elected to the National Academy of Sciences, and he went on to become a Fellow of the Royal Society and to receive the Penrose Medal.

Organized opposition to plate tectonics would arise and persist for more than a decade. Indeed, it still does. Specialists would work to be convinced by their special knowledge, and the converted would develop remarkably fruitful corollaries about continents and ocean basins. What was to come would not be less startling to new minds than what had gone before. The geologists who would make the discoveries that still continue would not find the contest less intense or the science less interesting, but something had happened that could not happen again. The revolution was over; the flowering of geology could begin.

EPILOGUE

In a history of science by a working scientist it has become customary to conclude with a chapter in which the writer attempts to link his material with the grander concepts of philosophy, history, and sociology. I, too, had such a chapter in mind, but that was when I expected to cover the 1970s and perhaps early 1980s in this volume. The present account terminates in 1968 after following the idea of continental drift from its conception to its proof. As a social phenomenon, however, the scientific revolution was not over at that time. The existence of a proof was yet to be accepted by most geologists. Moreover, the astonishingly fruitful quantitative corollaries of plate tectonics were still unknown. Geology in its way had reached the stage of physics with the Bohr atom, but quantum mechanics was in the future. Under the circumstances, generalizations about scientific revolutions seem inappropriate. Instead I shall merely venture a few summary comments on what has gone before.

MEMORY, PERCEPTION, AND EVOLVING HISTORY

When I began this book I thought that I knew about the events that it would record and that it was worthwhile to write them down before everyone involved was gone. I was certainly wrong. As I reread the old literature, including my own papers, and talked to friends and colleagues, many things emerged that surprised all of us. I had to rewrite the chapters on the discovery of sea-floor spreading, transform faults, magnetic symmetry, and plate tectonics after discussions based on the original drafts of the manuscript. None of my informants, I think it is fair to say, had a realistic perception of the role of multiple discoveries in this history. Certainly I did not. No one known to me except Talwani, Heirtzler, and John Ewing had any real perception of Maurice Ewing's initial response to sea-floor spreading. And who had heard of the spectacular paper by Coode or remembered the predictions by Backus? So many surprises appeared that it seems likely that others are still hidden and that this history will evolve.

MULTIPLE DISCOVERIES

Merton's thesis that multiple discoveries are normal in science is wholly confirmed. In the geographical discovery phase in the fifties, the same instruments and techniques were widely available, and inevitably new types of crustal structure and physiographic features were found independently. As to the conception of ideas, one must believe either in widespread plagiarism by eminent scientists or in independent multiple discoveries. The evidence supports the latter interpretation, although at times it is hard to understand how one scientist could have avoided knowing about another's work. As Einstein observed, however, the fact that an idea is known may seem less important to a creative scientist than the invention of a novel proof that it is correct. Ideas are cheap indeed to people who think of several a day. In any event, in every case I examined from personal knowledge or that of close and long-time associates, the discoveries were not only independent but viewed at the time as rather minor efforts—definitely not meriting a priority squabble.

The list of multiple is long. Continental drift was discussed casually in a talk by Bailey, casually in print by Taylor, and at great length by Wegener while he had time on his hands on convalescent leave—all at about the same time. The idea of an expanding earth was discovered independently time and again for decades. Sea-floor spreading was proposed by Holmes, Carey, Heezen, Hess, and Dietz, in that order, although only those with the right origin for the spreading received much credit. Morley, Vine, and Heirtzler's student all conceived of the origin of magnetic anomalies. Coode and Wilson thought of ridge-ridge transform faults almost simultaneously. Coode, Vine, Wilson, and Pitman all discovered magnetic symmetry, although at slightly different times. The rigidity of oceanic plates, and therefore of the global lithosphere, was discovered by Morgan, McKenzie, and Parker, and to a limited extent by Oliver and Isacks all at about the same time, but by independent types of evidence. Is "discovered" the right word, considering Backus's ideas on magnetic symmetry and rigid rotations of the sea floor?

GRASPING OPPORTUNITY

The fact that almost every discovery in this history was multiple makes it obvious that opportunities were equally available to many scientists. What distinguished the actual discoverers from their potential rivals? There is of course luck, but the perception of opportunity and the urge to grasp it are also important. Once Pitman perceived the magnetic symmetry, he plunged into elaborating it. When Sykes learned about it from Pitman he "dropped everything" and invented his own tests of spreading. After the NASA meeting, McKenzie "immediately started to work on the subject," and I informed Nierenberg that we had to change our whole geological research program. Ewing, much earlier, made similar changes as soon as he read Dietz's paper on sea-floor spreading. Heezen led his students in an invasion of the paleomagnetics laboratory within minutes of the discovery of the first reversal in sediment. Morgan dropped his scheduled talk and began to develop plate tectonics as soon as he read my paper. Wilson began to publish on sea-floor spreading within a few months of Dietz's paper. To do so he must have discarded a decade of his own research as soon as *Nature* arrived. To complete these examples, Wilson told Vine and Matthews to drop everything. Whatever they were doing could not be as important as helping him develop his new ideas on magnetic symmetry. In this revolution, the winners were the fastest starters.

DISCOVERY AND EXPLANATION

The young scientists who took to the seas in the fifties were adventuresome and inventive, and they made remarkable discoveries. Yet when the time came for the intellectual adventure of scientific theorizing, that inventiveness mainly failed. Why did the explanations come from those who were not involved in the discoveries? The dominant reason at Lamont may have been that so many of the explorers were physicists or geophysicists who did not understand and were not interested in geological problems. I think, however, that the main reason for marine explorers everywhere was merely a matter of style. Their efforts had been successful for a decade, so why change? The proper, proved approach was to seek answers at sea rather than manipulating a finite data set. It was the approach of generations of field geologists on land, equally logical, equally valid, equally limited. Moreover, success had become a trap. An astonishing fleet of research ships had been created, and it had hungry crews. Research laboratories were proliferating, and into them flocked eager graduate students who needed support. In Washington agencies were created to grant funding. The accepted way to finance a few kilobucks of thought by a senior scientist was to spend a megabuck at sea, and that took lots of time.

Even so, the finite data set was an adequate base for other people to pen geopoetry on an airplane or dash off a potboiler in La Jolla. Why was it not adequate for the field men when they could think about it, however briefly? Perhaps they knew too much, as Bullard surmised of Ewing. Much was known to the explorers that was not published. It was difficult not to be condescending in reviewing manuscripts by outsiders who were drawing wholly on published literature. Each new hypothesis in geology appears to have exceptions, and they were multiplied in the unpublished records that filled the minds of the explorers. Each hypothesis and then hypotheses in general became suspect. A little learning makes a man sanguine, too much promotes phlegm.

LOST OPPORTUNITIES

How could geologists have failed to realize that the magnetic anomalies off California deserved the same urgent study that physicists accorded to radioactivity when it, too, was accidentally discovered? Perhaps any equivalently speedy response was impossible because of the nature of the observations. Any physicists could duplicate the experiments with radioactivity; many did, and knowledge exploded. Only the survey off California existed, so it had to be studied in published literature, which was slow in coming. Perhaps, under the circumstances, explaining the magnetic anomalies in less than a decade *was* an intense effort. Still, many of us later had time to kick ourselves when earlier we had not had time to think.

AGE OF THE REVOLUTIONARIES

Age did not affect the creativity, imagination, or boldness of the inventors of sea-floor spreading and plate tectonics. The marine exploration and discovery were mainly the work of men in their early thirties, although the early leaders, Revelle and Ewing, were in their forties. That age relation, however, merely reflected the great influx of young men as oceanography began its rapid expansion. No such demographic effect determined the age of the creators of the major ideas that would follow. In 1960-1962 sea-floor spreading and its first-order corollaries were proposed. In 1960, Dietz was 46, Hess 54, and Wilson 52. The origin of marine magnetic anomalies was offered in 1963 by Vine, 24, and Matthews, 32. Wilson was 57 when he conceived of transform faults and magnetic symmetry, but Coode was half that age. McKenzie and Parker were in their mid-twenties when they invented their version of plate tectonics, and Backus and Morgan were in their early thirties, but Bullard had first applied Euler's theorem to tectonics a few years before when he was 57. The creative brilliance of young as opposed to old mathematicians and physicists has suggested that creativity itself is largely confined to the young. Possibly geologists, or physicists doing geology, are different. Indeed, it has been noted that geologists receive awards and distinctions 10 to 20 years later than physicists.

In the case under consideration, however, age was immaterial. Perhaps the vacuum in geological theory was so rarefied that other factors affecting creativity did not count. Everyone was sucked in. Perhaps that is why every discovery was multiple.

INVENTION AND PERSISTENCE

Credit and rewards for scientific discoveries are supposed to go to the first discoverer to publish. The necessity for publicizing and defending a hypothesis is recognized, but Darwin, not Huxley, received the credit for the theory of evolution. The matter is additionally complicated because Wallace proposed the theory of evolution independently and published jointly with Darwin, and yet Darwin alone got almost all the credit—even from Wallace. Other things being equal, the amount of effort expended in coming to a hypothesis is a major factor in the allotment of credit. This effect was strong in the present case. Morley dashed off his brilliant idea and had it rejected. Vine and Matthews showed that they had labored in the field, and they received the rewards. Coode encapsulated his lucid ideas in a two-page note, whereas Wilson fired a salvo of papers in *Nature* and *Science* on the same subject and got the credit. The effect was felt even in the exploratory phase. Raitt established the widespread occurrence of a standard oceanic crust but published sparingly while Ewing published a whole numbered series of papers based on a vast effort but scantier observations.

In the same vein, other things again being equal, credit for an idea may be related to the effort expended in publicizing it. Wilson spent much time and effort in invited lectures explaining transform faults. Vine made a prolonged and intense effort to prove and demonstrate magnetic symmetry and its cause. Dietz labored to publicize sea-floor spreading in invited lectures and popular publications. Other things, however, are not always perceived as equal, and only Wilson and Vine reaped rewards commensurate with their discoveries and efforts. There is some debate regarding the legitimacy of rewarding publicity beyond a natural appreciation that someone is willing to explain his ideas. Most of the creative scientists with whom I have discussed the point seem to feel that ideas are so cheap that the effort spent in demonstrating them should be a major factor in allotting credit. One scientist expressed her views rather forcefully in opposition to equal credit to all the multiple discoverers of an idea and in favor of credit to the one who worked to convince people about the idea's worth.

QUALITY OF SCHOLARSHIP IN A SCIENTIFIC REVOLUTION

Regrettable.

SCENARIOS FOR SCIENTIFIC CHANGE

It may be of interest to conclude by speculating on what could have happened to geology and particularly to tectonics if the marine discoveries in the 1950s had not occurred. Where, in short, would geology be if the thin, uniform oceanic crust, high heat flow, guyots, global median rift, linear magnetic anomalies, and fracture zones had not been discovered or had all been classified by an overzealous Navy?

It seems clear that further study of the stratigraphy and structure of continental rocks alone would not have led to acceptance of continental drift. In a series of Pyrrhic victories, geologists and especially paleontologists had successfully discredited the ability of their own data to discriminate between drifting and fixed continents.

The classical continental data plus paleomagnetism might have proved drift without the marine observations. The continental magnetic observations were generally viewed as inconclusive at best when the marine data settled the question of whether continents drift. Nevertheless, most of the perceived problems in paleomagnetism were already resolved. Cox, Doell, McDougall, and others were proving that the magnetic field reverses. Once that effect was removed and the proper ways to present the remaining data were developed, the paleomagnetic evidence for continental drift would have been much more compelling. Adding paleoclimatological data plus a new view of classical geology probably would have convinced a new generation of geologists of the reality of drift. Perhaps geology would have begun to flower in 1980 or so even without the quantitative predictive power of plate tectonics.

If there were no Navy and therefore no massive marine exploration in the fifties, there still would have been nuclear tests and the expansion of seismology for detection. Presumably, even without knowing about sea-floor spreading and transform faults, the not yet aged Sykes and McKenzie or their colleagues eventually would have plotted first motions of earthquakes in the right way and discovered ridge-ridge transform faults, the spreading rift, and the rigid motion of the North Pacific basin. It seems likely that these discoveries, plus the plunging lithosphere of trenches, would have come close together at some time within a decade of when they were discovered. Thus seismology would have demonstrated the existence of modern rigid plates. Possibly the congruence of the margins of the Atlantic would even have been enough to prove ancient rigid plates.

If this analysis is correct, even without the marine discoveries, evolution rather than revolution by this time would have led to a general acceptance of the hypothesis of continental drift and to rudimentary plate tectonics. The precise paths of plate motion would have been unknown, however, and so would all the phenomena associated with the aging and cooling of the lithosphere. Possibly even these somehow would have been discovered just as discoveries now inconceivable will inevitably occur.

ABSTRACT OF CHRONOLOGY
LOGARITHMIC SCALE

1870-1879 Daly and DuToit born.

1880-1889 Field, Vening Meinesz, and Wegener born.

1900-1909 Bullard, Ewing, Hess, Raitt, Revelle, and Wilson born. Reversely magnetized rocks found.

1910-1919 Dietz born. Taylor and Wegener publish on continental drift. Wegener predicts and confirms shift in longitude of Greenland.

1920-1929 Heezen, Heirtzler, Menard, and Runcorn born. Wegener's book issued in several editions and widely discussed. Deep earthquakes, negative gravity anomalies in trenches, and expanding universe discovered. Expanding earth proposed.

1930-1939 Wegener dies. Matthews, Morgan, Sykes, and Vine born. Median rift in Carlsberg Ridge. Ewing begins seismic refraction in shallow sea. DuToit confirms drift in southern hemisphere.

1940-1949 DuToit dies. McKenzie born. Guyots, seismicity of mid-ocean ridges, physiographic provinces in North Atlantic, and Benioff zones discovered. Ewing begins seismic refraction in deep sea. Holmes's second version of opening of the Atlantic.

1950-1954 Midpac and Capricorn expeditions. Thin pelagic sediment, volcanic layer, thin and uniform oceanic crust, high heat flow, abyssal hills, fracture zones, and abundance of submarine volcanism discovered. Guyots confirmed as drowned ancient islands. Tharp finds seismically active median rift in North Atlantic.

1955-1959 Ewing and Heezen find world-girdling rift. Mason and Raft discover magnetic stripes; Vacquier measures offsets on fracture zones. Hess finds Atlantic fracture zones. Heezen and Tharp publish physiographic diagrams. Runcorn's paleomagnetism supports continental drift. Mohole proposed. International Geophysical Year. First International Oceanographic Congress.

1960 Geophysics and history of East Pacific Rise. Hess circulates preprint on sea-floor spreading (SFS). Significance of paleomagnetic evidence debated.

1961   Dietz publishes SFS. Ewing begins field test of SFS. Wilson elab-
            orates SFS. Detailed maps of magnetic stripes published. Mohole
            test in deep water.

1962   Hess publishes SFS.

1963   Vine-Matthews hypothesis published. Morley rejected. Reliable
            reversal time scales. Fisher and Hess dispose of oceanic crust in
            trenches. Dietz and Wilson elaborate SFS. Bullard supports drift.

1964   Sea floor is tholeiitic basalt. Sequential development of ridges and
            expanding earth discussed. Royal Society symposium on conti-
            nental drift.

1965   Holmes dies. Transform faults and magnetic symmetry conceived
            by Wilson and Coode. Symmetry confirmed by Vine and Wilson.
            Hess elaborates sequential development of ridges. Wilson sepa-
            rates effects of relative and absolute motion.

1966   Vening Meinesz dies. Heirtzler and Pitman discover global mag-
            netic anomalies. Sykes confirms transform faults by earthquake
            motion. Ewing accepts SFS. Mohole project killed. Second Inter-
            national Oceanographic Congress in Moscow. November: Meet-
            ing of the National Aeronautics and Space Agency. December:
            SFS papers at the Geological Society of America (GSA) meeting;
            Hess awarded Penrose Medal; Vine, Pitman, and Heirtzler publish
            magnetic profiles and models.

1967   Spring: Oliver and Isacks submit paper on sinking lithosphere in
            Tonga region. April: Morgan talks about rigid plate motions at the
            American Geophysical Union (AGU) meeting. Fall: McKenzie
            and Parker confirm rigid plate motions by seismology. SFS talks
            increase at AGU and GSA.

1968   March: *Journal of Geophysical Research* publishes first detailed
            papers on regional magnetic anomalies by Lamont group and first
            paper on ancient plate tectonics by Morgan.

NOTES

CHAPTER 1

1. "Paleomagicians" is a convenient term for people who investigate paleomagnetics.

2. Merton (1961), p. 478.

3. Merton (1968).

4. Quoted in Pais (1982), p. 165.

5. Hess told me he wrote papers on airplanes and (like Einstein) sometimes lacked access to a library. Of course, I was interested in his papers on the evolution of the sea floor, which involved few data, and they were not even his own. His detailed papers on petrology and mineralogy were another matter.

6. Hess (1946), p. 785.

7. E. L. Hamilton (1956).

8. Published by the University of California Press at Berkeley in 1962 and 1976.

9. Merton (1961).

10. Quoted in ibid., p. 482.

11. Ibid.

12. Birch (1956); Frances Birch succeeded Daly as Sturgis Hooper Professor. Billings (1959); Marland Billings succeeded Daly as Chairman of the Department at Harvard.

13. Billings (1959).

14. Ruedemann (1926).

15. Price (1961).

16. Menard (1971).

17. Among other things Gilbert declined to be Director of the Geological Survey.

18. Pyne (1980), p. 255.

19. Watson (1968).

20. Dicke (1959), p. 25. He was delivering the 27th Joseph Henry Lecture of the Philosophical Society of Washington.

21. Richard Hey informs me that the same story is told at Princeton regarding Einstein.

22. Menard (1971).

CHAPTER 2

1. Carey (1958).

2. For a wealth of detail see Marvin (1973).

3. Greene (1982).

4. F. Darwin (1887), 3:230.

5. Coleman (1916), p. 190.

6. See Georgi (1962), Hallam (1975), Marvin (1973). I have interpreted his biography in the light of my own experiences.

7. Marvin (1973), p. 67.

8. Georgi (1962), p. 317.

9. Ibid., p. 322. The curious phraseology "fully fit and with rested dogs" I take as meaning that Wegener had done everything that a prudent, experienced explorer could do to prepare for a safe trip to the coast. Good scientific explorers believe that the "adventures" of adventurers are largely the consequences of poor planning. An exciting story after a safe return with the expected data, however, is expected of an explorer.

10. Marvin (1973), p. 106.

11. *Geologische Rundschau* 22 (1931): 272.

12. Bullard (1975b), p. 16.

13. Pickering (1907).

14. Taylor (1910).

15. Leverett (1939), p. 193. The quotation is from a letter from Taylor to Leverett written in 1932.

16. Georgi (1962), p. 309.

17. Wegener (1922), p. 5.

18. Watson (1968). The article in *Nature* appeared in April 1953, the resulting Nobel Prize in 1962.

19. Wegener (1929), p. 23.

20. Wegener (1929), p. 98.

21. Wegener (1929), p. 170.

22. Quoted in Hallam (1983), p. 128.

23. Willis (1928), p. 76.

24. Willis (1944) refers to drift as a fable. The quote is from my own memory of a professional meeting in Los Angeles in 1946 or early 1947. Willis was in the process of deriding my professor, John Maxon, for having the temerity to ask how Willis's paper on global tectonics related to continental drift.

25. A. G. Fischer informs me that an incident is "strangely engraved" in his memory in which Gutenberg gave a talk at a GSA meeting and in the following discussion Andy Lawson shook a long bony finger at him and said, "and, Mr. Gutenstein or whatever your name is. . . ." It seems likely that we two witnessed the same scene in which one of the distinguished, old American scientists was confused about Gutenberg's name.

26. Wright (1923), Jeffreys (1962).

27. Lake (1922), p. 338.

28. Schuchert (1928), p. 134.

29. Berry (1928), p. 194.

30. Daly (1923), p. 448.

31. Longwell (1958), p. 2.

32. Discussion, p. 61, accompanying Daly (1923).

33. Termier (1925), p. 236.

34. Longwell (1944), p. 221.

CHAPTER 3

1. Bullard (1975a), p. 272.
2. Ibid.
3. Hess (1962a), p. 86.
4. Hess (1962b), p. 2.
5. Field (1933), p. 9.
6. Ibid.
7. Ibid., p. 14.
8. Bullard (1975a), p. 269.
9. Ibid., p. 271. I once wrote to Teddy to warn him that his memorials were so beautifully written as to cause his colleagues to long for an early death.
10. Wertenbaker (1974), p. 21.
11. Bullard (1975a), p. 271. Teddy thought that it was probably the worst advice ever given by a professor to a student.
12. Quoted in Wertenbaker (1974), p. 22.
13. Bullard (1975a), p. 272.
14. Ibid., p. 273.
15. M. Ewing, Crary, and Rutherford (1937).
16. Bullard (1975a), p. 273.
17. Indeed, Ewing had hardly anything to do with this paper. Apparently, they struck a bargain whereby Heezen's name appeared on another paper with which he had hardly anything to do.
18. Leet (1937), p. 354.
19. For forgotten reasons Ewing called John L. Worzel "Joe" when he was an undergraduate. So he remained in our small world.
20. The citation for the Day Medal that was awarded to Ewing in 1949 states that he was commissioned as a Lieutenant in the Naval Reserve from 1938 to 1941. I find no other mention of this.
21. A friend told me, three decades ago, that when she was a young girl in Pasadena and Roger would visit home, the neighborhood children would stare in awe and say, "That's Roger Revelle—he's a genius."
22. Revelle (1944). The interval of 15 years between collection of the samples and publication is illustrative of old-style geology and oceanography. Things would shortly change.
23. Incredibly enough, the old ship made it back into the movies. She was the schooner that had her superstructure dismantled and burned to fuel the hectic Atlantic passage in the movie "Around the World in 80 Days." Even then she was not finished. Somehow she got to Tahiti and sank in Papeete harbor, where Scripps oceanographers used to salute the stubs of her masts.
24. It had been renamed "Navy Electronics Laboratory" when I went there to work under Dietz in 1949.
25. Letter to Andrew Hamilton, Public Information Office, UCLA, 20 May 1948. SIO Archives.

26. The reader should be alert to a potential source of bias. As a recipient of several million dollars from the NSF, mainly to run ships, I tend to view it fondly.

27. "Effective Use of the Sea," a report of the Panel of Oceanography, President's Science Advisory Committee, The White House, June 1966. Gordon MacDonald was Chairman, and members included the oceanographers Walter Munk and Henry Stommel. I was then on the White House science staff (Office of Science and Technology) and served as technical assistant. This report was but one in a lengthy series by which the Executive Branch and the Congress monitor the outpouring of Federal money. We visited 9 different oceanographic centers, polled 13 for data, interviewed 15 Federal agencies, and consulted countless organizations and individuals.

28. Memorandum to A. N. Richards and Detlev Bronk, 1 September 1948. Revelle papers, SIO Archives.

29. Letter to Harald Sverdrup, 6 January 1948. Revelle papers, SIO Archives. The letter was featured in a library display of the history of Scripps Institution.

30. Ibid.

31. "The research ship *Horzion*," SIO Reference 74-3 (1974).

32. Bullard (1975a), p. 292.

33. Pyne (1980).

34. Gray (1956).

35. Menard (1964), p. 152.

36. Dedicatory epistle to Lord Howard of Effingham, in R. Hakluyt, *The principal navigations and voyages, traffiques and discoveries of the English nation*, vol. 1 (1599); p. 16 in edition by J. M. Dent and Sons, Ltd. (London, 1927).

37. The reader will note that I am giving the same kind of defensive detail as did Georgi when he described Wegener's death. Personally, I was doing everything that a prudent, experienced explorer could do. So, I assume, was Ewing.

CHAPTER 4

1. Pettersson (1953).

2. Koczy (1954).

3. Kuenen (1950).

4. Gutenberg and Richter (1954). It is a commentary on the nature of fame that Charles Richter of the "Richter scale" is now far better known than his illustrious collaborator—or any other seismologist.

5. Ibid.

6. Hess (1946). The evolution of ideas on guyots appeared recently in Menard (1984a).

7. M. Ewing et al. (1949).

8. On *Horizon* on our first expedition across the Atlantic, Edward Goldberg, Harmon Craig, and I were on board with a number of students delighted to sail from Martinique to Monte Carlo. As I recall, I was expedition leader and planned the heat flow. The measurements were taken routinely by students including R. D. Nason and W.H.K. Lee. The profile showed high heat flow at the ridge crest and low on the flanks—just what Maxwell and von Herzen had found on the East Pacific Rise. After some debate among the senior scientists it was decided that the heat-flow profile should be published

in *Nature* with the students as sole authors. We wanted to needle Ewing with the idea that the heat-flow measurements that had eluded him in the Atlantic were hardly noteworthy at Scripps. Ewing complained to me when I next saw him that his efforts at Lamont were frustrated by incompetents. That very year his program began to succeed.

9. Teddy seldom appeared to take his honors seriously, although I remember once when we went ashore somewhere in the Fiji Islands he made a point of signing the guest book at Government House and was offended when he was not speedily invited to dinner. He also took umbrage at the name of our mini-schnauzer, which is "Sir Roderick Impey Murchison" after the discoverer of the Silurian System. The real Teddy of memory, however, was the man on the dock as we prepared to leave Easter Island. The young Chilean naval governor and his wife were kindly seeing us off when a native offered Teddy a last souvenir. We had all exhausted our money, but Teddy proposed to trade his trousers, was accepted, removed them, boarded the skiff, and joined us in waving as we pulled away from the dock. The governor's reaction was one of surprise, but his wife's was one of delight.

10. Or so Teddy told me. If so, both Ewing and Bullard received discouraging advice as they began their careers.

11. Shor (1984).

12. R. W. Raitt (1956).

13. Ewing, Sutton, and Officer (1954).

14. Deitz, Menard, and Hamilton (1954).

15. Menard (1956).

16. Menard (1964).

17. Revelle and Maxwell (1952) with commentary by Bullard (1952).

18. Hess (1951).

19. Griggs (1939).

20. E. L. Hamilton (1956).

21. Hess (1946).

22. Revelle (1953) and others produced the "Shipboard Report." Helen Raitt (1956), Russ's wife, wrote her own account, which had a second edition in 1964, was translated into several languages, and had a significant effect in popularizing oceanography.

23. R. W. Raitt (1956).

CHAPTER 5

1. I was born in Fresno, California in 1920, obtained a B.S. from Caltech in 1942, became a Naval Reserve Ensign immediately after Pearl Harbor, and was trained as a photo interpreter. I served initially in the Solomon Islands, went to England at the time of the Normandy invasion, and spent the last year of the war with the fast carrier task force in the western Pacific. I was not advanced enough to do research in geology, but I managed a little field mapping here and there and carried a scientific library on shipboard. After the war I returned to Caltech for an M.S. in geology. I married Gifford Merrill in 1946, and we went on to Harvard where I received a Ph.D. in geology in 1949. My thesis, under Henry Stetson, was an experimental study of sediment transport in a flume at Woods Hole Oceanographic Institution.

2. As civil servants we had "Secret" clearances. Because Midpac and Capricorn involved nuclear tests, the participants had "Q" clearances from the Atomic Energy Commission. As a naval officer I had "Top Secret" and "Ultra Secret" clearances.

3. See also Menard (1964), pp. 1-10. The U.S. Navy Hydrographic Office became the Oceanographic Office, but at first many of the people were the same.

4. Murray (1939).

5. Shepard and Emery (1941).

6. Cordilleran Section of the Geological Society of America, March 1951.

7. Merton (1968).

8. Wooster (1951).

9. Wooster (1952).

10. "Shirley Trough" west of central Baja California is named after her.

11. Vening Meinesz (1947).

12. This is the jargon that came after the revolution, as discussed in Chapter 21.

13. Gutenberg (1951), p. 178.

14. Bijlaard (1951). Discussion on p. 519.

15. Menard (1971).

16. Brooks (1941).

17. Hess (1955a).

18. Heezen et al. (1964).

CHAPTER 6

1. Bullard (1968), p. 492.

2. Hospers (1951).

3. Nagata (1951).

4. Bullard (1968), p. 483.

5. R. L. Fisher provided details about the expedition.

6. By the mid-1950s I was running a casual but successful seamount discovery system under the auspices of the InterAmerican Tuna Commission and with the cooperation of the San Diego tuna fleet. My assistant, Tom Chase, distributed our bathymetric maps to the fishermen, and they told us what they found instead of keeping trade secrets. They were particularly useful in disproving the existence of the many spurious shoals that appeared on Hydrographic Office charts of the region. The "shoals" were echoes off the plankton in the newly discovered deep scattering layer.

7. According to Roger Revelle, who supplied the funds. This section is based on several conversations with Raff and Revelle on the occasion of the latter's 75th birthday symposium in 1984.

8. Being surveyors, the C and GS people ran the Precision Depth Sounder in accordance with the instructions of the manufacturer. Thus they obtained quite-small-scale records in deep water. Oceanographers modified the recorder so the depth could be read as accurately (± 1 fathom) in deep as in shallow water.

9. College administrations shuddered at the news that Roger Revelle was on campus to recruit faculty. At his best, he probably could have lured Vacquier, or the rest of us, from Scripps *to* the New Mexico School of Mines with promises of clear desert air and the opportunity to create something unique.

10. Curiously enough, a wave of rationality swept through the Navy classifiers about 1960, and I obtained a release for a very detailed chart off central California. Bruce Heezen and I were writing a joint paper for Volume 3 of *The Sea*, so I included the chart in it. The volume was delayed—Hess withdrew his essay on geopoetry as a consequence—and the Navy restored the Confidential classification. I honestly forgot about my chart in the joint paper, so it was a breach of security when it was at last published. No one else ever noticed.

11. Menard and Vacquier (1958). Naturally we would have been embarrassed if someone later had proposed that we had priority in Mason's discovery.

12. Mason (1958).

13. Vacquier (1959).

14. See Pais (1982).

15. Vacquier, Raff, and Warren (1961); Mason and Raff (1961); Raff and Mason (1961).

16. Menard (1960).

17. Menard (1979).

18. Vacquier, Raff, and Warren (1961), p. 1257.

19. Menard (1962).

20. Raff (1962).

21. Ibid., p. 471.

CHAPTER 7

1. Coleman (1933).

2. Coleman (1924), p. 398.

3. Bowie (1935), p. 447.

4. Jonas (1935).

5. Gutenberg (1936).

6. Gunn (1936), Watts (1935), Bucher (1933), Chaney (1940).

7. Du Toit (1927).

8. Du Toit (1937).

9. Ibid., p. vii.

10. Simpson (1943), p. 2.

11. Du Toit (1944), p. 146.

12. Willis (1944), p. 509.

13. Holmes (1928).

14. Holmes (1929).

15. Hess (1962a).

16. Holmes (1944), p. 509.

17. Longwell (1944), p. 220.

18. Hedberg (1957), p. 69.

19. Ibid., p. 71.

20. Quoted in Pyne (1980), p. 94.

21. Bullard (1975b), p. 14.

22. Runcorn (1962), p. 9.

23. Ibid., pp. 17-18.

24. Bullard (1975b).
25. Runcorn (1956a, 1956b).
26. Munk and MacDonald (1960), p. 259.
27. Blackett, Clegg, and Stubbs (1960).
28. Hess (1962a), p. 608.
29. Heezen (1960), p. 10.
30. See Carey (1958).
31. Longwell (1958).
32. King (1958a).
33. King (1958b).
34. Carey (1958).
35. Ibid., p. 178.
36. Ibid., p. 186.
37. Ibid., p. 191.
38. Ibid., pp. 197-198.
39. Ibid., pp. 196-197.
40. Merton (1961).

CHAPTER 8

1. President Johnson hoped for a similar solution during the exponential expansion of the Vietnam war. The U.S. Geological Survey's "heavy metals" project developed new geochemical techniques to map the distribution of gold in extremely low concentrations in the United States.
2. Stocks and Wüst (1933).
3. Wiseman and Sewell (1937), p. 220.
4. Ibid., p. 222.
5. See Sullivan (1974), p. 53.
6. Tolstoy and Ewing (1949).
7. Ibid., p. 1528.
8. Ibid., p. 1530.
9. Tolstoy (1951).
10. See Bullard (1967).
11. Ibid., p. 202.
12. Hill (1956).
13. Hill (1957).
14. Hill (1960).
15. Heck (1935).
16. Gutenberg and Richter (1945), p. 642.
17. Rothé (1954), p. 387. Cited from the abstract. The paper is in French. Presumably the talk was as well.
18. Gutenberg and Richter (1954).
19. Ibid., p. 77.
20. Ibid., p. 74.
21. Youngquist and Heezen (1948).
22. Talwani (1982), p. 2.

23. Wertenbaker (1974), p. 144.

24. Tharp (1982), p. 22.

25. Heezen, Ewing, and Miller (1953).

26. Heezen (1956).

27. Ewing and Heezen (1956a).

28. Heezen and Tharp (1956).

29. M. Ewing and Heezen (1956b), p. 80.

30. About that time, their distinguished senior colleague, Walter Bucher, told A. G. Fischer that he doubted the existence of such a feature and considered it a mistake in interpretation. Personally, I often compared the rift to the Midgard serpent—world-girdling but a myth.

31. Menard (1958), p. 1182.

32. Heezen and Ewing (1961), p. 628.

33. Wertenbaker (1974), p. 148.

34. Ibid.

35. Heezen (1959b). English abstract. Bruce and I, among others, wrote in English, and our papers were then translated into French.

36. Heezen, Tharp, and Ewing (1959).

37. Ewing and Heezen (1960), Heezen (1960), Heezen and Ewing (1961).

38. Heezen (1959b); Heezen, Tharp, and Ewing (1959).

39. Talwani (1982); Bullard (1975a).

40. Heezen and Ewing (1961), p. 640.

CHAPTER 9

1. See James (1973).

2. Rubey (1968), p. 85.

3. Ibid.

4. Hess (1939), p. 263.

5. See also Bascom (1961), pp. 148-150.

6. Rubey (1968), p. 84.

7. Hess (1954).

8. Ibid., p. 341.

9. Ibid., pp. 341-342.

10. Ibid., p. 344.

11. Revelle and Maxwell (1952).

12. Rubey (1951).

13. Letter to Menard, 10 October 1955.

14. Hess (1955a), p. 403.

15. Maxwell and Revelle (1954).

16. Salisbury and Christensen (1978).

17. Estabrook (1956), p. 686.

18. Hess (1959a), Bascom (1961). Walter Munk, who attended one and hosted the other gathering, agrees with Hess.

19. Hess (1959a), p. 340.

20. Hedberg (1962).

21. Greenberg (1964).

22. I consulted with my colleague, Herbert York, before taking this job. As former Director of Defense Research and Engineering he was on intimate terms with the highest levels in government. "It doesn't sound like much," he said, "being an advisor to an advisor to someone." He paused, "But you have to remember who that someone is."

23. Shakespeare, *Henry IV*, Pt. 1, Act 3, Sc. 1.

CHAPTER 10

1. Kuenen (1935), p. 79.
2. See Daly (1940), pp. 253-255.
3. Ibid., p. 252.
4. See Longwell (1945).
5. Hess (1957), p. 183.
6. Daly (1940), p. 256.
7. Vening Meinesz (1934).
8. Bijlaard (1936).
9. Hess (1948), p. 422.
10. Bucher (1933), p. 461.
11. Ibid., p. 240.
12. Ibid., p. 75.
13. Richter (1958).
14. Benioff (1949), p. 1838.
15. Benioff (1954).
16. Eponymy, the naming of features after their discoverer, can be a mixed blessing. It doesn't matter so much if the features are rare, but if by chance they become the subject of an extensive literature, the situation changes. For a while the oceanic fracture zones were called "Menard zones" by some scientists, so I thought about the matter. By this time, 35 years after his discovery, everyone in the world except Benioff would be referring to his "great faults" as "Benioff zones."

17. Fisher and Revelle (1955), p. 37.

18. Later, when the bathyscaphe was developed, it became a challenge to be the first person to go to the deepest point. The bathyscaphe *Trieste* was built while Jacques Piccard lived at Lausanne, Switzerland, and the depth gauge was tested in the deep Lake Lucerne. When Piccard and Navy Lt. Don Walsh dove in the Marianas Trench in 1960, they were over the deepest known point—35,800 feet of water. Imagine their excitement when the depth gauge showed them below 36,000, then 37,000, and finally on bottom at 37,800 feet! Of course it was all a mistake; no one had properly corrected the depth gauge for the increased density of sea water compared with fresh.

19. Fisher and Revelle (1955).
20. Ibid., pp. 39-40.
21. Ewing and Worzel (1954), p. 165.
22. Worzel and Shurbet (1955), p. 93.
23. Ewing and Heezen (1955), p. 266.
24. Ewing, Worzel, and Shurbet (1957).

25. Hess (1957), p. 187.
26. Officer et al. (1957), p. 366.
27. Talwani, Sutton, and Worzel (1959).
28. Talwani, Worzel, and Ewing (1961).
29. Fisher (1961), Shor and Fisher (1961), Fisher and Raitt (1962).

CHAPTER 11

1. Fisher (1958).
2. I had not been idle since Capricorn. For a few years after that expedition I rarely went into the deep sea. With five partners, including Bob Dietz and Ed Hamilton, I was moonlighting on weekends as a diving geologist consulting for oil companies. I made more than 1,000 dives to depths as great as 150 feet in the cold murky waters off Southern California. Curiously enough, even after 30 years none of us shows any physical deterioration from the diving. Radiation damage from the sun is another matter. In 1956, I left the Navy Electronics Laboratory to become an Associate Professor in the University of California's Institute of Marine Resources and Scripps Institution of Oceanography. Rather by coincidence, I then resumed cruising in the deep sea.
3. Menard (1960, 1961).
4. Ewing and Ewing (1959).
5. Ibid., pp. 307-308.
6. Raitt and Shor (pp. 36-38), quoted in Fisher (1958), p. 37.
7. von Herzen (1959).
8. The best known variable-into-constant was Einstein's definition of the speed of light in a vacuum as a constant ''C.'' A reverse change, of interest here, is Dirac's redefinition of the gravity constant ''G'' as a variable that decreases with time.
9. Menard (1960), p. 1741.
10. Ibid., p. 1744.
11. Ibid., p. 1745.
12. Ibid.
13. Ibid., p. 1746. I mentioned in the Introduction that scientists tend to erase hypotheses from their minds when they are no longer useful. This quotation is an example. I had completely fogotten my own work after accepting the reality of sea-floor spreading. I thought I had proposed no more deformation than that indicated by the stretching of the crest of the East Pacific Rise. Thus I have been surprised to find that I accepted continental drift without reservation in 1960.

CHAPTER 12

1. Greene (1982).
2. S. W. Carey, *The Expanding Earth: An Essay Review*, 56 pp., privately issued. Undated, the most recent references are three in 1973. The early references he cites are rare, and I have not attempted to obtain them. The brief, unidentified quotes in this chapter are from this valuable but unpublished review.
3. B. Lindemann. 1927. *Kettengebirge, Kontintale Zerspaltung und Erdexpansion*. Jena: Gustav Fischer Publ. Not seen.

4. M. Bogolepow. 1930. Die Dehnung der Lithosphäre. *Zeit. dt. geol. Ges.* 82:206-228. Not seen.

5. Not seen by Carey or me.

6. Not noted by Carey in his review.

7. Some astrophysicists still question the meaning of the red-shift.

8. O. C. Hilgenberg. 1933. *Vom wachsenden Erdball.* Berlin. Not seen.

9. J. K. Halm. 1935. An astronomical aspect of the evolution of the earth. *Astron. Soc. So. Africa* 4:1-28. Not seen.

10. Dirac (1937).

11. Dicke (1957).

12. Translated as P. Jordan. 1966. *The expanding earth.* Oxford: Pergamon. The fact of the translation suggests the intensity of interest in the expansion hypothesis during the plate tectonics revolution.

13. Joksch (1955).

14. Egyed (1956), p. 534.

15. Carey (1973?).

16. Egyed (1957), p. 108.

17. Ibid., p. 109.

18. Merton (1961), p. 474.

19. Egyed (1957), p. 113.

20. Jones (1954).

21. I have an otherwise unidentified reprint of Carey's *Tectonic approach to the origin of the Indian Ocean* from the Comptes Rendus of the third "Congrès de la P.I.O.S.A., Tanarive 1957, section C." It contains many of the figures and analyses he later published in 1958.

22. Carey (1958), p. 318.

23. Dicke (1957).

24. The accepted age has been stable for the last two decades at 4.55 aeons.

25. Heezen (1959b), p. 295.

26. Tharp (1982), p. 22.

27. Heezen (1959a), p. 28.

28. Wilson (1960).

29. Ibid., p. 880.

30. Heezen (1960), p. 12.

31. Sullivan (1974), p. 57.

CHAPTER 13

1. Hess (1959a).

2. Hess (1959b).

3. Hess (1959a), p. 343.

4. Incredibly enough, the first paper he submitted for the Buddington volume was unanimously rejected by the three editors, or so I am assured by two of them, A.E.J. Engel and Harold James. The paper addressed one of his specialties, namely, mineralogy. The editors thought that Hess's geopoetry was just that. He "had a good track record" for wild ideas, however, so they accepted it.

5. Dietz (1961a).

6. Hess (1962a).

7. Piccard and Dietz (1961).

8. Dietz (1946).

9. Menard (1971).

10. Weariness can be extreme. At a ceremony honoring the economist Seymour Harris at UCSD, Clark Kerr, one of his former students, remarked that when he realized that Harris was writing books faster than he could read them, he abandoned economics and became President of the University of California.

11. Dietz (1961a).

12. Phone conversation, February 1984.

13. Menard (1959), p. 215.

14. Hess (1962a), p. 599.

15. Ibid.

16. Dietz (1961a), p. 854.

17. Ibid.

18. Hess (1962a), p. 614.

19. Hess (1962a), pp. 608-609.

20. Dietz (1961a), reprint p. 4.

21. Holmes (1944), p. 508.

22. Dietz (1968), p. 6567.

23. Hess (1968), p. 6569.

24. Dietz (1962a), p. 12.

25. Phone conversation, February 1984.

26. Phone conversation, February 1984.

27. Hess (1962a), p. 611.

28. Letter to Menard, 12 June 1961.

29. Letter to Menard, 6 September 1966.

30. Dietz (1961a), reprint p. 4.

31. Ibid., p. 7.

CHAPTER 14

1. *Journal of Geology, Bulletin of the Geological Society of America, Journal of Geophysical Research, Bulletin of the American Association of Petroleum Geologists, Quarterly Journal of the Geological Society of London*, and *Geotimes*. Many of the journals include letters and commentaries.

2. Cox and Doell (1960), p. 763.

3. Ibid., p. 645.

4. E. L. Hamilton (1959, 1960).

5. Billings (1960), p. 389.

6. Ibid., p. 394.

7. I am indebted to Allan Cox for providing me with copies of the autobiographical information he solicited in connection with his book on *Plate Tectonics and Geomagnetic Reversals* (1973). Tuzo Wilson and I have corresponded and met at professional

meetings for 30 years, and he has recently given me additional autobiographical information.

8. Letter to Allan Cox, 21 June 1972.

9. Wilson (1951), p. 85.

10. Ibid., p. 86.

11. Letter to Allan Cox, 21 June 1972.

12. Wilson (1949, 1951, 1952, 1953, 1954, 1957, 1959). During this period his
basic concepts hardly changed; he merely modified them in accordance with the latest
results from geophysical research. In retrospect, the papers seem remarkably similar,
but at the time we tended to emphasize the differences.

13. Wilson (1959), p. 15.

14. Ibid.

15. Ibid., p. 5.

16. Wilson (1951), Fig. 8 and others.

17. Wilson (1957), p. 228.

18. Wilson (1960), p. 881.

19. Letter to Allan Cox, 21 June 1972.

CHAPTER 15

1. Dietz (1961b).

2. According to conversations on 31 May 1984.

3. Glen (1982), pp. 312 and 313.

4. According to conversations with Heirtzler, 5 December 1983; Talwani, 5 April
and 1 May 1984; and J. Ewing (by telephone), 4 June 1984.

5. Bernal (1961), p. 123.

6. Dietz (1961b), p. 124.

7. Wilson (1961), p. 126.

8. Ibid., p. 128.

9. Heezen and Ewing (1961), p. 640.

10. W. Hamilton (1961). Is it purely coincidence that, after being standard for most
of a century, the practice of printing the Director's authorization on each paper by a
USGS geologist was soon discontinued?

11. Letter to Menard, 24 May 1984.

12. Dietz (1962a, 1962b, 1962c).

13. Dietz (1962b), pp. 7-8.

14. It includes six of my published illustrations.

15. Heezen (1962), p. 285.

16. Ibid., p. 278.

17. Ibid.

18. Carey (1976), p. 57.

19. Ibid., p. 54.

20. Beloussov (1962), p. 746.

21. Ibid., p. 750.

22. Ibid., p. 753.

23. Ibid., p. 747.

24. Ibid., p. 679.
25. Ibid., p. 682.
26. Wilson (1962), p. 5.

CHAPTER 16

1. Hodgson told amusing anecdotes at that time regarding his experiences with other seismologists. Depending on the first motion, an earthquake radiates compressive waves in some directions and rarefactions in others. By mapping the occurrence of first motions at seismographic stations around the world it is possible to determine the nature of the offset on a distant fault even if it is under water or within the interior of the earth. There is, however, some ambiguity in the solution, and either of two orientations of the fault is possible. Hodgson's problem was that he could not make consistent maps of first motion. There were always anomalous compressions mixed in with the rarefactions and vice versa. Hodgson sought an explanation. What a seismometer records is not a high or low pressure but a wiggly line that initially is either above or below normal and then oscillates as more waves are recorded. Hodgson wrote to the seismologists whose results were consistently anomalous and suggested that they check the polarity of their electrical connections. After sometimes strained correspondence, the anomalies disappeared.
2. Wilson (1963a, 1963b, 1963c).
3. Stearns (1961).
4. Wilson (1963b).
5. Ibid., pp. 536-537.
6. Menard (1983).
7. Wilson (1963b) p. 537.
8. Dietz (1963a, 1963b, 1964).
9. Dietz (1963b), p. 950.
10. Ibid.
11. Dietz (1963a), p. 314.
12. Ibid., p. 326.
13. Menard and Ladd (1963), p. 384.
14. Conversation with Marie Tharp, 4 June 1984.
15. According to conversation with D. Hayes, November 1983.
16. Draft of a letter from Heezen to President Kirk of Columbia University apparently in 1966. It may not have been sent. My copy courtesy of Marie Tharp.
17. Conversation with Manik Talwani, 5 April 1984.
18. Conversation with Marie Tharp, 31 May 1984.
19. Ewing (1963), p. 42.
20. Ibid.
21. Ibid., p. 44.
22. Conversation with Manik Talwani, 5 April 1984.
23. Ewing (1963), p. 55.
24. Ewing, Ewing, and Talwani (1964).
25. Ewing and Ewing (1964).
26. Ibid., p. 528.

27. Ibid., p. 535.

28. Drake and Woodward (1963).

29. Drake et al. (1963).

30. Drake and Girdler (1964).

31. Sykes (1963).

32. Letter to Allan Cox in 1972.

33. Sykes and Landisman (1964).

34. Rusnak and Fisher (1964).

35. There are two stories from Princeton. One comes from Engel about taking a laboratory course in crystallography from Harry Hess. Hess walked into the laboratory for the first meeting, waved a hand toward the goniometer and other instruments, told the students to learn how to use them, and was never seen again. The second story is about Engel. He listened to the opening of a professorial lecture, stood up, commented ''With that statement you mark yourself as incompetent in the field,'' and departed. It is possible that the two stories are not unrelated.

36. Engel and Engel (1964). The Engel papers ignored a century of tradition in which samples were identified by their expedition sample numbers. They created their own numbers, which made it difficult, for example, to compare heat flow or sediment type with petrography or rock chemistry. Land geologists have gradually complicated the lives of simple sailors for the two decades since. As early as 1972, stratigraphers began to feel the necessity to apply their special terminology. Stratigraphers, like petrographers, name rocks after a type location, so we were cursed by the ''application of classical North American stratigraphic principles'' with the names ''Clipperton Oceanic Formation'' and ''Marquesas Oceanic Formation.'' They were applied to muck, from Deep-Sea Drilling holes, that could hardly stand under its own weight.

37. Bullard (1975b).

38. Menard (1964), p. 103.

39. Ibid., p. 113.

40. Menard (1984b); the idea that this rise existed turned out to have a complex history. I abandoned it at one time as no longer being fruitful, but it is now having something of a vogue.

41. I tried to persuade Harold Urey to write something in this vein, assuming that the master of the origin of the planets could end the discussion. He thought, however, that a concept involving such densities was too absurd for discussion.

42. Menard (1964), p. 169.

CHAPTER 17

1. Glen (1982).

2. Bullard (1968).

3. Letter from McDougall to Allan Cox, 1973.

4. Balsley was Assistant Director for Research when I was Director. He was also head of a special group attempting to translate Survey data into innovative, integrated reports. Most Survey scientists and the Congress seemed to prefer traditional ways.

5. Cox, Doell, and Dalrymple (1963a, 1963b); McDougall and Tarling (1963).

6. Krause (1960).

7. Cox and Doell (1962).

8. Raff (1963), p. 955.

9. Matthews (1961).

10. Matthews (1963).

11. Letter from Morley to Cox, 22 November 1971.

12. Morley and Larochelle (1964).

13. Bullard (1975b), p. 19.

14. Letter from Morley to Cox, 22 November 1971.

15. Frankel (1982), Glen (1982).

16. Letter from Morley to Cox, 22 November 1971; Glen (1982); Frankel (1982).

17. Frankel (1982). Glen (1982) gives the date of this talk as June 1964 and says (p. 299) that it was ''Following that second rejection. . . .'' Frankel's quote (p. 17) ''at that time [the paper] was still pending acceptance'' seems to make more sense. The talk was published in 1964.

18. Letter from Morley to Cox, 22 November 1971; Glen (1982).

19. Frankel (1982), p. 12.

20. This may have been the same text that my son used at the same age. The difference was that at Cambridgeshire High School in 1962 the teacher indicated that geologists were no longer doubtful about drift.

21. Frankel (1982), p. 12.

22. Vine's response upon receiving the Arthur L. Day Medal for 1968 from the Geological Society of America.

23. Letter from Sclater to Cox, 1972.

24. Glen (1982), p. 280.

25. Ibid., p. 279.

26. Backus (1964), p. 591.

CHAPTER 18

1. Eaton (1964).

2. Gilliland (1964).

3. Gilluly (1963). First it was Vine talking to the Sedgwick Club, now Gilluly delivering the William Smith Lecture. The great names of the nineteenth century were being linked to continental drift.

4. Letter from Hamilton to Menard, 24 May 1984.

5. W. Hamilton (1963a), p. 14.

6. W. Hamilton (1963b).

7. A ''grumpy aside'' in Hamilton's letter to me, 24 May 1984.

8. Axelrod (1963).

9. W. Hamilton (1964).

10. Ibid., p. 1667.

11. Northrup and Meyerhoff (1963).

12. Northrup was particularly active at one time in utilizing SOFAR data to locate small oceanic earthquakes.

13. MacDonald (1964), pp. 928-929.

14. Runcorn (1963), p. 629. I always advise students against sarcasm in print be-

cause it is almost impossible not to appear heavy-handed. I take the quotation to be a successful exception.

15. Bullard (1964), p. 15.

16. Lee and MacDonald (1963). This paper achieved a certain notoriety for the bizarre consequences of the spherical harmonic analysis.

17. Bullard (1964), p. 22.

18. Teddy knew a great deal of poetry. I once had occasion to quote part of a line from "The Burial of Sir John Moore at Corunna" to him in a context in which it was not obviously a quotation. He remembered the entire poem from the occasion when he had won a school prize for reciting it. In the sentence quoted from the Geological Society meeting he seems to be paraphrasing "Alice in Wonderland."

19. Bullard (1975b), p. 19.

20. Blackett (1965), p. ix.

21. Bullard et al. (1965), p. 42.

22. Worzel (1965), p. 137.

23. Vacquier (1965).

24. Heezen and Tharp (1965).

25. MacDonald (1965), p. 215.

26. Bullard et al. (1965), p. 323.

CHAPTER 19

1. Hess (1965), p. 318.

2. Glen (1982), p. 304. Harry was very encouraging to young scientists in particular. And, of course, everyone thought the hypothesis was fantastic, but most meant it literally.

3. Ibid., p. 330.

4. Dick Hey informed me of the existence of this paper while we were aboard R/V *Thomas Washington* near the end of the Mendocino escarpment in August 1984.

5. Coode (1965).

6. Ibid., p. 400.

7. Unless otherwise credited this account is based on conversations with Hess and Wilson and an informal interview with Wilson in March 1984.

8. Wilson (1965a).

9. Ibid., p. 343.

10. Glen (1982), p. 304.

11. Ibid., p. 305.

12. Letter from Sclater to Allan Cox, 1972.

13. Glen (1982), p. 303.

14. Wilson (1965b).

15. Wilson (1965c), Vine and Wilson (1965).

16. See, for example, Glen (1982), p. 306.

17. Wilson (1965c), p. 482.

18. Ibid., p. 485.

19. Menard and Dietz (1952).

20. Vine and Wilson (1965).

21. Ibid., p. 485.

22. Glen (1982), p. 310.

23. LePichon et al. (1965), p. 331.

24. Talwani, LePichon, and Ewing (1965).

25. Heirtzler and LePichon (1965).

26. Ibid., p. 4028.

27. Talwani, LePichon, and Heirtzler (1965), p. 1109.

28. Ibid., p. 1114.

CHAPTER 20

1. Menard (1971).

2. W. Hamilton (1966a), p. 178.

3. Ibid., p. 186.

4. Hamilton and Myers (1966).

5. Menard (1971).

6. Dr. Peter kindly sent me copies of his correspondence with Hess.

7. Peter (1966).

8. Christoffel and Ross (1965), p. 2857.

9. M. Ewing, LePichon, and Ewing (1966), p. 1624.

10. Ibid., p. 1627.

11. Ibid., p. 1633.

12. Ibid., p. 1635.

13. Frankel (1982), p. 36. I am citing Frankel for my own correspondence because I do not have copies. Hess kept his, however, and Frankel found them at Princeton.

14. Saito, Ewing, and Burckle (1966).

15. We had flown together across the Atlantic to the Royal Society meeting on continental drift. Orowan pointed out how the rivets in the wings were moving and cheerfully estimated the probability that a wing would fall off before we reached London.

16. Orowan (1966).

17. Letter from Orowan to Menard, 9 May 1966.

18. Dietz and Holden (1966a).

19. Dietz and Holden (1966b).

20. Gilluly (1966), p. 2.

21. W. Hamilton (1966b), p. 348.

22. As of 1984 the data acquisition and processing systems begin with the same analog records as in 1965. In addition, the data are acquired digitally, the computer automatically removes the earth's field, the remaining value is merged with the satellite navigation, and the anomalies appear in analog form beside the ship's track on a flatbed plotter. All this is in principle. At the time when I was writing during the Zed expedition, the computer was stubbornly rejecting the satellite fixes, and without them it would not plot a corrected track or the anomalies. The only systems still working are like those from 1965.

23. What follows in the text is a combination of my memory, conversations with Heirtzler in 1983 and with Pitman and Hayes in 1984, and published sources that are cited.

24. Wertenbaker (1974), p. 203.

25. Glen (1982), p. 312.

26. Ibid., p. 313.

27. Ibid., p. 312.

28. Bullard (1975a), p. 288.

29. Ibid., p. 289.

30. Glen (1982), p. 357.

31. Ibid., p. 356.

32. Conversation at Lamont, May 1984.

33. Bullard (1975a), p. 288.

34. Ibid., p. 286.

35. Glen (1982), pp. 329-330. His account agrees, as well as I can remember, with what Bruce told me in Moscow. Certainly Bruce was bitter, but he often was so when describing events at Lamont in those years. These particular events did not loom as large then as they would shortly.

36. Ibid., p. 330.

37. Glen (1982), p. 321.

38. *Anatomy of an Expedition* published by McGraw-Hill in 1969, and written largely while at sea.

39. Phinney (1968).

40. Letter from McKenzie to Allan Cox, 1972.

41. Phinney (1968), p. 87. In a footnote to his published paper Vine had said that "Rutherford once stated that if he obtained results which required statistics to interpret them he would throw them away."

42. We had been surveying back and forth between Heirtzler #21 and #24 for several days when I wrote this.

43. Quoted in Wertenbaker (1974), p. 218.

44. Discussion following Menard (1968), pp. 116-117.

45. Phinney (1968), pp. 226-228.

46. Rubey (1968), p. 83.

47. J. Ewing et al. (1966), Pitman and Heirtzler (1966), Vine (1966).

48. J. Ewing et al. (1966), p. 1131.

49. Glen (1982), p. 337.

CHAPTER 21

1. Tuzo sent me a copy of the letter.

2. McManus (1967).

3. I remember this talk very well because it was the only time that Ewing ever visited my office.

4. Menard (1967a).

5. Menard (1966), p. 74.

6. To the best of my knowledge I coined the terms "Euler pole" and "Euler latitude" when it grew tedious to refer to "pole of relative motion" while explaining Euler's theorem to one audience after another.

7. Many people, who had barely heard the word before, were about to become tectonophysicists.

8. Menard (1967b).

9. Katsumata quoted, from the English abstract of the Japanese article, by Oliver and Isacks (1967) in a "Note added in proof."

10. Utsu cited in Oliver and Isacks (1967).

11. Oliver and Isacks (1967), p. 4259.

12. Letter from Oliver to Allan Cox, 10 May 1972.

13. Oliver and Isacks (1967), p. 4259.

14. I introduce this merely to acknowledge the coup by Leon Knopoff in writing a paper titled with record brevity "Q."

15. Oliver and Isacks (1967), p. 4273.

16. McKenzie (1967), p. 6261.

17. Letter from McKenzie to Cox, 1972.

18. Letter from Sclater to Cox, 1972.

19. Conversation with Parker in August 1984.

20. McKenzie and Parker (1967).

21. Draft of Morgan (1968), p.1.

22. Morgan (1968), p. 1959.

23. Heirtzler et al. (1968), p. 2119.

BIBLIOGRAPHY

Axelrod, D. I. 1963. Fossil floras suggest stable, not drifting, continents. *J. Geophys. Res.* 68:3257-3263.

Backus, G. E. 1964. Magnetic anomalies over oceanic ridges. *Nature* 201, no. 4919:591-592.

Bascom, Willard. 1961. *A hole in the bottom of the sea.* Garden City, NY: Doubleday.

Beloussov, V. V. 1962. *Basic problems in geotectonics.* New York: McGraw-Hill.

Benioff, H. 1949. Seismic evidence for the fault origin of oceanic deeps. *Bull. Geol. Soc. Am.* 60:1837-1856.

————. 1954. Orogenesis and deep crustal structure—additional evidence from seismology. *Bull. Geol. Soc. Am.* 65:385-400.

Bernal, J. D. 1961. Continental and oceanic differentiation. *Nature* 192, no. 4798:123-125.

Berry, E. W. 1928. Comments on the Wegener hypothesis. In *Theory of continental drift*, 194-196. Tulsa: Am. Assoc. Petrol. Geol.

Bijlaard, P. P. 1936. Théorie des deformations plastiques et locales par rapport aux anomalies negatives de la gravitation. *Rept. Int. Union Geodesy and Geophysics, Congress at Edinburgh*, pp. 3-23.

————. 1951. On the origin of geosynclines, mountain formation, and volcanism. *Trans. Am. Geophys. Un.* 32:518-519.

Billings, M. P. 1959. Memorial to Reginald Aldworth Daly. *Proc. Geol. Soc. Am. for 1958*, pp. 115-122.

————. 1960. Diastrophism and mountain building. *Bull. Geol. Soc. Am.* 71:363-398.

Birch, F. 1956. Reginald Daly. *Nat. Acad. Sci. Biogr. Mem.* 34:31-64.

Blackett, P. M. 1965. Introduction. *Phil. Trans. Roy. Soc.* 258:vii-x.

————, S. Clegg, and P. H. Stubbs. 1960. An analysis of rock magnetic data. *Proc. Roy. Soc.* (ser. A) 256: 291-322.

Bott, M. H. 1965. The upper mantle in relation to the origin of movements at the earth's surface. In *The upper mantle symposium, New Delhi*, 20-28. Copenhagen: Int. Un. Geodesy Geophys.

Bowie, W. 1935. The origin of continents and oceans. *Sci. Monthly* 41:444-449.

Brooks, H. 1941. Cyclic convection currents. *Trans. Am. Geophys. Un.* 22:548-551.

Bucher, W. H. 1933. *The deformation of the earth's crust.* Princeton: Princeton Univ. Press.

Bullard, E. C. 1952. Commentary. *Nature* 170:199.

————. 1964. Continental drift. *Quart. Jour. Geol. Soc. London* 120:1-34.

————. 1965. Concluding remarks. *Phil. Trans. Roy. Soc.* 258:322-323.

Bullard, E. C. 1967. Maurice Neville Hill. *Biogr. Mem. Fellows Roy. Soc.* 13:193-203.
———. 1968. Reversals of the earth's magnetic field. *Phil. Trans. Roy. Soc.* 263:481-524.
———. 1975a. William Maurice Ewing. *Biogr. Mem. Fellows Roy. Soc.* 21:269-311.
———. 1975b. The emergence of plate tectonics: A personal view. *Ann. Rev. Earth Planet. Sci.* 3:1-30.
Bullard, E. C. et al. [J. E. Everett and A. G. Smith]. 1965. The fit of the continents around the Atlantic. *Phil. Trans. Roy. Soc.* 258:41-51.
Carey, S. Warren. 1958. A tectonic approach to continental drift. In *Continental drift*, S. Warren Carey, convener, 177-355. A Symposium of the Geol. Dept., Univ. of Tasmania.
———. 1973? *The expanding earth: An essay review*. Privately circulated.
———. 1976. *The expanding earth*. Amsterdam: Elsevier Scientific Publ. Co.
Chamberlin, R. T. 1938. Review of our wandering continents. *J. Geol.* 46:791-792.
Chaney, R. W. 1940. Bearing of forests on the theory of continental drift. *Sci. Monthly* 51:489-499.
Christoffel, D. A., and D. I. Ross. 1965. Magnetic anomalies south of the New Zealand plateau. *J. Geophys. Res.* 70:2857-2861.
Coleman, A. P. 1916. Dry land in geology. *Bull. Geol. Soc. Am.* 27:171-204.
———. 1924. Ice ages and the drift of continents. *Am. J. Sci.* (5th ser.) 7:398-404.
———. 1933. Ice ages and the drift of continents. *J. Geol.* 41:409-417.
Collinson, D. W., and S. K. Runcorn. 1960. Polar wandering and continental drift: Evidence from paleomagnetic observations in the United States. *Bull. Geol. Soc. Am.* 71:915-958.
Coode, A. M. 1965. A note on oceanic transcurrent faults. *Can. J. Earth Sci.* 2:400-401.
Cook, K. L. 1962. The problem of the mantle-crust mix: Lateral inhomogeneity in the uppermost part of the earth's mantle. *Advances Geophys.* 9:295-360.
———. 1965. Rift system in the Basin and Range province. In *The world rift system*, edited by T. N. Irvine, 246-279. Ottawa: Geol. Surv. Can.
Cox, A. 1973. *Plate tectonics and geomagnetic reversals*. San Francisco: W. H. Freeman.
———, and R. R. Doell. 1960. Review of paleomagnetism. *Bull. Geol. Soc. Am.* 71:645-768.
———. 162. Magnetic properties of the basalt in hole EM7, Mohole Project. *J. Geophys. Res.* 67:3997-4004.
Cox, A., R. R. Doell, and G. B. Dalrymple. 1963a. Geomagnetic polarity epochs and Pleistocene geochronometry. *Nature* 198, no. 4885:1049-1051.
———. 1963b. Geomagnetic polarity epochs: Sierra Nevada II. *Science* 142:382-385.
———. 1964. Reversals of the earth's magnetic field. *Science* 144:1537-1543.
Daly, R. A. 1923. A critical view of the Taylor-Wegener hypothesis. *J. Wash. Acad. Sci.* 13:445-450.
———. 1940. *Strength and structure of the earth*. New York: Prentice-Hall.
Darwin, Francis. 1887. *The life and letters of Charles Darwin*. 3 vols. London: John Murray.

Dicke, R. H. 1957. Principles of equivalence and the weak interactions. *Rev. Mod. Phys.* 29:355-362.

———. 1959. Gravitation—an enigma. *Am. Sci.* 47:25-40.

Dietz, R. S. 1961a. Continent and ocean basin evolution by spreading of the sea floor. *Nature* 190:854-857.

———. 1961b. Note. *Nature* 192, no. 4798:124.

———. 1962a. Ocean-basin evolution by sea-floor spreading. In *The crust of the Pacific Ocean*, edited by G. A. Macdonald and H. Kuno. Am. Geophys. Un. Geophy. Monogr. 6:11-12.

———. 1962b. Ocean-basin evolution by sea-floor spreading. *J. Oceanogr. Soc. Jap.*, 20th Anniv. Vol., pp. 4-14.

———. 1962c. Ocean-basin evolution by sea-floor spreading. In *Continental drift*, edited by K. Runcorn, 289-298. New York: Academic Press.

———. 1963a. Collapsing continental rises: An actualistic concept of geosynclines and mountain building. *J. Geol.* 71:314-333.

———. 1963b. Alpine serpentines as oceanic rind fragments. *Bull. Geol. Soc. Am.* 74:947-952.

———. 1964. Origin of continental slopes. *Am. Sci.* 52:50-69.

———. 1965a. Collapsing continental rises: A reply. *J. Geol.* 73:901-906.

———. 1965b. Colston symposium: Marine geology and geophysics. *Science* 149:94-95.

———. 1966. Passive continental margins, spreading sea floors, and collapsing continental rises. *Am. J. Sci.* 264:177-193.

———. 1968. Reply. *J. Geophys. Res.* 73:6567.

Dietz, R. S., and J. C. Holden. 1966a. Mioclines (miogeosynclines) in space and time. *J. Geol.* 74:566-583.

———. 1966b. Deep-sea deposits in but not on the continents. *Bull. Am. Assoc. Petrol. Geol.* 50:351-362.

Dietz, R. S., H. W. Menard, and E. L. Hamilton. 1954. Echograms of the Mid-Pacific Expedition. *Deep-Sea Res.* 1:258-272.

Dirac, P.A.M. 1937. The cosmological constants. *Nature* 139:323.

Drake, C. L. 1964. World rift system. *Trans. Am. Geophys. Un.* 45:435-440.

Drake, C. L., N. J. Campbell, G. Sander, and J. E. Nafe. 1963. A mid-Labrador sea ridge. *Nature* 200, no. 4911:1085-1086.

Drake, C. L., and R. W. Girdler. 1964. A geophysical study of the Red Sea. *Geophys. J.* 8:473-495.

Drake, C. L., and H. P. Woodward. 1963. Appalachian curvature, wrench faulting, and offshore structures. *Trans. New York Acad. Sci.* 26:48-63.

Du Toit, A. L. 1927. *A geological comparison of South America with South Africa.* Carnegie Inst. Wash. Publ. 381.

———. 1937. *Our wandering continents.* Edinburgh: Oliver and Boyd.

———. 1944. Tertiary mammals and continental drift. *Am. J. Sci.* 242:145-163.

Eaton, G. P. 1964. Windborne volcanic ash: A possible index to polar wandering. *J. Geol.* 72:1-35.

Egyed, L. 1956. Determination of changes in the dimensions of the earth from paleo-geographical data. *Nature* 178:534.

———. 1957. A new dynamic conception of the internal constitution of the earth. *Geol. Rund.* 46:101-121.

Engel, A. E., and C. G. Engel. 1964. Igneous rocks of the East Pacific Rise. *Science* 146:477-485, and cover photo.

Estabrook, F. B. 1956. Geophysical research shaft. *Science* 124:686.

Ewing, J., and M. Ewing. 1967. Sediment distribution on the mid-ocean ridges with respect to spreading of the sea floor. *Science* 156:1590-1592.

Ewing, J., J. L. Worzel, M. Ewing, and C. Windisch. 1966. Ages of horizon A and the oldest Atlantic sediments. *Science* 154:1125-1132.

Ewing, M. 1963. Sediments of ocean basins. In *man, science, learning and education*, 41-59. Houston: Rice Univ.

Ewing, M., A. P. Crary, and H. M. Rutherford. 1937. Geophysical investigations in the emerged and submerged Atlantic coastal plain. Part I: Methods and results. *Bull. Geol. Soc. Am.* 48:753-802.

Ewing, M., and J. Ewing. 1959. Seismic-refraction measurements in the Atlantic Ocean basins, in the Mediterranean Sea, on the Mid-Atlantic Ridge, and in the Norwegian Sea. *Bull. Geol. Soc. Am.* 70:291-318.

———. 1964. Distribution of oceanic sediments. In *Studies in Oceanography*, edited by K. Yoshida, 525-537. Tokyo: Tokyo Univ. Press.

Ewing, M., J. Ewing, and M. Talwani. 1964. Sediment distribution in the oceans: The Mid-Atlantic Ridge. *Bull. Geol. Soc. Am.* 75:17-36.

Ewing, M., and B. C. Heezen. 1955. Puerto Rico Trench topographic and geophysical data. *Geol. Soc. Am. Spec. Paper* 62:255-268.

———. 1956a. Mid-Atlantic Ridge seismic belt. *Trans. Am. Geophys. Un.* 37:343.

———. 1956b. Some problems of Antarctic submarine geology. In *Antarctica in the I.G.Y.* Am. Geophys. Un. Geophys. Monogr. 1:75-81.

———. 1960. Continuity of mid-oceanic ridge and rift valley in the southwestern Indian Ocean confirmed. *Science* 131:1677-1679.

Ewing, M., X. LePichon, and J. Ewing. 1966. Crustal structure of mid-ocean ridges, 4. *J. Geophys. Res.* 71:1611-1636.

Ewing, M., G. H. Sutton, and C. B. Officer, Jr. 1954. Seismic refraction measurements in the Atlantic Ocean. Pt. 6: Typical deep stations, North America Basin. *Bull. Seism. Soc. Am.* 44:21-38.

Ewing, M., and J. L. Worzel. 1954. Gravity anomalies and structure of the West Indies, Pt. 1. *Bull. Geol. Soc. Am.* 65:165-174.

Ewing, M., J. L. Worzel, J. B. Hersey, F. Press, and G. R. Hamilton. 1949. Seismic refraction measurements in the Atlantic Ocean Basin. *Bull. Geol. Soc. Am.* 60:1303.

———. 1950. Seismic refraction measurements in the Atlantic Ocean Basin, Pt. 1. *Bull. Seism. Soc. Am.* 40:233-242.

Ewing, M., J. L. Worzel, and G. L. Shurbet. 1957. Gravity observations at sea in U.S. submarines. *Kon. Nederl. Geol.-Mijnb. Gen. Verh. Geol. Ser.* 18:49-115.

Field, R. M. 1933. Report of committee on geophysical and geological study of ocean basins. *Trans. Am. Geophys. Un.*, 14th Ann. Meet., pp. 9-16.

Fisher, R. L. 1958. *Preliminary report on Expedition Downwind*. IGY Gen. Rept. Series 2. Washington, D.C.: Nat. Acad. Sci.-Nat. Res. Coun.

———. 1961. Middle America Trench: Topography and structure. *Bull. Geol. Soc. Am.* 72:703-720.

———. 1962. Pacific Ocean. In *McGraw-Hill yearbook of science and technology*, New York: McGraw-Hill.

———, and H. H. Hess. 1963. Trenches. In *The sea*, edited by M. Hill, 3:411-436. New York: Interscience Publ.

Fisher, R. L., and R. R. Revelle, 1955. The trenches of The Pacific. *Sci. Am.* 193:36-41.

Fisher, R. L., and R. W. Raitt. 1962. Topography and structure of the Peru-Chile trench. *Deep-Sea Res.* 9:423-443.

Frankel, H. 1982. The development, reception, and acceptance of the Vine-Matthews-Morley hypothesis. *Hist. Studies Phys. Sci.* 13 (pt. 2):1-39.

Georgi, J. 1962. Memories of Alfred Wegener. In *Continental drift*, edited by K. Runcorn, 309-324. New York: Academic Press.

Gilliland, W. N. 1962. Possible continental continuation of the Mendocino fracture zone. *Science* 137:685-686.

———. 1964. Extension of the theory of zonal rotation to explain global fracturing. *Nature* 202, no. 4939:1276-1278.

Gilluly, J. 1963. The tectonic evolution of the western United States. *Quart. J. Geol. Soc. London* 119:133-174.

———. 1966. *Volcanism, tectonism and plutonism in the western United States*. Geol. Soc. Am. Spec. Paper 80.

Girdler, R. W. 1962. Initiation of continental drift. *Nature* 194, no. 4828:521-524.

Glen, W. 1982. *The road to Jaramillo*. Stanford: Stanford Univ. Press.

Gray, G. W. 1956. The Lamont Geological Observatory. *Sci. Am.* 195:83-94.

Greenberg, D. S. 1964. Mohole: The project that went awry. Pts. 1-3. *Science* 143:115-119, 223-227, and 334-337.

Greene, M. T. 1982. *Geology in the nineteenth century* Ithaca: Cornell Univ. Press.

Griggs, D. L. 1939. A theory of mountain-building. *Am. J. Sci.* 237:611-650.

Gunn, R. 1936. On the origin of continents and their motions. *J. Franklin Inst.* 222:475-492.

Gutenberg, B. 1936. Structure of the earth's crust and the spreading of continents. *Bull. Geol. Soc. Am.* 47:1587-1610.

———. 1951. Hypotheses on the development of the earth. In his *Internal Constitution of the Earth*, 178-226. New York: Dover.

Gutenberg, B., and C. F. Richter. 1945. Seismicity of the earth. *Bull. Geol. Soc. Am.* 56:603-668.

———. 1954. *Seismicity of the earth and associated phenomena*. Princeton: Princeton Univ. Press.

Hallam, A. 1975. Alfred Wegener and the hypothesis of continental drift. *Sci. Am.* 232:88-97.

Hallam, A. 1983. *Great geological controversies*. Oxford: Oxford Univ. Press.

Hamilton, E. L. 1956. Sunken islands of the Mid-Pacific Mountains. *Geol. Soc. Am. Mem.* 64.

———. 1959. Thickness and consolidation of deep-sea sediments. *Bull. Geol. Soc. Am.* 70:1399-1424.

———. 1960. Ocean basin ages and amounts of original sediments. *J. Sed. Petrol.* 30:370-379.

Hamilton, W. 1961. Origin of the Gulf of California. *Bull. Geol. Soc. Am.* 72:1307-1318.

———. 1963a. Tectonics of Antarctica. *Am. Assoc. Petrol. Geol. Mem.* 2:4-15.

———. 1963b. *Metamorphism in the Riggins region, western Idaho*. U.S. Geol. Surv. Prof. Paper 436.

———. 1964. Discussion of paper by D. I. Axelrod. *J. Geophys. Res.* 69:1666-1667.

———. 1966a. Formation of the Scotia and Caribbean arcs. In *Continental margins and island arcs*, edited by W. H. Poole, 178-185. Ottawa: Geol. Surv. Can.

———. 1966b. *Origin of the volcanic rocks of eugeosynclines and island arcs*. Geol. Surv. Can. Paper 66-15, pp. 348-356.

———. 1967. Tectonics of Antarctica. *Tectonophys.* 4:555-568.

Hamilton, W., and W. B. Myers. 1966. Cenozoic tectonics of the western United States. *Rev. Geophys.* 4:509-549.

Heck, N. H. 1935. A new map of earthquake distribution. *Geogr. Rev.* 25:125-130.

Hedberg, H. 1957. Presentation of the Penrose Medal to Arthur Holmes. *Proc. Geol. Soc. Am. for 1956*, 69-74.

———. 1962. AMSOC oceanic deep drilling project. *Geotimes* 7:8-13.

Heezen, B. C. 1956. Outline of North Atlantic deep-sea geomorphology. *Bull. Geol. Soc. Am.* 67:1703.

———. 1959a. Paleomagnetism, continental displacements, and the origin of submarine topography. *International Oceanographic Congress, Reprints of Abstracts*, 26-28. Washington, D.C.: Am. Assoc. Advance. Sci.

———. 1959b. Géologie sous-marine et déplacements des continents. In *La Topographie et la Géologie des Profondeurs Océaniques*, 83:295-304. Paris: Colloq. Int. Cent. Nat. Res. Sci.

———. 1960. The rift in the ocean floor. *Sci. Am.* 203:98-110.

———. 1962. The deep sea floor. In *Continental drift*, edited by K. Runcorn, 235-286. New York: Academic Press.

Heezen, B. C., E. T. Bunce, J. B. Hersey, and M. Tharp. 1964. Chain and Romanche fracture zones. *Deep-Sea Res.* 11:11-33.

Heezen, B. C., and C. L. Drake. 1964. Grand Banks slump. *Bull. Am. Assoc. Petrol. Geol.* 48:221-225.

Heezen, B., and M. Ewing. 1961. The mid-oceanic ridge and its extension through the Arctic Basin. In *Geology of the Arctic*, 622-642. Toronto: Univ. of Toronto Press.

Heezen, B. C., M. Ewing, and E. T. Miller. 1953. Trans-Atlantic profile of total magnetic intensity and topography, Dakar to Barbados. *Deep-Sea Res.* 1:25-33.

Heezen, B., and M. Tharp. 1956. Physiographic diagram of the North Atlantic. *Bull. Geol. Soc. Am.* 67:1704.

————. 1965. Tectonic fabric of the Atlantic and Indian oceans and continental drift. *Phil. Trans. Roy. Soc.* 258:90-106.

Heezen, B., M. Tharp, and M. Ewing. 1959. *The floors of the oceans. I. The North Atlantic.* Geol. Soc. Am. Spec. Paper 65.

Heirtzler, J. R., and D. E. Hayes. 1967. Magnetic boundaries in the North Atlantic Ocean. *Science* 157:185-187.

Heirtzler, J. R., and X. LePichon. 1965. Crustal structure of the mid-ocean ridges, 3. *J. Geophys. Res.* 70:4013-4033.

Heirtzler, J. R., G. O. Dickson, E. M. Herron, W. C. Pitman, III, and X. LePichon. 1968. Marine magnetic anomalies, geomagnetic field reversals, and motions of the ocean floor and continents. *J. Geophys. Res.* 73:2119-2136.

Hess, H. H. 1939. Island arcs, gravity anomalies and serpentine intrusions. *International Geological Congress, Moscow, 1937* 2:263-283.

————. 1946. Drowned ancient islands of the Pacific Basin. *Am. J. Sci.* 244:772-791.

————. 1948. Major structural features of the western North Pacific. *Bull. Geol. Soc. Am.* 59:417-446.

————. 1951. Comment on mountain building. *Trans. Am. Geophys. Un.* 32:528-531.

————. 1954. Geological hypotheses and the earth's crust under the oceans. *Proc. Roy. Soc.* (ser. A) 222:341-348.

————. 1955a. Serpentines, orogeny and epeirogeny. *Geol. Soc. Am. Spec. Paper* 62:391-408.

————. 1955b. The oceanic crust. *J. Mar. Res.* 14:423-439.

————. 1957. The Vening Meinesz negative gravity anomaly belt of island arcs 1926-1956. *Kon. Nederl. Geol.-Mijnb. Gen. Verh. Geol. Ser.* 18:183-188.

————. 1959a. The AMSOC hole to the earth's mantle. *Trans. Am. Geophys. Un.* 40:340-345.

————. 1959b. Nature of the great oceanic ridges. *Internat. Oceanog. Cong. Preprints of Abstracts,* 33-34. Washington, D.C.: Am. Assoc. Advance. Sci.

————. 1962a. History of ocean basins. In *Petrologic Studies: A volume to honor A. F. Buddington,* edited by A.E.J. Engel, H. L. James and B. F. Leonard, 599-620. New York: Geol. Soc. Am.

————. 1962b. Richard Montgomery Field. *Trans. Am. Geophys. Un.* 43:1-3.

————. 1965. Mid-oceanic ridges and tectonics of the sea-floor. In *Submarine Geology and Geophysics,* edited by W. F. Whittard and R. Bradshaw, 317-332. London: Butterworths.

————. 1968. Reply. *J. Geophys. Res.* 73:6569.

Hill, M. N. 1956. Notes on the bathymetric chart of the N. E. Atlantic. *Deep-Sea Res.* 3:229-231.

————. 1957. Geophysical investigations on the floor of the Atlantic Ocean in R.R.S. *Discovery* II, 1956. *Nature* 180:10-13.

————. 1960. A median valley of the Mid-Atlantic Ridge. *Deep-Sea Res.* 6:193-205.

Holmes, Arthur. 1928. Continental drift: A review. *Nature* 122:431-433.

————. 1929. Radioactivity and earth movements. *Trans. Geol. Soc. Glasgow* 18:559-606.

Holmes, Arthur. 1944. *Principles of physical geology*. New York: The Ronald Press. (2d ed., 1965).

Hospers, J. 1951. Remanent magnetism of rocks and the history of the geomagnetic field. *Nature* 168:1111-1112.

James, H. L. 1973. Harry Hammond Hess. *Nat. Acad. Sci. Biogr. Mem.* 43:108-128.

Jeffreys, H. 1962. *The earth, its origin, history and physical constitution*. 4th ed. Cambridge: Cambridge Univ. Press.

Joksch, H. C. 1955. Statistische Analyse der hypsometrischen Kurve der Erde. *Zeit. Geophys.* 21:109-112.

Jonas, A. I. 1935. Pre-Devonian structural zones in Scotland and eastern North America. *J. Wash. Acad. Sci.* 25:166-173.

Jones, H. S. 1954. Dimensions and rotation. In *The earth as a planet*, edited by G. P. Kuiper, 1-41. Chicago: Univ. of Chicago Press.

King, L. 1958a. A new reconstruction of Laurasia. In *Continental drift*, S. Warren Carey, convener, 13-23. A Symposium of the Geol. Dept., Univ. of Tasmania.

————. 1958b. The origin and significance of the great sub-oceanic ridges. In *Continental drift*, S. Warren Carey, convener, 62-102. A Symposium of the Geol. Dept., Univ. of Tasmania.

Koczy, F. F. 1954. A survey on deep-sea features taken during the Swedish deep-sea expedition. *Deep-Sea Res.* 1:176-184.

Krause, D. C. 1960. Geology of sea floor east of Guadalupe Island. *Deep-Sea Res.* 8:28-38.

Kuenen, Ph. H. 1935. *Geological results of the Snellius Expedition, Pt. 1*. Ulrecht: Kemink en Zoon N.V.

————. 1950. *Marine Geology*. New York: John Wiley and Sons.

Lake, P. 1922. Wegener's displacement theory. *Geol. Mag.* 59:338-346.

Lee, W. H., and G. J. MacDonald. 1963. The global variations of terrestrial heat flow. *J. Geophys. Res.* 68:6481-6492.

Leet, L. D. 1937. Review of geophysical investigations. *Bull. Seis. Soc. Am.* 27:353-354.

LePichon, X., R. E. Houtz, C. L. Drake, and J. E. Nafe. 1965. Crustal structure of the mid-ocean ridges, 1. *J. Geophys. Res.* 70:319-339.

Leverett, F. 1939. Memorial to Frank Bursley Taylor. *Proc. Geol. Soc. Am. for 1938*, pp. 191-200.

Longwell, C. 1938. Review of our wandering continents. *Econ. Geol.* 33:358-359.

————. 1944. Some thoughts on the evidence for continental drift. *Am. J. Sci.* 242:218-231.

————. 1945. Presentation of Penrose Medal to Felix Andies Vening Meinesz. *Proc. Geol. Soc. Am. for 1945*, pp. 121-127.

————. 1958. My estimate of the continental drift concept. In *Continental drift*, S. Warren Carey, convener, 1-12. A Symposium of the Geol. Dept., Univ. of Tasmania.

MacDonald, G. J. 1964. The deep structure of continents. *Science* 143:921-929.

————. 1965. Continental structure and drift. *Phil. Trans. Roy. Soc.* 258:215-227.

Marvin, U. B. 1973. *Continental drift*. Washington, D.C.: Smithsonian Inst. Press.

Mason, R. G. 1958. A magnetic survey off the west coast of the United States. *Geophys. J.* 1:320-329.

Mason, R. G., and A. D. Raff. 1961. Magnetic survey off the west coast of North America, 32°N latitude to 42°N latitude. *Bull. Geol. Soc. Am.* 72:1259-1266.

Matthews, D. H. 1961. Lavas from an abyssal hill on the floor of the North Atlantic Ocean. *Nature* 190, no. 4771:158-159.

————. 1963. A major fault scarp under the Arabian Sea displacing the Carlsberg Ridge near Socotra. *Nature* 198, no. 4884:950-952.

Maxwell, A. E., and R. Revelle. 1954. Heat flow through the Pacific Ocean floor. *Bureau Central Seismo. Internat.* (sec. A) 19:395-405.

McDougall, I., and D. H. Tarling. 1963. Dating of polarity zones in the Hawaiian Islands. *Nature* 200:171-172.

McKenzie, D. P. 1967. Some remarks on heat flow and gravity anomalies. *J. Geophys. Res.* 72:6261-6273.

McKenzie, D. P., and R. L. Parker. 1967. The North Pacific: An example of tectonics on a sphere. *Nature* 216, no. 5122:1276-1280.

McManus, D. A. 1967. Physiography of Cobb and Gorda rises, northeast Pacific Ocean. *Bull. Geol. Soc. Am.* 78:527-546.

Menard, H. W. 1955. Deformation of the northeastern Pacific and the west coast of North America. *Bull. Geol. Soc. Am.* 66:1149-1198.

————. 1956. Archipelagic aprons. *Bull. Am. Assoc. Petrol. Geol.* 40:2195-2210.

————. 1958. Development of median elevations in ocean basins. *Bull. Geol. Soc. Am.* 69:1179-1186.

————. 1959. Geology of the Pacific sea floor. *Experentia* 15:205-213.

————. 1960. The East Pacific Rise. *Science* 132:1737-1746.

————. 1961. The East Pacific Rise. *Sci. Am.* 205:52-61.

————. 1962. Correlation between length and offset on very large wrench faults. *J. Geophys. Res.* 67:4096-4098.

————. 1964. *Marine geology of the Pacific.* New York: McGraw-Hill.

————. 1965. The world-wide rise-ridge system. *Phil. Trans. Roy. Soc.* 258:109-122.

————. 1966. Fracture zones and offsets of the East Pacific Rise. *J. Geophys. Res.* 71:682-685.

————. 1967a. Extension of northeastern Pacific fracture zones. *Science* 155:72-74.

————. 1967b. Sea-floor spreading, topography, and the second layer. *Science* 157:923-924.

————. 1968. Some remaining problems in sea-floor spreading. In *The history of the earth's crust*, edited by R. A. Phinney, 109-118. Princeton: Princeton Univ. Press.

————. 1971. *Science, growth and change.* Cambridge: Harvard Univ. Press.

————. 1979. Very like a spear. In *Two-hundred years of geology in America*, edited by C. J. Schneer, 19-30. Hanover: Univ. Press of New England.

————. 1983. Insular erosion, isostasy and subsidence. *Science* 220:913-918.

————. 1984a. The origin of guyots: The Beagle to Seabeam. *J. Geophys. Res.* 89:11,117-11,123.

————. 1984b. Darwin reprise. *J. Geophys. Res.* 89:9960-9968.

Menard, H. W., and R. S. Dietz. 1952. Mendocino submarine escarpment. *J. Geol.* 60:266-278.

Menard, H. W., and H. S. Ladd. 1963. Oceanic islands, seamounts, guyots and atolls. In *The sea*, edited by M. N. Hill, 3:365-388. New York: Interscience Publ.

Menard, H. W., and V. V. Vacquier. 1958. Magnetic survey of part of the deep sea floor off the coast of California. *Res. Rev.* (June):1-5.

Merton, R. K. 1961. Singles and multiples in scientific discovery: A chapter in the sociology of science. *Proc. Am. Philos. Soc.* 105:470-486.

———. 1968. The Matthew effect in science. *Science* 159:56-63.

Morgan, W. J. 1968. Rises, trenches, great faults and crustal blocks. *J. Geophys. Res.* 73:1959-1982.

Morley, L. W., and A. Larochelle. 1964. Paleomagnetism as a means of dating geological events. *Roy. Soc. Can. Spec. Publ.* 8:39-50.

Munk, W. H., and G. J. MacDonald. 1960. *The rotation of the earth*. Cambridge: Cambridge Univ. Press.

Murray, H. W. 1939. Submarine scarp off Mendocino, California. *Field Engineers Bull.* 13:27-33.

Nagata, T. 1951. Reverse thermo remnant magnetism. *Nature* 169:704-705.

Nairn, A. E. 1960. Paleomagnetic results from Europe. *J. Geol.* 68:285-306.

Nichols, J. M. 1932. *Bibliography of North American geology, 1919-1928*. U.S.G.S. Bull. 823.

Northrup, J. W., and A. A. Meyerhoff. 1963. Validity of polar and continental movement hypotheses based on paleomagnetic studies. *Bull. Am. Assoc. Petrl. Geol.* 47:575-585.

Officer, C. B., J. I. Ewing, R. S. Edwards, and H. R. Johnson. 1957. Geophysical investigations in the eastern Caribbean. *Bull. Geol. Soc. Am.* 68:359-378.

Oliver, J., and B. Isacks. 1967. Deep earthquake zones, anomalous structures in the upper mantle, and the lithosphere. *J. Geophys. Res.* 72:4259-4275.

Opdyke, N. D., and S. K. Runcorn. 1960. Wind direction in the western United States in the late Paleozoic. *Bull. Geol. Soc. Am.* 71:959-972.

Oppenheim, V. 1967. Critique of the hypothesis of continental drift. *Bull. Am. Assoc. Petrol. Geol.* 51:1354-1360.

Orowan, E. 1966. Age of the ocean floor. *Science* 154:413-416.

Pais, A. 1982. *"Subtle is the Lord. . . ."* New York: Oxford Univ. Press.

Peter, G. 1966. Magnetic anomalies and fracture pattern in the northeast Pacific Ocean. *J. Geophys. Res.* 71:5365-5374.

Peter, G. et al. 1965. Structure of the Aleutian Trench. *J. Geophys. Res.* 70:353-366.

Pettersson, H. 1953. *Westward ho with the albatross*. New York: E. P. Dutton.

Phinney, R. A. (ed.). 1968. *The history of the earth's crust*. Princeton: Princeton Univ. Press.

Piccard, J., and R. S. Dietz. 1961. *Seven miles down*. New York: G. P. Putnam's Sons.

Pickering, W. A. 1907. The place of origin of the moon—the volcanic problem. *J. Geol.* 15:23-38.

Pitman, W. C., III, and W. C. Heirtzler. 1966. Magnetic anomalies over the Pacific-Antarctic Ridge. *Science* 154:1164-1171.

Price, D. J. de Solla. 1961. *Science since Babylon*. New Haven: Yale Univ. Press.

Pyne, S. J. 1980. *Grove Karl Gilbert*. Austin: Univ. of Texas Press.

Raff, A. D. 1962. Further magnetic measurements along the Murray fault. *J. Geophys. Res.* 67:417-418.

——. 1963. Magnetic anomaly over Mohole drill hole EM7. *J. Geophys. Res.* 68:955-956.

——, and R. G. Mason. 1961. Magnetic survey off the west coast of North America, 40°N latitude to 52°N latitude. *Bull. Geol. Soc. Am.* 72:1267-1270.

Raitt, H. 1956. *Exploring the deep Pacific*. Denver: Sage Books.

Raitt, R. W. 1956. Seismic-refraction studies of the Pacific Ocean Basin. *Bull. Geol. Soc. Am.* 67:1623-1640.

——, R. L. Fisher, and R. G. Mason. Tonga Trench. *Geol. Soc. Am. Spec. Paper* 62:237-254.

Reitan, P. H. 1960. The earth's volume change and its significance for orogenesis. *J. Geol.* 68:678-680.

Revelle, R. R. 1944. *Marine bottom samples collected in the Pacific Ocean by the* Carnegie *on its seventh cruise*. Carnegie Inst. Wash. Publ. 556.

——. 1953. *Shipboard report, Capricorn expedition*. SIO 53-15, Scripps Inst. Ocean. Misc. Papers.

Revelle, R. R., and A. E. Maxwell. 1952. Heat flow through the floor of the eastern North Pacific Ocean. *Nature* 170:199.

Rice, W. A. 1938. Our wandering continents (Review). *Am. J. Sci.* 35:391-393.

Richter, C. F. 1958. *Elementary seismology*. San Francisco: W. H. Freeman.

Rothé, J. P. 1954. La zone seismique médiane Indo-Atlantique. *Proc. Roy. Soc. London* 222:387-392.

Rubey, W. W. 1951. Geologic history of sea water. *Bull. Geol. Soc. Am.* 62:1111-1148.

——. 1968. Presentation of the 1966 Penrose Medal to Harry Hammond Hess. *Proc. Geol. Soc. Am.* 1966:83-86.

Ruedemann, R. 1926. Neuere amerikanische Theorien über die Entstehung der Kontinente und Ozeane. *Geol. Rundschau* 17a:49-61.

Runcorn, S. K. 1956a. Paleomagnetic comparisons between Europe and North America. *Proc. Geol. Assoc. Can.* 8:77-85.

——. 1956b. Paleomagnetic survey in Arizona and Utah: Preliminary results. *Bull. Geol. Soc. Am.* 67:301-316.

——. 1962. Paleomagnetic evidence for continental drift and its geophysical cause. In *Continental Drift*, edited by S. K. Runcorn, 1-39. New York: Academic Press.

——. 1963. Satellite gravity measurements and convection in the mantle. *Nature* 200, no. 4907:628-630.

——. 1966. Corals as paleontological clocks. *Sci. Am.* 215:26-33.

Rusnak, G. A., and R. L. Fisher. 1964. Structural history and evolution of the Gulf of California. In *Marine geology of the Gulf of California—A symposium, Mem. 3*, edited by T. H. van Andel and G. G. Shor, Jr., 144-156. Tulsa: Am. Assoc. Petrol. Geol.

Saito, T., M. Ewing, and L. H. Burckle. 1966. Tertiary sediment from the Mid-Atlantic Ridge. *Science* 151:1075-1079.

Salisbury, M. H., and N. I. Christensen. 1978. The seismic velocity structure of a traverse through the Bay of Islands ophiolite complex, Newfoundland, an exposure of oceanic crust and upper and mantle. *J. Geophys. Res.* 83:805-817.

Schuchert, C. 1928. The hypothesis of continental displacement. In *The Theory of Continental Drift*, edited by W.A.J.M. van Waterschoot van der Gracht, 104-144. Tulsa: Am. Assoc. Petrol. Geol.

Shand, S. J. 1937. Review of our wandering continents. *J. Geomorph.* 1:250-251.

Shepard, F. P., and K. O. Emery. 1941. *Submarine topography off the California coast.* Geol. Soc. Am. Spec. Paper 31.

Shneiderov, A. J. 1943. The exponential law of gravitation and its effects on seismological and tectonic phenomena: A preliminary exposition. *Trans. Am. Geophys. Un.* 3:61-88.

———. 1944. Earthquakes on an expanding earth. *Trans. Am. Geophys. Un.* 25:282-288.

Shor, E. N. 1984. E. C. Bullard's first heat probe. *EOS* 65:73.

Shor, G. G., Jr., and R. L. Fisher. 1961. Middle America Trench: Seismic-refraction studies. *Bull. Geol. Soc. Am.* 72:721-730.

Shor, G. G., Jr., H. W. Menard, and R. W. Raitt. 1970. Structure of the Pacific Basin. In *The sea*, edited by A. E. Maxwell, 4:3-28. New York: Interscience Publ.

Simpson, G. G. 1943. Mammals and the nature of continents. *Am. J. Sci.* 241:1-31.

Stearns, H. T. 1961. Eustatic shorelines on Pacific islands. *Zeitschrift für Geomorphologie* 3:3-16.

Stocks, T., and G. Wüst. 1933. Die Tiefenverhaltnisse des offenen Atlantischen Ozeans. *Wiss. Erg. Deutschen Atlantischen Exped. Meteor.* Vol. 3, pt. 1. Berlin: Walter de Gruyter and Co.

Sullivan, W. 1974. *Continents in motion.* New York: McGraw-Hill.

Sykes, L. R. 1963. Seismicity of the South Pacific Ocean. *J. Geophys. Res.* 68:5999-6006.

Sykes, L. R., and M. Landisman. 1964. The seismicity of east Africa, the Gulf of Aden and the Arabian and Red Seas. *Bull. Seism. Soc. Am.* 54:1927-1940.

Talwani, M. 1982. Bruce Heezen—an appreciation. *The ocean floor*, edited by R. A. Scrutton and M. Talwani, 1-2. New York: John Wiley and Sons.

———, X. LePichon, and M. Ewing. 1965. Crustal structure of the mid-ocean ridges, 2. *J. Geophys. Res.* 70:341-352.

Talwani, M., X. LePichon, and J. R. Heirtzler. 1965. East Pacific Rise: The magnetic pattern and the fracture zones. *Science* 150:1109-1115.

Talwani, M., G. H. Sutton, and J. L. Worzel. 1959. Crustal section across the Puerto Rico Trench. *J. Geophys. Res.* 64:1545-1555.

Talwani, M., J. L. Worzel, and M. Ewing. 1961. Gravity anomalies and crustal section across the Tonga Trench. *J. Geophys. Res.* 66:1265-1278.

Taylor, F. B. 1910. Bearing of the Tertiary mountain belt on the origin of the earth's plan. *Bull. Geol. Soc. Am.* 21:179-226.

Teller, E. 1948. On the change of physical constants. *Phys. Rev.* 73:801-802.

Termier, P. 1925. The drifting of continents. *Ann. Rept. Smithsonian Inst., 1924*, pp. 219-236.

Tharp, M. 1982. Mapping the ocean floor—1947 to 1977. In *The ocean floor*, edited by R. A. Scrutton and M. Talwani, 19-32. New York: John Wiley and Sons.

Tolstoy, I. 1951. Submarine topography in the North Atlantic. *Bull. Geol. Soc. Am.* 62:441-450.

Tolstoy, I., and M. Ewing. 1949. North Atlantic hydrography and the Mid-Atlantic Ridge. *Bull. Geol. Soc. Am.* 60:1527-1540.

Udintsev, G. B. 1965. Results of upper mantle project studies in the Indian Ocean by the research vessel "Vitiaz." In *The world rift system*, edited by T. N. Irvine, 148-172. Ottawa: Geol. Surv. Can.

Vacquier, V. V. 1959. Measurements of horizontal displacement along faults in the ocean floor. *Nature* 183:452-453.

———. 1965. Transcurrent faulting in the ocean floor. *Phil. Trans. Roy. Soc.* 258:77-81.

Vacquier, V. V., A. D. Raff, and R. E. Warren. 1961. Horizontal displacements in the floor of the northeastern Pacific Ocean. *Bull. Geol. Soc. Am.* 72:1251-1258.

Vening Meinesz, F. A. 1934. *Gravity expeditions at sea, 1923-32.* Vol. 2: *Interpretation of the Results. Publ. Neth. Geod. Comm.* Delft: Waltman.

———. 1947. Shear patterns of the earth's crust. *Trans. Am. Geophys. Un.* 28:1-61.

Vine, F. J. 1966. Spreading of the ocean floor: New evidence. *Science* 154:1405-1415.

———, and D. H. Matthews, 1963. Magnetic anomalies over oceanic ridges. *Nature* 199:947-949.

Vine, F. J., and J. T. Wilson. 1965. Magnetic anomalies over a young oceanic ridge off Vancouver Island. *Science* 150:485-489.

von Herzen, R. P. 1959. Heat-flow values from the southeastern Pacific. *Nature* 183:882-883.

Walker, R. T., and W. J. Walker. 1954. *The origin and history of the earth*. The Walker Corp., Colorado.

Watson, J. D. 1968. *The double helix*. New York: Atheneum.

Watts, W. W. 1935. Form, drift and rhythm of the continents. *Science* 82:203-213.

Wegener, A. L. 1922. *The origin of continents and oceans*. Translation of the third German edition by J. G. Skerl. New York: E. P. Dutton, 1924. First edition published in 1915.

———. 1929. *The origin of continents and oceans*. Translation of the fourth revised German edition by J. Biram. New York: Dover, 1966.

Wertenbaker, W. 1974. *The floor of the sea*. Boston: Little, Brown.

Willis, B. 1928. Continental drift. In *The theory of continental drift*, edited by W.A.J.M. van Waterschoot van der Gracht, 76-82. Tulsa: Am. Assoc. Petrol. Geol.

———. 1944. Continental drift, ein Märchen. *Am. J. Sci.* 242:509-513.

Wilson, J. T. 1949. The origin of continents and Precambrian history. *Trans. Roy. Soc. Can.* 43:157-184.

———. 1951. On the growth of continents. *Proc. Roy. Soc. Tasmania for 1950*, pp. 85-111.

Wilson, J. T. 1952. Orogenesis as the fundamental geological process. *Trans. Am. Geophys. Un.* 33: 444-449.

———. 1953. On grabens, rifts, major wrench faults, and straight chains of islands. *Trans. Am. Geophys. Un.* 34:350.

———. 1954. The changing worlds of geology and geophysics. *Trans. Roy. Soc. Can.* 48:87-91.

———. 1957. Origin of the earth's crust. *Nature* 179:228-230.

———. 1959. Geophysics and continental growth. *Am. Sci.* 47:1-24.

———. 1960. Some consequences of expansion of the earth. *Nature* 185:880-882.

———. 1961. Note. *Nature* 192, no. 4798:125-128.

———. 1962. Cabot fault. *Nature* 195, no. 4837:135-138.

———. 1963a. Pattern of uplifted islands in the main ocean basins. *Science* 139:592-594.

———. 1963b. Evidence from islands on the spreading of ocean floors. *Nature* 197, no. 4867:536-538.

———. 1963c. A possible origin of the Hawaiian Islands. *Can. J. Phys.* 41:863-870.

———. 1965a. A new class of faults and their bearing on continental drift. *Nature* 207, no. 4995:343-347.

———. 1965b. Submarine fracture zones, aseismic ridges and the International Council of Scientific Unions Line: Proposed western margin of the East Pacific Ridge. *Nature* 207, no. 5000:907-911.

———. 1965c. Transform faults, oceanic ridges and magnetic anomalies southwest of Vancouver Island. *Science* 150:482-485.

Wiseman, J.D.H., and R.B.S. Sewell. 1937. The floor of the Arabian Sea. *Geol. Mag.* 74:219-230.

Wooster, W. S. 1951. *Operation Northern Holiday.* SIO 51-46, Scripps Inst. Ocean. Occasional Papers.

———. 1952. *Preliminary report, Shellback expedition.* SIO 52-47, Scripps Inst. Ocean. Occasional Papers.

Worzel, J. L. 1965. Discussion. *Phil. Trans. Roy. Soc.* 258:137.

———, and J. C. Harrison. 1963. Gravity at sea. In *The sea*, 3:134-154. New York: Interscience Publ.

Worzel, J. L., and G. L. Shurbet. 1955. *Gravity interpretations from standard oceanic and continental crustal sections.* Geol. Soc. Am. Spec. Paper 62.

Wright, W. B. 1923. The Wegener hypothesis. Discussion at the British Association, Hull. *Nature* 111:30-31.

Youngquist, W., and B. C. Heezen. 1948. Some Pennsylvanian conodonts from Iowa. *J. Paleo.* 22:767-773.

INDEX

abandoned ridge, 79
abstracts, 7
abyssal hills, 51-52, 209
abyssal plain, 100
Adams, R. D., 189
Aden, Gulf of, 137, 206
Agassiz, Alexander, 134, 211
Albatross, 48, 134
Albatross Plateau, 44, 134
Aleutian Trench, 259
Allen, Clarence, 234
American Geophysical Union (AGU), 31, 285-286
American Miscellaneous Society (AMSOC), 76, 115-116, 117
Anderson, Don, 274
Antarctica, 179, 183
Antarctic plate, 196
Arabian Sea, 95
Arctic Basin, 180
Arden, D. D., Jr., 168
Arnold, Jim, 13
Arrhenius, Gus, 52
aseismic ridges, 92, 247
astroblemes, 154, 196
Atlantic: age of, 282-283; crust of, 51; fracture zones in, 68-69
Atlantis, 33-34, 96-97
Atlantis fracture zone, 106
Atwater, Tanya, 77, 281, 290
aulocogens, 148
Austin, T. S., 37
Australian National University, 213-214
awards, 11, 13, 15
Axelrod, D. I., 225, 230, 231

Backus, George, 190, 221-222, 285
Bacon, Francis, 145
Bahamas, 194
Bailey, Sir Edward, 22, 230
Baird: on Capricorn expedition, 63; on Down-

wind expedition, 104, 133, 135; Vacquier on, 75-76; on Yo-yo expedition, 71
Balsley, James, 214, 320n4
Barnes, C. A., 37
Barracuda, 68
Barracuda fracture zone, 68
basalt, 51, 207
basaltification, 184-187
Bascom, Willard (Bill), 115, 116, 216
Bass, 153
Bateman, Alan, 108
bathyscaphe, 314n18
Beloussov, Vladimir Vladimirovich, 167, 180, 184-187, 254
Benioff, Hugo, 65, 125-126, 172
Benioff zones, 125-126, 191, 289-290
Bernal, J. D., 175, 178-179
Berry, E. W., 27, 28
Bigelow, Henry, 33
Bijlaard, P. P., 65, 122
Bikini, 36-37
Billings, Marland, 59, 166-167, 172, 305n12
biographical memoirs, 8-10
Birch, Frances, 65, 305n12
Blackett, Patrick, 89-90, 166, 232, 233
Blanco fracture zone, 246, 249, 253
Blow Me Down complex, 113
Bogolepow, M., 143
BOMM, 208
Boucot, Arthur, 276
Bowie, William, 27, 30, 31, 33, 80-81
Bowin, Carl, 153
Bradley, W. H., 108
Bramlette, M. N., 108
branching line, 284
Bretz, J. Harlan, 13-14
Bronk, Detlev, 308n28
Brooks, Harvey, 66
Brown and Root, Inc., 117
Brunhes, Bernard, 70, 88
Bryan, Kirk, 113, 212

Krause, Dale, 215-216, 268
Kuenen, Philip: on deep-sea floor, 45, 46; on
    marine geology, 196-197; on trenches and
    island arcs, 119, 122, 123
Kunaratnam, V., 219

Labrador Sea, 205
Ladd, H. S., 37
LaFond, Eugene C., 37, 56
Lake, Arthur, 123
Lake, Philip, 28
Lambert, W. D., 26, 27
Lamont Geological Observatory (LGO): data
    collection by, 41-42; Ewing at, 35, 39; fleet
    of, 40; heat flow studies at, 48; magnetic re-
    versals at, 215; marine geology at, 199-206;
    median rift at, 99-102; research by, 41; sea-
    floor spreading at, 176-178, 251-254, 263-
    268; seismology at, 288-290; tectonics at,
    176-178
Lamplugh, G. W., 27, 230
land bridges, 18-19, 24
Laubscher, H. P., 255
lava, 52
Lawson, A. C. (Andy), 123, 171, 306n25
Lee, W.H.K., 308n8
Lee, Willie, 229
Leet, L. Don, 34
Le Pichon, Xavier, 177, 251-253
Libby, Willard, 13
Licht, A. L., 167
Lill, Gordon, 115
limestones, 51
Lindemann, B., 143
lithosphere, 176, 290, 291, 292
Longwell, Chester, 27, 29, 83, 86, 90, 108
LORAN-C, 72
low-K tholeiites, 207
Lüder's lines, 62, 65
Lyman, John, 37

MacDonald, Gordon James Fraser: on conti-
    nental drift, 227-228, 229, 236; early life
    of, 226-227; on magnetics, 89, 166
Macelwane, J. B., 31
Madingley Rise, 208, 219, 243
magnetic anomalies: age of, 293-294; at
    American Geophysical Union meeting,
    285-286; in classroom, 281-282; on East
    Pacific Rise, 264; flank, 253, 259; on Juan

de Fuca Ridge, 250; at Lamont, 215, 252-
    253; Mason and, 71, 73-75; Menard on,
    277-278; offsets of, 75-76; Peter on, 258-
    259; in Pioneer Survey, 71-73; Pitman on,
    264-265; publications on, 76-79; at Scripps,
    215-216; Vine on, 274; in Vine-Matthews
    hypothesis, 219-221, 250
magnetic quiet zone, 288
magnetic reversals: dating of, 212-215; early
    findings on, 70; hypothesis of, 190; at La-
    mont, 215; Mason on, 71; on Mendocino
    escarpment, 77; Morley on, 216-218; Pit-
    man on, 264-265; Runcorn on, 89; at
    Scripps, 215-216; Vine on, 274; and Vine-
    Matthews hypothesis, 219-221
magnetic stratigraphy, 268-269, 274
magnetic stripes: discovery of, 72, 73-74; Ma-
    son on, 72, 73-74; Morley on, 217; publica-
    tions on, 76-79; symmetry of, 212; Vac-
    quier on, 75-76; in Vine-Matthews
    hypothesis, 219, 221
magnetic symmetry: discovery of, 162, 212;
    on East Pacific Rise, 264, 273; Menard on,
    272-273, 275; and sea-floor spreading, 238,
    243, 250
magnetometer, 72-73, 74
Main Range, 96
mantle: composition of, 111-114; definition
    of, 137; seismicity of, 289-290
mantle convection. See convection
Marianas Trench, 128, 209
marine geology: Dietz on, 196-198; at La-
    mont, 199-206; as published during 1960s,
    258-262; at Scripps, 206-211; in The Sea,
    198-199
Marquesas fracture zone, 63
Mason, Ronald: on Capricorn expedition, 55;
    on global tectonics, 180; on magnetic
    anomalies, 70, 71, 72, 73-74, 76-77; on
    sea-floor spreading, 160-161; on trenches,
    126
mass spectrometer, 213
Matthew effect, 5, 67
Matthews, Drummond Hoyle, 212, 216, 219-
    220, 246
Maury trough, 61, 62
Maxon, John, 306n24
Maxwell, Art, 49, 52-53, 55, 116
McConnell, R. B., 231
McDougall, Ian, 213-214

sedimentation rates, 165-166

seismology, 34-35, 51, 288-290. *See also* earthquakes

sequential hypothesis, 132-141

serpentinization, 112-113, 159-160, 182, 210

Sewell, Seymour, 95-96

Seychelles, 192

Shackleton, R. M., 234

shales, 263

Shand, S. J., 83

shatter cones, 154

shear net, 259

Shellback expedition, 61

Shepard, Francis Parker: C and GS soundings used by, 58; on Committee on Geophysical and Geological Study of Oceanic Basins, 31; and MacDonald, 226; Menard and, 57; on *Scripps* expedition, 36; on submarine canyons, 187

Shipek, 134

ships: acquisition of, 39-40; dangers aboard, 42-43

Shor, George Gershon, Jr.: on Downwind expedition, 134, 137; early life of, 130; on Mohole project, 115, 116; on seismicity of trenches, 130

Shumway, George, 68, 72

Shurbet, Lynn, 128

sial, 51, 82, 92

Silverman, Max, 99

Simpson, George Gaylord, 83-84, 167

sinuosities, 69

Smith, Stuart, 251

Snodgrass, J. M., 49

Society Islands, 195

Sonstadt, E., 94

soundings, 51-52, 56-57

sphenochasm, 91, 92

Spiess, Fred, 278, 290

spreading centers, 194-196

Stearns, Harold, 193

Steenland, N. C., 74

Stehli, F. G., 287

Stetson, Henry, 42, 57, 309n1

Stewart, Harris B., Jr., 59-61, 258, 262

Stocks, Theodor, 95

Stommel, Henry, 308n27

Stone, Lewis, 36

Stoner, J. O., 284

stratigraphy, 268-269, 274, 286, 320n36

strike-slip faults: Billings on, 166-167; Buwalda on, 167; Dietz on, 60; Hess on, 59; Hodgson on, 191; Mason on, 74; Menard on, 60, 78; Sykes on, 276

structural geologists, 231-232

Strutt, R. J., 84

Stubbs, P. H., 89-90

subduction, 202-203, 283-284

submarine canyons, 187

submarines, 71-72

Suess, Edward, 181

Sverdrup, Harald, 38-39, 308n29

Swartz, D. H., 168

Sweepstakes hypothesis, 167-168

Sykes, Lynn: early life of, 205-206; on earthquakes, 206, 276, 286, 289; on magnetic anomalies, 265

Tahiti, 195

Talwani, Manik: on continental drift, 268; on gravity anomaly, 129; on Heezen, 100, 102, 200; on magnetic anomalies, 252-253; on marine geology, 201, 202; on Mid-Atlantic Ridge, 180; on sea-floor spreading, 177

Tarling, Don, 213-214

Taylor, F. B., 22, 27

Taylor-Wegener hypothesis, 22

tectogene, 122-123, 125, 127-128

tectonic lines, 62

tectonics. *See* global tectonics; plate tectonics

Teller, Edward, 144, 148

temperature-gradient probe, 48-49

tension: in expansion hypothesis, 143; in trench formation, 122-123, 125, 127-128

Terman, L. M., 36

Termier, Pierre, 17, 21, 27, 29

terminology, 10-11

Tharp, Marie: bathymetric charts by, 64; on continental drift, 235; on Heezen, 149, 199, 201, 271; on median rift, 101, 105-106

tholeiites, 207

Thompson, G. A., 255

Thomson, William, 87

thrust faults, 167

Timor, 122

Tofua Trough, 127

Tolstoy, Ivan, 96-97, 202

Tonga arc, 289

Tonga Trench, 120, 122, 125, 126-127, 191

Library of Congress Cataloging-in-Publication Data

Menard, Henry W. (Henry William), 1920-1986.
   The ocean of truth.
   (Princeton series in geology and paleontology)
   Bibliography: p.
   Includes index.
   1. Geology, Structural—History.   2. Plate tectonics—
History.   I. Title.   II. Series.
QE501.3.M45   1986   551.1′36′09   85-43300
ISBN 0-691-08414-9 (alk. paper)

                                        Rev.